RENEWALS 458-4574
DATE DUE

WITHDRAWN
UTSA LIBRARIES

THE POLITICS OF IRRIGATION REFORM

Global Environmental Governance Series

Series Editors: John J. Kirton and Konrad von Moltke

Global Environmental Governance addresses the new generation of twenty-first century environmental problems and the challenges they pose for management and governance at the local, national, and global levels. Centred on the relationships among environmental change, economic forces, and political governance, the series explores the role of international institutions and instruments, national and sub-federal governments, private sector firms, scientists, and civil society, and provides a comprehensive body of progressive analyses on one of the world's most contentious international issues.

Also in the series

Hard Choices, Soft Law
Voluntary Standards in Global Trade, Environment and Social Governance
Edited by John J. Kirton and Michael J. Trebilcock
ISBN 0 7546 0966 9

Agricultural Policy Reform:
Politics and Process in the EU and US in the 1990s
Wayne Moyer and Tim Josling
ISBN 0 7546 3050 1

Governing Global Biodiversity:
The Evolution and Implementation
of the Convention on Biological Diversity
Edited by Philippe G. Le Prestre
ISBN 0 7546 1744 0

Linking Trade, Environment, and Social Cohesion:
NAFTA Experiences, Global Challenges
Edited by John J. Kirton and Virginia W. Maclaren
ISBN 0 7546 1934 6

The Politics of Irrigation Reform
Contested Policy Formulation and Implementation
in Asia, Africa and Latin America

Edited by

PETER P. MOLLINGA and ALEX BOLDING

ASHGATE

© Peter P. Mollinga and Alex Bolding 2004

All rights reserved. No part of this publication may be reproduced, stored in a retrieval system or transmitted in any form or by any means, electronic, mechanical, photocopying, recording or otherwise without the prior permission of the publisher.

Peter P. Mollinga and Alex Bolding have asserted their right under the Copyright, Designs and Patents Act, 1988, to be identified as the editors of this work.

Published by
Ashgate Publishing Limited
Gower House
Croft Road
Aldershot
Hants GU11 3HR
England

Ashgate Publishing Company
Suite 420
101 Cherry Street
Burlington, VT 05401-4405
USA

Ashgate website: http://www.ashgate.com

British Library Cataloguing in Publication Data
The politics of irrigation reform : contested policy
　formulation and implementation in Asia, Africa and Latin
　America. - (Global environmental governance)
　1. Irrigation projects - Political aspects - Case studies
　2. Irrigation - Management - Case studies 3. Irrigation -
　Government policy - Case studies 4. Irrigation - Social
　aspects - Case studies 5. Water resources development -
　Government policy - Case studies
　I. Mollinga, Peter P. II. Bolding, Alex
　333.9'14

Library of Congress Cataloging-in-Publication Data
The politics of irrigation reform : contested policy formulation and implementation in Asia,
　Africa and Latin America / edited by Peter P. Mollinga and Alex Bolding.
　　p. cm. -- (Global environmental governance)
　Includes bibliographical references and index.
　ISBN 0-7546-3515-5
　　1. Irrigation--Government policy--Asia. 2. Irrigation--Government policy--Africa. 3. Irrigation--Government policy--Latin America. I. Mollinga, Peter P. II. Bolding, Alex. III. Global environmental governance series.

HD1741.A78P65 2003
333.91'3--dc22

ISBN 0 7546 3515 5

2003058365

Printed and bound in Great Britain by MPG Books Ltd, Bodmin, Cornwall

Contents

List of Figures, Diagrams and Maps	vii
List of Tables	viii
List of Boxes	ix
List of Contributors	x
Preface	xii
Foreword	xiii
List of Abbreviations	xv

1. Introduction
   Peter P. Mollinga and Alex Bolding — 1

2. Water Policy and Law Review Process in South Africa with a Focus on the Agricultural Sector
   Marna de Lange — 11

3. The Politics of Creating Commitment: Irrigation Reforms and the Reconstitution of the Hydraulic Bureaucracy in Mexico
   Edwin Rap, Philippus Wester and Luz Nereida Pérez-Prado — 57

4. Irrigation Development and Management Reform in the Philippines: Stakeholder Interests and Implementation
   Thomas Panella — 95

5. From Voice to Empowerment: Rerouting Irrigation Reform in Indonesia
   Bryan Bruns — 145

6. Irrigation Policy Discourse and Practice: Two Cases of Irrigation Management Transfer in Zimbabwe
   Alex Bolding, Emmanuel Manzungu and Conrade Zawe — 166

7. Irrigation Policy Reforms in Pakistan: Who's Getting the Process Right?
   Edward J. van der Velde and Jamshed Tirmizi — 207

8. Capture and Transformation: Participatory Irrigation Management in Andhra Pradesh, India
   Peter P. Mollinga, R. Doraiswamy and Kim Engbersen — 240

9. The Politics of Irrigation Policy Implementation: Networks of Votes, Bribes and Coca-Cola in the Philippines
   Joost Oorthuizen   263

10. The Politics of Irrigation Reform: Research for Strategic Action
    Peter P. Mollinga and Alex Bolding   291

Index   *319*

List of Figures, Diagrams and Maps

Figures

9.1	The Penaranda River irrigation scheme	266
9.2	Layout of the PRIS system, indicating municipal boundaries of Zone 1, as well as the boundaries of Baritan village	274
9.3	Layout of PRIS, indicating San Anton village as part of the municipality of Baliwag	282

Diagrams

2.1	Main components of the institutional framework for water law review in South Africa	30
2.2	Overview of the elements of the water policy and law review in South Africa	32

Maps

8.1	Andhra Pradesh and research site in India	243

List of Tables

3.1	Development of CNA's income from water rights and fees (millions of constant 1994 pesos)	85
3.2	Financial self-sufficiency levels of the CNA (millions of constant 1994 pesos)	85
4.1	NIA foreign loans and capital expenditures from 1969-1986	99
4.2	Foreign lending to NIA by funding agency	102
4.3	Growth in national and communal irrigation systems (1000 ha)	103
4.4	Irrigation rate change 1968 to 1974	107
4.5	NIA administrators	115
4.6	Collection efficiencies and ISF as percentage of O&M expenses	122
4.7	NIS irrigation performance 1993 to 1998	123
4.8	NIA functionality rating of irrigator associations 1994 to 1998	126
4.9	Socialized ISF from AO 17	134
5.1	Sizes of public irrigation areas	147
5.2	Project activities to implement the 1987 Irrigation Operation & Maintenance Policies	149
6.1	Structure of Zimbabwe's irrigation sector (1997)	169
6.2	Evolution of smallholder irrigation policy in Zimbabwe: 1912-1980	171
6.3	Institutions in smallholder irrigation	172
6.4	Nyanyadzi blocks: official data compared with *de facto* use of the scheme (1997)	181
6.5	Musengezi irrigation schemes size, number of plot holders and funding source	191
6.6	Summary of the phases and source of funds for operation and maintenance	193
6.7	Summary of the crop contracts during phase 3	195
7.1	Distributary farmer organization pilot projects in Pakistan, 1998	219
8.1	Characteristics of distributaries studied	247
8.2	WUA characteristics in the study area	248
8.3	Irrigated area of different crops (hectares)	252

List of Boxes

2.1	Extract from the National Constitution: the process	13
2.2	Extracts from the Constitution: clauses of particular significance for water legislation	15
2.3	Some differences between the first Draft Bill (May 1997) and the final National Water Act (Act 36 of 1998) of relevance to the commercial irrigation sector	25
2.4	Some differences between the first Draft Bill (May 1997) and the final National Water Act (Act 36 of 1998) of particular relevance to the small-scale irrigation sector	26
4.1	Powers and provisions for NIA under Presidential Decree 552	106
6.1	Aims of Irrigation Management Committees	177
6.2	Characteristics of Nyanyadzi irrigation scheme	182

List of Contributors

Alex Bolding is a Ph.D. researcher and former faculty at the Irrigation and Water Engineering Group, Wageningen University, the Netherlands.

Bryan Bruns (Ph.D. in Development Sociology, Cornell University, USA) works as a consulting sociologist specializing in improving participation in irrigation and water resources management. He has worked extensively in Indonesia and elsewhere in Southeast Asia.

R. Doraiswamy is a Postgraduate in Sociology, Bangalore University, India. His fields of interests are to promote users participation in water policy formulation and implementation. He is presently with Pragathi, a non-governmental organisation established by farmers with its headquarters in Bangalore, India.

Kim Engbersen is a M.Sc. graduate from the Irrigation and Water Engineering Group at Wageningen University, the Netherlands.

Marna de Lange is a civil engineer and representative in South Africa for the International Water Management Institute. She was a consultant to the Department of Water Affairs and Forestry during the water law review and national irrigation policy development, and is presently involved in the establishment of catchment management agencies.

Emmanuel Manzungu holds a Ph.D. in irrigation management from Wageningen University, the Netherlands. He is currently a research associate in the Department of Soil Science and Agriculture, University of Zimbabwe where he conducts research in irrigation and water resources management.

Peter P. Mollinga is Associate Professor at the Irrigation and Water Engineering Group at Wageningen University, the Netherlands, and Convenor of SaciWATERs (South Asia Consortium for Interdisciplinary Water Resources Studies), based in Hyderabad, India.

Luz Nereida Pérez-Prado is a Visiting Research Fellow at the Center for U.S.-Mexican Studies, University of California, San Diego and holds a Ph.D. in Anthropology from the University of Manchester. She is knowledgeable on issues related to the irrigation water reform, state-civil society relations and the politics of gender relations in Mexico.

List of Contributors

Joost Oorthuizen completed his Ph.D. thesis on irrigation reform in the Philippines in 2003, at Wageningen University. He was a faculty at the Irrigation and Water Engineering Group, Wageningen University, the Netherlands from 1997 to 1999 and is presently working as a freelance management consultant.

Thomas Panella is currently with the Asian Development Bank in Manila and finishing his doctoral dissertation on the political economy of irrigation reform in the Philippines at the Goldman School of Public Policy at the University of California at Berkeley.

Edwin Rap is a Ph.D. researcher studying irrigation reform in Mexico at the Rural Development Sociology Group, Wageningen University, the Netherlands.

Jamshed Tirmizi is the founder Director and principal Social and Institutional Analyst at SEER, a Pakistan-based development management firm. He completed his doctorate and masters in Sociology and Anthropology from Heidelberg University, Germany.

Edward J. van der Velde (Ph.D. in Geography, University of Michigan, USA) works as an independent consultant in irrigation management, specializing on institutional issues. He was a senior irrigation management specialist with the International Irrigation Management Institute's branch in Pakistan from 1986 to 1994.

Philippus Wester is Assistant Professor Water Reforms and Ph.D. candidate at the Irrigation and Water Engineering Group, Wageningen University, the Netherlands. He has worked as a water researcher in Senegal, Pakistan, the Netherlands, Bangladesh and Mexico in the past ten years.

Conrade Zawe is an Agricultural Extension Officer (irrigation), in the Department of Agricultural and Extension (AREX) of Zimbabwe. He completed the MSc programme on irrigation management at the Irrigation and Water Engineering Group, Wageningen University, the Netherlands, where he is presently doing his Ph.D. research.

Preface

As is not uncommon, this edited volume is very much a joint effort – of the editors and authors, but also of a considerably larger set of people. The chapters were written as part of a research project called *The long road to commitment. A socio-political perspective on irrigation reform*. The project was initiated by Geert Diemer (World Bank and INPIM), L.K. Joshi (IndiaNPIM) and Peter P. Mollinga. The authors met to discuss their draft papers in Hyderabad, India in December 1999. This researchers' workshop immediately preceded *the Fifth International Conference on Participatory Irrigation Management* organised by INPIM (the International Network on Participatory Irrigation Management) together with the Government of Andhra Pradesh and the World Bank. The discussions in the researchers' workshop were reported in this larger conference. The 'long road' research project, part of the Collaborative Work Programme between the Rural Development Department of the World Bank and the Irrigation and Water Engineering Group at Wageningen University, was initially coordinated by Peter P. Mollinga and Joost Oorthuizen. When the latter left Wageningen University, Alex Bolding took over.

We thank the Directorate General for International Cooperation of the Dutch Ministry of Development Cooperation for its financial support of the research, and IndiaNPIM (India Network on Participatory Irrigation Management) and the Government of Andhra Pradesh for their support and co-operation in the organisation of the researchers' workshop. For their assistance in the emergence of this volume we would like to thank, apart from the authors and those already mentioned above, Mr. Raymond Peter, Irrigation Department, Government of Andhra Pradesh at the time of the workshop, for overall guidance and practical assistance, Ms. Aparna Karve (Hyderabad) for being an excellent organizing consultant, Dr. M.V.K. Sivamohan and Ms. Letty Christian at the Administrative Staff College of India (ASCI), Hyderabad, for hosting and facilitating the workshop, and Ms. Trudy Freriks and Ms. Gerda de Fauw, secretaries at the Irrigation and Water Engineering Group and Ms. Theresa d'Souza and Mr. Vijayasekhar, secretaries at ASCI for a lot of logistic, editing and other practical support. We also thank the referees for their comments on the manuscript.

To return to commonalities, it is also not uncommon that the publication of an edited volume takes too long. One would have wanted this book out immediately after the workshop. We can only apologize for our failure to keep to the agreed time schedule, and hope the content of the book will compensate for the delay.

Alex Bolding and Peter P. Mollinga
Wageningen and Hyderabad, August 2001

Foreword

Between 1900 and 2000 the area under irrigation sextupled to 260 million hectares. It may be estimated that governments contributed around half. Their gargantuan achievement now poses the formidable challenge of keeping the public schemes intact. This is vital because all irrigated areas together produce more than 40 percent of all food. But sustainability of many public multi-user schemes is not assured; managers of many of them do not get the resources to keep canals and structures operational. Numerous public schemes are even shrinking.

Behind this challenge looms another one: responding to water scarcity. In developing countries, irrigation systems take in 70 to 90 percent of the available fresh water. As people grow in number and riches, and environmental services produced by rivers and wetlands gain recognition, demand for fresh water will increase. Irrigation managers and farmers will need to take in less but can do so only if schemes are in good condition.

Achieving the sustainability of public schemes is an organizational and technical issue as well as a political one. It supposes shifting management responsibilities from government staff to private users, and shifting funding for operation and maintenance from public to non-public sources.

What does this book tell us about this political dimension? It shows that foreign funding cannot buy change. It can however help shape minds, stakeholder coalitions and reform packages. The second lesson is that reform advocates need to be prepared to use unexpected opportunities.

What action does the book imply? Two main ones stand out: engaging irrigation agencies in debate with civil society and greater selectivity in signing loans. Irrigation bureaucracies react weakly to user feedback on performance. Building explicit mechanisms for accountability into the operation and maintenance of irrigation systems is therefore needed to help trigger improvements. User satisfaction surveys may be held among the irrigators. NGOs and farmer leaders can discuss the outcomes with administrative and political management, summarize them in report cards and disseminate them through mass media. User groups may also track budget allocations for maintenance and report sizable deviations in the press.

Donors and governments may spark public debate also by increasing transparency. Think of newsletters communicating the budget and progress of a loan and of annual reports by irrigation departments reporting their use of the budget and their hydraulic performance.

Last but not least, donors may invite borrowers to report on the status of recent rehabilitations. If canals and structures are again in poor operational shape, the cause may be institutional and lending be held up until the borrower rewires itself for sustainability.

Hopefully this book will bring these changes nearer.

Geert Diemer
Senior Water Resources Management Specialist
Rural Development Department, World Bank, Washington, DC, June 2001

List of Abbreviations

ADB	Asian Development Bank
ADCC	Agricultural Development Coordinating Council (Philippines)
AFC	Agricultural Finance Corporation (Zimbabwe)
AFMA	Agricultural and Fisheries Modernization Act (Philippines)
AGRITEX	Department of Agricultural, Technical and Extension Services (Zimbabwe)
ANC	African National Congress (South Africa)
AO	Administrative Order (Philippines)
AOU	Advanced Operation Unit (Indonesia)
APFMIS Act	Andhra Pradesh Farmers Management of Irrigation Systems Act (Andhra Pradesh, India)
ARDA	Agricultural and Rural Development Agency (Zimbabwe)
ASCI	Administrative Staff College of India
AWB	Area Water Board (Pakistan)
AWFPC	Agriculture, Water Affairs and Forestry Portfolio Committee (South Africa)
BAPPENAS	National Planning Board (Indonesia)
BMZ	Bundesministerium für Wirtschaftliche Zusammenarbeit und Entwicklung (Germany)
BOD	Boards of Directors (Philippines)
BPWID	Bureau of Public Works Irrigation Department (Philippines)
CARP	Comprehensive Agrarian Reform Programme (Philippines)
CBZ	Commercial Bank of Zimbabwe
CDA	Canal and Drainage Act (Pakistan)
CIA	Council of Irrigator Associations (Philippines)
CIC	Communal Irrigation Committee (Philippines)
CIESAS	*Centro de Investigaciones y Estudios Superiores en Antropología Social*; Centre for Investigation and Higher Education in Social Anthropology
CIS	Communal Irrigation System (Philippines)
CMA	Catchment Management Agency (South Africa)
CMB	Cotton Marketing Board (Zimbabwe)
CNA	*Comisión Nacional del Agua*; National Water Commission (Mexico)
CNI	*Comisión Nacional de Irrigación*; National Irrigation Commission (Mexico)
CO	Community Organizer (Philippines)
COSAB	Council of South African Banks
COSATU	Council of South African Trade Unions

DA	Department of Agriculture (Philippines)
DA	District Administrator (Zimbabwe)
DANIDA	Danish International Development Assistance (Zimbabwe)
DBM	Department of Budget and Management (Philippines)
DC	Distributary Committee (Andhra Pradesh, India)
DERUDE	Department of Rural Development (Zimbabwe)
DEVAG	Department of Agricultural Development (Zimbabwe)
DG	Director General (South Africa)
DGWRD	Directorate General of Water Resources Development (Indonesia)
DILG	Department of the Interior and Local Government (Philippines)
DWAF	Department of Water Affairs and Forestry (South Africa)
DWD	Department of Water Development (Zimbabwe)
EOM	Efficient Operation and Maintenance Programme (Indonesia)
ESAP	Economic Structural Adjustment Programme (Zimbabwe)
FAO	Food and Agriculture Organization
FARMESA	Farm-level Applied Research Methods Programme for East and Southern Africa (Zimbabwe)
FIG	Farmers Irrigator Group (Philippines)
FIO	Farmers Irrigator Organizers (Philippines)
FO	Farmer Organizations (Pakistan)
GDP	Gross Domestic Product
GMB	Grain Marketing Board (Zimbabwe)
GOAP	Government of Andhra Pradesh (Andhra Pradesh, India)
GOI	Government of India
GOP	Government of Pakistan
GTZ	Deutsche Gesellschaft für Technische Zusammenarbeit GmBH
IA	Irrigation Association/ Irrigator Association (Philippines)
IBP	Indus Basin Plan (Pakistan)
ID	Irrigation Department (Pakistan)
IDD	Institutional Development Department (Philippines)
IDO	Institutional Development Officer (Philippines)
IEPES	Instituto de Estudios Políticos, Económicos y Sociales; Institute of Political, Economic, and Social Studies (Mexico)
IFAD	International Fund for Agricultural Development
IFPRI	International Food Policy Research Institute
IM	Irrigation Manager (Zimbabwe)
IMC	Irrigation Management Committee (Zimbabwe)
IMT	Irrigation Management Transfer
IndiaNPIM	India Network on Participatory Irrigation Management
INEGI	Instituto Nacional de Estadística, Geografía e Informática; The National Institute of Statistics, Geography and Informatics (Mexico)
INPIM	International Network for Participatory Irrigation Management
IOMP	Irrigation Operation and Maintenance Policy (Indonesia)

IOSP	Irrigation Operations Support Project (Philippines)
IRA	Internal Revenue Allotment (Philippines)
IRRI	International Rice Research Institute
ISF	Irrigation Service Fee (Philippines)
ISIP	Irrigation Systems Improvement Project (Philippines)
ISSP	Irrigation Subsector Project (Indonesia)
IWMI	International Water Management Institute
JICA	Japan International Cooperation Agency (Philippines)
JIWMP	Java Irrigation and Water Resources Management Project (Indonesia)
JSM	Joint System Management
KBP	Kissan Board of Pakistan
LBDC	Lower Bari Doab Canal (Pakistan)
LGU	Local Government Unit (Philippines)
MCM	million cubic meters
MLA	Member of the Legislative Assembly (Andhra Pradesh, India)
MOA	Memorandum of Agreement (Philippines)
MPW	Ministry of Public Works (Indonesia)
MRIIS	Magat River Integrated Irrigation System (Philippines)
MRMP	Magat River Multipurpose Project (Philippines)
NACOD	Nyanyadzi Advisory Committee on Development (Zimbabwe)
NAFU	National African Farmers Union (South Africa)
NCIA	The National Confederation of Irrigator Associations (Philippines)
NDP	National Drainage Programme (Pakistan)
NEDA	National Economic and Development Authority (Philippines)
NEDLAC	National Economic Development and Labour Council (South Africa)
NFA	National Food Authority (Philippines)
NFIF	National Farm Irrigation Fund (Zimbabwe)
NGO	Non-Government Organization
NIA	National Irrigation Administration (Philippines)
NIACONSULT	A Subsidiary Corporation of the National Irrigation Administration (Philippines)
NIS	National Irrigation System (Philippines)
NPA	New People's Army (Philippines)
NWA	National Water Act (South Africa)
NWFP	North West Frontier Province (Pakistan)
NWMP	National Water Management Programme (India)
O&M	Operation and Maintenance
OECF	Overseas Economic Cooperation Fund (Philippines)
OFWM	On-Farm Water Management (Pakistan)
OPEC	Oil Producing Exporting Countries (Philippines)
PAD	Provincial Agriculture Department (Pakistan)
PD	Presidential Decree (Philippines)

PDR	Process Documentation Research
PID	Provincial Irrigation Department (Pakistan)
PIDA	Provincial Irrigation and Drainage Authority (Pakistan)
PIM	Participatory Irrigation Management
PNH	*Plan Nacional Hidráulico*; National Water Plan (Mexico)
PRI	*Partido Revolucionario Institucional;* Institutional Revolutionary Party (Mexico)
PRIS	Penaranda River Irrigation System (Philippines)
PRODERITH	*Programa de Desarrollo Rural Integrado del Trópico Húmedo*; Programme for the Integrated Rural Development of the Humid Tropics (Mexico)
SAAU	South African Agricultural Union
SADC	Southern African Development Community
SAG	*Secretaría de Agricultura y Ganaderia*; Ministry of Agriculture and Livestock (Mexico)
SAMWU	South African Municipal Workers' Union
SARH	*Secretaría de Agricultura y Recursos Hidráulicos*; Ministry of Agriculture and Water Resources (Mexico)
SCARP	Salinity Control and Reclamation Project (Pakistan)
SFRA	Stream Flow Reduction Activities (South Africa)
SHINO	*Sistema Hidráulico Interconectado del Noroeste*; Northwestern Interconnected Hydraulic System (Mexico)
SIDA	Swedish International Development Assistance
SISP	Smallholder Irrigation Support Programme (Zimbabwe)
SOP	Standard Operating Procedure (Philippines)
SPP	*Secretaría de Programación y Presupuesto*; Ministry of Programming and Budget (Mexico)
SRH	*Secretaría de Recursos Hidráulicos*; Ministry of Water Resources (Mexico)
TC	Territorial Constituency (Andhra Pradesh, India)
TSA	Turnout Service Area (Philippines)
UCT	University of Cape Town (South Africa)
UNAM	National Autonomous University of Mexico (Mexico)
UNDP	United Nations Development Programme
UNHCR	United Nationals High Commission of Refugees
UPRIIS	Upper Pampanga River Integrated Irrigation System (Philippines)
UPRP	Upper Pampanga River Project (Philippines)
USAID	United States Agency for International Development
USBR	United States Bureau of Reclamation
VIDCO	Village Development Committee (Zimbabwe)
WAPDA	Water and Power Development Authority of Pakistan (Pakistan)
WARDCO	Ward Development Committee (Zimbabwe)
WATSAL	Water Resources Sector Adjustment Loan (Indonesia)
WB	World Bank

WOTRO	Netherlands Foundation for the Advancement of Tropical Research
WRDP	Water Resources Development Project (Philippines)
WRFT	Water Resources Field Technician (Philippines)
WUA	Water Users Association
WUF	Water Users Formation (Pakistan)
WUTP	Water Users Training Project (Indonesia)
ZANU (PF)	Zimbabwe African National Union Patriotic Front (Zimbabwe)
ZESA	Zimbabwe Electricity Supply Authority (Zimbabwe)
ZINWA	Zimbabwe National Water Authority

For Kees van Straaten

Chapter 1

Introduction

Peter P. Mollinga and Alex Bolding

Irrigation is a form of land and water management to enhance agricultural production by manipulating the availability of water in time and space for better crop growth. By diverting and storing water and applying it to agricultural fields at appropriate times and in appropriate quantities, irrigation allows stabilisation of yields, expansion of cultivated areas, and intensification of land use through double and triple cropping and higher input use. It has been a leading input (Ishikawa, 1967) in agricultural development in those parts of the world that have deficit, seasonal and erratic rainfall.

Historically irrigation has been strongly associated with state formation and state governance. This association has been perceived to be so strong that it gave rise to theories that explained 'oriental despotism' from the need of centralised control of irrigation systems in 'hydraulic societies' (Wittfogel, 1957). Less grandly, irrigation has been an important asset for rulers in pre-colonial, colonial as well as modern times for a number of reasons. For British colonial canal irrigation Stone (1984) summarises that

> [i]t was intended to serve the perceived interests of its masters (...). In its design, modes of operation, and intended effects, canal irrigation was ultimately a cultural expression, representing the priorities and aspirations of its western architects, and was inextricably bound up with some of the most vital aspects of colonial rule. (p.8) (...) on a policy level it was simultaneously linked with famine prevention, revenue stability, the settling of unruly tribes, expansion of cultivation, extended cultivation of cash crops, enhanced taxable capacity, improved cultivation practices, and political stability. (p.9)

While pre-19th century rulers often financed the construction of irrigation, levied tax and exerted varying degrees of political control over it, they rarely directly managed the systems on a day-to-day basis. Management was left to local users of the systems. The 19th century witnessed the emergence of irrigation bureaucracies as government departments. These agencies not only designed, built and governed the systems, they also assumed responsibility for day-to-day operation and maintenance. Recent history has seen enormous expansion of irrigated area, mostly through the construction of large-scale canal irrigation systems. This happened in the period from, roughly, the 1950s to the 1980s, particularly in the newly independent nation states of the South, and was strongly supported by international development loans. There was a concomitant expansion in size, importance and power of the government agencies that designed, built and managed these systems, the Irrigation Departments.

Contemporary irrigation professionals and irrigators live in an age of irrigation reform. There is worldwide debate about and practical efforts at institutional reform in the irrigation sector. In the 1990s the thinking about improving performance of the irrigation sector has moved beyond improvement of the performance of individual systems. The present focus is the reform of the irrigation bureaucracies that manage the systems and a redefinition of the relationships between these agencies and irrigators/water users. Central themes are the transfer of management (and sometimes governance) functions to user organisations, attaining financial sustainability through cost recovery, the introduction of financial and economic incentives for better performance, and down-sizing of the irrigation bureaucracies. A recent theme is the need for an 'integrated' perspective: seeing irrigation as a part of the larger water resources sector.

This book investigates the politics of these irrigation reform processes in a variety of settings, and looks at different phases and elements of the policy formulation and implementation processes. The meaning and scope of 'reform' varies widely across the cases: from broadly conceived water sector reform to the narrowly defined establishment of water users associations. We use reform to refer to any process of purposive transformation of the institutional features of irrigation agencies, the laws and regulations that constitute them and irrigation water use, and the relationships of these agencies with the waters users and other relevant actors. The country case studies are, in order of appearance, South Africa, the Philippines, Indonesia, Zimbabwe, Pakistan, one Indian State, and finally the Philippines again. Together they form a reasonable cross-section of irrigation reform processes.[1]

This chapter describes the objectives, perspective and context of the book. It discusses the reasons for its compilation by presenting the sources of the present focus on reform in the irrigation sector, and by commenting on the nature of the irrigation reform discourse and practice, in which an acknowledgement of its political dimensions is felt wanting. It then discusses the features of the 'politics' perspective adopted, gives the objectives of the book, its major conclusions and finally outlines its structure.

The Focus on Irrigation Reform

The sources of the emphasis on irrigation *reform* in irrigation policy in the 1990s are several. Firstly, it has been partly generated within the sector itself, based on experience. Since the mid-1960s there has been recognition of performance problems in particularly the larger canal irrigation systems: not all area created was actually irrigated, yields were below projection, maintenance was below standard and rehabilitation frequently required, widespread waterlogging and salinity started to occur, and costs were not recovered. This recognition was followed by a series of intervention programmes to improve performance. These programmes have gradually increased their scope. Efforts started at plot or tertiary unit, that is, the local level, in the system. On Farm Development at individual farms, making the canal system reach every farm and improving irrigation methods was one approach. The focus in this early period was to improve the use of water at farmer level.

Soon it was discovered, realised or admitted that farmer level problems were partly caused by main system management. In the 1980s the scope of debate and intervention gradually moved up the canal system, to include the secondary and primary canals. In India an example was the NWMP, National Water Management Programme, which included the making of operational plans for main system management.

In the 1990s debate and intervention shifted one more level up. It is now felt that the irrigation sector requires reorganisation at the agency, policy and legal levels, to enable the solution of performance problems at system and farmer level. The newest *avatar* of this view is a call for establishing river basin management organisations for integrated water resources management.[2]

Apart from an enlargement of scope in terms of levels, the scope also enlarged in terms of the type of intervention. In the beginning the emphasis was almost exclusively on technical improvements. Soon the topic of farmer participation was added to the agenda, and at present self-governance and accountability are central concerns.

The first source of the irrigation reform drive of the 1990s has thus been a learning process within the irrigation sector, which has produced an increasingly comprehensive problem analysis of the issues in the sector.

However, this process does not explain the interest in irrigation reform fully. A second source for the interest in the reform agenda of the 1990s is that of external pressure by development funding agencies. Organisations like the World Bank and the Asian Development Bank have made their loans conditional on particular reform packages. Their agenda is strongly flavoured by the neo-liberal development paradigm. It emphasises a reduced role of the state and a larger one for the private sector, economic pricing of water, financial autonomy of irrigation agencies, and devolution of management responsibilities to lower levels.

A third source for reform is the internal developments in the nations and states concerned. Many suffer from fiscal/budgetary crises, and find the subsidies to irrigation increasingly difficult to continue and justify. In some countries there has been increasing public criticism of particularly large-scale canal irrigation. National and global debate on the environmental and social problems associated with large dams and large-scale infrastructure development generally has considerably reduced the status that irrigation enjoys. Finally, and more simply, the problem of decaying systems, and the danger of total loss of the investment in the infrastructure, has induced some governments to opt for more fundamental reforms.

There are thus three different forces that underpin the 1990s emphasis on more fundamental reform in the sector: developments and learning within the sector, external and international pressure, and domestic fiscal, economic and political concerns. How important each of these is, and whether or not they come together and translate into actual reform, and what kind of reform, differs from place to place. This book investigates how this happens.

Observations on the Irrigation Reform Debate and Practice

After this general sketch of the background of the irrigation reform exercises of the 1990s we offer some observations on the discussion on reform and its relation to practice. These observations will explain the rationale of the composition of the book.

The first observation is that there is a strong tendency to think in terms of reform 'models'. Models are success stories to be replicated elsewhere.[3] In the 1970s and 1980s the Philippines, and the NIA (National Irrigation Administration) reform, was the model the irrigation world was worshipping. In the 1990s the Mexico model appeared, and many irrigation professionals were invited to visit this country to see the successes and replicate the experience at home. In the coming years Turkey, Andhra Pradesh (India) and China have a good chance of becoming the new success stories that the international irrigation community portrays to other nations as paths to follow and routes to take.

One could speculate where this fascination with 'models' comes from. There are probably several sources. The first is that it fits the engineering mind. It is an example of blueprint or prototype thinking: once you've got it right and working, it can be done anywhere. Secondly, this social engineering approach also fits the conventional planner's mind: a standardised solution that promises to fit all the different irrigation systems that need to be dealt with. And thirdly and finally, promising success is a structural element of the donor world. Consultants tendering for project contracts, or development funding agencies designing the conditionalities of a loan, need to suggest that within a given time frame, say three or five years, problems can be solved. The existence of successful models helps to convince those involved of the feasibility of the undertaking.

This leads to the second observation on the debate on and practice of irrigation reform. There is a substantial discrepancy between theory and practice. Quite consistently the models do not turn out to be so wonderful when practice is studied more closely. Sometimes much less has been achieved than suggested, and always experience is much more diverse and complex than the models lead us to believe. Policy gets transformed during implementation in different ways in different localities, leading to a variety of outcomes, some successful, others much less so. Also, transformation rarely happens in three or five years. The processes are much more gradual, curvy and complex than the rational planner would like.

This in its turn leads to the third observation. Notwithstanding this consistent finding that practice is much more diverse and complex than theory, and models often much less successful and applicable elsewhere than claimed, there is very little space and attention for the debate of such experience. Serious reflection on partial results and complexities in implementation is rare. The tendency is the flight forward: if one model doesn't work, jump to the next promise.[4]

A fourth observation is that the word 'politics' is virtually absent in the formal policy discourse on irrigation reform. It sometimes appears in euphemistic terms and sometimes appears as a black box. An example of the latter is the frequently used phrase of the need for 'political will' to implement reform. However, explicit analysis of the political dimensions of irrigation and irrigation reform is rare.

This doesn't mean that it is not talked about. It is talked about very intensively, but informally, in the corridors of meetings, over a cup of coffee, or a drink at the bar. Endless numbers of anecdotes exist on the political dimensions of irrigation, but very little explicit debate. Again, one could speculate about the reasons for this absence and discrepancy. And again there are probably several.

The conventional policy and planning model doesn't allow much space for politics. Politicians decide on particular policies, but after that it is a matter of implementation by expert planners and sector professionals. Politics is often considered as the opposite of a rational and scientific approach. When one would go into the field in any State in India, and talk to engineers they would literally speak of 'unscientific water management', referring to the 'political interference' in their work. In the irrigation sector politics is often seen as a nuisance and as undesirable, something to be kept out.

A second reason is that the irrigation sector is a very closed sector. In many countries, particularly the large irrigation countries, it is fully dominated by one discipline: civil engineering. It is not very accessible to non-engineers. Political scientists, who might want to study the politics of irrigation, may not feel very confident to enter into a field that seemingly requires a lot of technical expertise, and may choose sectors where access is easier.

A third reason for the absence of the word politics in the debate is that a lot of the research done on irrigation is funded by government agencies, or depends on the co-operation of governments and government agencies. These are often not too keen on the investigation of their internal workings. Self-evaluation and self-criticism are usually professed to be desirable, but rarely practised. And because politics, as we shall see, is about the mediation of social power and about strategic action, it may not be a popular subject for vested interests. Because most research in the irrigation sector is commissioned by governments and other 'involved' institutions the principle that one doesn't bite the hand that feeds you seems to find some application. In the organisation of the conference for which most papers in this book were commissioned and in the composition of the book itself we also experienced that the level of sensitivity is indeed very high (see Chapter 10).

One of the main intentions of this book follows directly from these observations. A major objective of the book is to make a socio-political perspective on irrigation reform a legitimate subject for discussion.[5]

The Meaning of 'Politics'

Given the importance we attach to politics and the socio-political perspective, it is relevant to explain in some detail what we mean with that word. In colloquial use 'politics' often refers to official politics, that is, state and party politics. This one also finds in the dictionary definition of politics: "the art and science of directing and administering states and other political units" (*New Collins Concise English Dictionary*, 1982). But the description continues as follows. Politics is also "the complex or aggregate of relationships of men in society, especially those relationships involving authority or power", "any activity concerned with the acquisition of power", and "manoeuvres or factors leading up to or influencing (something)". In short, politics in this broader sense is about the mediation of social power, and the strategic action related to that mediation, that is, the process through which the social relations of power are constituted, negotiated, reproduced, transformed or otherwise shaped.

Kerkvliet gives a similar broad definition of politics in relation to resource use. He says that politics is "the debates, conflicts, decisions, and cooperation among individuals, groups and organisations regarding the control, allocation, and use of resources and the values and ideas underlying these activities" (Kerkvliet, 1990:11). Using this broad concept of politics we distinguish four levels.

At the lowest level there is what Kerkvliet has called the 'everyday politics' of resource use. This refers to the daily struggle over access to and use of a resource, in this case irrigation water. The politics of day-to-day water distribution at irrigation system or at the village level provide classical examples of this type of politics.

Grindle has labelled the second level as 'the politics of policy' (Grindle, 1977). This refers to the political nature of policy formulation and implementation processes, policy as contested by different interest groups at all stages of its existence, and with all interest groups trying to shape it in particular ways. This book is devoted to such analysis of irrigation reform policies.

The third level is the level of the 'official' state and party politics. In the case of water resources the inter-state politics regarding water allocation in a shared basin, could also be included. The term 'hydropolitics' has been coined to refer to this, and it is the one level where there is an extensive literature on water and politics (see Waterbury, 1979; Ohlsson, 1995).

The fourth level is what could be called the emerging global politics of water. This refers to institutions, agreements and conventions at global level regarding water resources use. Examples are the Rio and Dublin conferences on water, environment and development held in 1992 where governments have signed documents accepting certain general principles for water resources management, and the subsequent emergence of organisations like the World Water Council and the Global Water Partnership.

It is the second level that particularly interests us in this book: a socio-political perspective on the process of irrigation reform policy formulation and implementation. We want to look at policy formulation and implementation as social processes in which different interest groups try to shape the content and effects of these policies by their strategic action.

Objectives

The book has the following immediate objectives. The first is to make available to the general public detailed accounts of irrigation reform processes from a socio-political perspective. However, the publication is not only a form of information provision. We also hope to overcome some of the resistance to explicit discussion on the politics of irrigation. The legitimization of the topic in public discourse is the second objective. The third objective is agenda setting. Which agenda for further research and analysis can be formulated on the basis of the cases presented, and how can this be undertaken? We thus want to initiate a further research programme on the politics of irrigation reform.

The practical value of more explicit discussion of the politics of irrigation reform can summarised as follows. Firstly, it can help to avoid wastage of development funds by making more realistic assessments of reform models and processes. Such assessments would avoid over-enthusiastic and large-scale investments based on the perceived success of particular 'models'.

Secondly, socio-political analysis can inform the strategic action of reform advocates (and obviously also of those opposing it). It can help to make more grounded choices for courses of action and approaches to take. It is an essential element in a reform process that wants to learn from experience, and develop and improve itself on the basis of that experience.

And thirdly, socio-political analysis of reform processes can be an important source of information for public debate and awareness-raising on the need for reform and on the desired content of that reform.

The broad objective of the book is – thus – to contribute to advancing processes of irrigation reform in the direction of democratic governance and management, efficient and equitable water distribution, and sustainable use of land and water. We also hope that the book will help to increase the number of 'reflective practitioners' (Schön, 1983) in the irrigation sector and will induce more political scientists and policy studies scholars to study irrigation reform.

Main Conclusions

The first conclusion that can be drawn from the case studies presented in this book is that there is sufficient evidence to warrant very careful use of 'model' approaches to irrigation reform. It is shown that in cases that have been declared 'models' like the Philippines, the achievements can be much less than is claimed. Apart from this need for caution, the important points are that both 'success' and 'failure' require explanation, that irrigation reform processes are complex processes in which many transformation mechanisms, issues and interests play a role, and that they have a history and need to be understood in context.

The second main conclusion that the case studies suggest is that for understanding the dynamics of irrigation reform processes a close look at the 'alignment strategies' may be a useful entry point. With the alignment of interests we refer to the process in which different policy actors (government departments, irrigators, politicians, international development agencies and others) negotiate (or fail to do so) the content as well as the process of reform. We suggest that a study of the way social power 'works' in irrigation reform processes should be part of this analysis. We identify three ways in which social power manifests itself: 1) as the social relations of power in negotiation and other forms of interaction in irrigation reform policy formulation and implementation; 2) as discursive power in debates on reform, where it defines the terms of debate and the way the issues are framed; and 3) as the role of politicians in irrigation reform.

We identify three themes for further research, which also summarise some of the main empirical findings of the book: a) the resilience of irrigation bureaucracies in resisting reform; b) the role of international development agencies; and c) the capture and re-shaping of reform policy in the implementation process.

Lastly we draw some conclusions on the politics of research on irrigation reform. We conclude that despite its sensitivity as a topic, research on the politics of irrigation/water sector reform needs to be expanded and strengthened. This would benefit water sector practitioners and policy makers, as well as the academic community. Intellectual interfaces need to be created where reflection on concrete irrigation/water sector reform processes can take place, combining contributions of different types of 'insiders' with those of independent scholars.

Structure

The book contains eight case studies of irrigation reform processes, situated in seven different countries in Asia, Southern Africa and Latin America. The inclusion in the book is in an approximate order going from a focus on policy formulation to a focus on policy implementation. We thus follow the journey of policies from 'top' to 'bottom'. But the chapters can be read independently, and in any order.

Chapter 2 starts the journey with an account of the public consultation process that took place in South Africa as part of the formulation of a new water law and irrigation policy. It is a fascinating example that shows that productive consultation processes are possible in highly polarized and politicized situations, and partly because of that politicization.

Chapter 3 is a detailed account of the history of the Mexican irrigation reform process that culminated in the early 1990s with massive turnover of irrigation systems to user organisations. Contrary to common assumptions this was not a sudden process triggered by financial crisis of the state, but a much longer and circumlocutory process revolving around the autonomy of the irrigation agency.

Chapter 4 is the first of two chapters on the Philippines, the first and most famous of the irrigation reform models. It documents the trajectory of the National Irrigation Administration irrigation reform exercise, outlines the role of international development funding agencies in it, and shows that, like in the Mexican case, the maintenance of institutional autonomy and construction-orientation is a major motivation in the strategic action of irrigation agencies.

Chapter 5 looks at the 1987-1999 period of irrigation reform in Indonesia. It provides a good case study of how an irrigation bureaucracy was able to transform an irrigation development programme with institutional reform conditionalities attached, into a conventional 'hardware' oriented programme. It also discusses the new opportunities and initiatives for irrigation reform after the end of the authoritarian Suharto regime.

Chapter 6 takes us to Zimbabwe, where one of the issues is the disjunction between policy discourse and what happens on the ground. There has been a lot of debate on irrigation reform policy, but no such policy has been consolidated. At the same time the financial bankruptcy of the state has induced 'reform' processes at field level.

Chapter 7 reports the efforts of particular international development funding agencies to induce or enforce an institutional reform process in the irrigation sector in Pakistan in the 1990s. It analyses, among other things, how the institutional structure of the irrigation sector and the resistances it produces, contributes to the extreme tardiness of the reform.

Chapter 8 is a case study of the much acclaimed irrigation reform in Andhra Pradesh, India. It shows how, in one system, the policy is captured and transformed during implementation by the local rural elite and irrigation agency staff.

Chapter 9 is a return to the Philippines. It presents a case study of the politics of policy implementation in a large irrigation system. It shows that both 'success' and 'failure' of policy interventions can only be understood by analysing their embeddedness in local, regional and national political processes. It argues the case for more process documentation research as part of reform exercises.

Chapter 10 sums up the themes and issues raised by the preceding chapters and outlines an agenda for further research on the politics of irrigation reform.

Not all papers presented at the workshop in December 1999 are included. We included the papers focussing on the *process* dimensions of irrigation reform, and not included those focussing on the *impact* of reform processes. Not all papers we wanted to include are in the book because some authors unfortunately were unable to submit a revised version of their paper. Two papers, chapters 8 and 9 were not presented at the workshop, as they were still unfinished at the time it was held.

Notes

1   The major regions missing are that of the former Soviet Union and Eastern Europe, and the People's Republic of China where reform processes from highly centralised state controlled irrigation to more decentralised and user-controlled systems are ongoing.
2   There are many reviews available of the evolution of the irrigation management debate. An authoritative one is Chambers (1988). Also see Merrey (1997).
3   A more detailed discussion of the meaning of 'model' as used here is provided in the concluding chapter.
4   That in the rural development sector promises of 'success' can co-exist with repeated 'failure' in implementation, and that there may be a symbiotic relation between the two, is discussed in Ferguson (1994).
5   There are some shifts happening in the water policy discourse that may allow more explicit engagement with the political dimensions of irrigation reform to become easier. One is the increasing emphasis on governance. A second is the increasing interest in the issue of allocation (in addition to management or distribution). These concepts lead to questions on representation, legitimacy, accountability, rights, decision-making procedures and the like. These can also be understood in non-political ways, but a socio-political perspective may be found acceptable more easily as one possible approach than in the 'technical' domains of management and operation.

References

Chambers, Robert (1988), *Managing Canal Irrigation. Practical Analysis from South Asia*, Oxford and IBH, New Delhi.
Ferguson, James (1994), *The Anti-Politics Machine. 'Development', Depoliticization and Bureaucratic Power in Lesotho*, University of Minnesota Press, Minneapolis and London (orig. Cambridge University Press, 1990).
Grindle, Merilee S. (1977), *Bureaucrats, Politicians and Peasants in Mexico. A Case study in Public Policy*, University of California Press, Berkeley.
Ishikawa, Shigeru (1967), *Economic Development in Asian Perspective*, Economic Research Series No. 8, The Institute of Economic Research, Hitotsubashi University, Kinokuniya Bookstore, Tokyo.
Kerkvliet, Benedict J. (1990), *Everyday Politics in the Philippines. Class and Status Relations in a Central Luzon Village*, University of California Press, Berkeley.
Merrey, Douglas J. (1997), *Expanding the Frontiers of Irrigation Management Research. Results of Research and Development at the International Irrigation Management Institute 1984 to 1995*, IIMI, Colombo.
Ohlsson, L. (1995), *Hydropolitics: Conflicts over Water as a Development Constraint*, Zed Books, London.
Ostrom, Elinor (1990), *Governing the Commons. The Evolution of Institutions for Collective Action*, Cambridge University Press, New York.
Ostrom, Elinor (1992), *Crafting Institutions for Self-Governing Irrigation Systems*, Institute for Contemporary Studies Press, San Francisco.
Schön, Donald A. (1983), *The Reflective Practitioner. How Professionals Think in Action*, Basic Books, New York.
Stone, Ian (1984), *Canal Irrigation in British India. Perspectives on Technological Change in a Peasant Economy*, Cambridge University Press, Cambridge.
Waterbury, J. (1979), *Hydropolitics of the Nile valley*, Syracuse University Press, New York.
Wittfogel, Karl (1957), *Oriental Despotism. A Comparative Study of Social Power*, Vintage Books, New York (1981 edition).

Chapter 2

Water Policy and Law Review Process in South Africa with a Focus on the Agricultural Sector

Marna de Lange

Introduction

This paper is about process. It seems like only yesterday when South Africa started to become aware of the importance of the 'process', and not just the 'product', and now this has become a science in itself. I share with you something of the experience with the major policy and legislative reform process in South Africa in our transition to democracy, and refer to content issues mainly where it serves to explain the process and its outcomes.

The paper reflects my own viewpoint from participation in two parallel and related processes: the water law review and irrigation policy development. I was mainly involved with the agricultural sector and was responsible for the irrigation policy consultation process as a consultant to the Department of Water Affairs and Forestry and the National Department of Agriculture.

I had to ensure interaction between the irrigation policy process and the drafting of the National Water Bill, mainly through regular participation in the meetings of the Water Law Drafting Team. Late in the process I had the unique opportunity to participate in a meeting between the Minister and his officials, when the near-final Bill was discussed line-by-line and some significant changes were introduced. However, I never participated in the parliamentary processes that followed after the final National Water Bill was produced. Information on this part of the process I extracted from the minutes and proceedings of the parliamentary committees.[1]

So naturally, there will be many other views from people who observed the process from other vantage points. Mine remains only one viewpoint of an outsider to the Department of Water Affairs and Forestry (DWAF) and it is certainly not intended to reflect any official stand. Wherever I am critical, it is with the advantage of hindsight and as much a criticism of myself as it may be of anyone else who was involved. Any such criticism is in the interest of doing it better next time, which is now, in the ongoing democratisation of South Africa.

The paper starts with an overview of water issues in South Africa. It discusses how the development of South Africa's National Constitution set the scene for comprehensive water law reform with substantial public participation.

The major driving forces in the water policy reform range from a real need to prepare the country for increasing water scarcity and the political imperative to redress the results of past racial and gender discrimination. The role that some key individuals and groups played in the review process is discussed next, followed by an analysis of the irrigation sector and its main concerns about the water law review. The institutional framework and some elements of the review process is described and finally the water law review is analysed according to some of the questions often raised.

Water Issues in South Africa

The fact that South Africa had a policy of *apartheid* or separate development is well known. Less well appreciated is the dualistic nature of both the society and its economy. While South Africa is ranked as a middle-income country, 40% of its rural population lives in poverty. In 1994, out of a population of 40 million, 12 million did not have access to basic water requirements.

The irrigation sector accounts for half of South Africa's water use on 1.3 million hectares of irrigated land and produces 25% of the agricultural output on 10% of the cultivated land. Approximately one third of the irrigated area is planted to high value horticultural crops, much of which is exported.

South Africa has the largest economy in the SADC[2] region, but agriculture contributes only about 5% to the GDP compared to up to 80% in the economies of some of the other SADC countries. South Africa has about 80% of the total irrigation development in SADC, but has virtually reached its ceiling. There is only water left for a further 200,000 hectare of irrigation development, and only in some parts of the country. However, South Africa represents less than 10% of the total irrigation potential in the SADC region (DWAF, 1997).

South Africa is a water scarce country, and expected to be among the most water scarce countries in the world by 2025.[3] One of the main characteristics of the region, and particularly of South Africa, is its highly erratic rainfall, evident in the unpredictable cycles of droughts and floods. This has resulted in an emphasis on water resource development and particularly dam building. South Africa's water resources are almost fully developed, so that the need for trade-offs and transfers between water user sectors is becoming more and more of a reality.

Approximately 15% of the water use comes from groundwater resources, and the balance from surface water. Water quality issues related to the highly developed and growing industrial and mining sectors are on the increase.

The country's electricity supply is generated mainly by coal-fired power stations, which are supplied by opencast coalmines. The coal mining operations in the upper Olifants river basin pollutes the water to such an extent that clean water is imported from neighbouring river basins for the cooling towers of the power stations. Hydropower is generated mainly to supply in peak demands, and this often leads to conflict with irrigation users. The water releases for power generation often don't coincide with irrigation needs, so water is lost from storage and becomes inaccessible for irrigation and other uses.

Box 2.1 Extract from the National Constitution: the process

Constitution of the Republic of South Africa 1996

As adopted on 8 May 1996 and amended
on 11 October 1996 by the Constitutional Assembly

One law for One nation

Act 108 of 1996 ISBN 0-620-20214-9

EXPLANATORY MEMORANDUM

This Constitution was drafted in terms of Chapter 5 of the interim Constitution (Act 200 of 1993) and was first adopted by the Constitutional Assembly on 8 May 1996. In terms of a judgement of the Constitutional Court, delivered on 6 September 1996, the text was referred back to the Constitutional Assembly for reconsideration. The text was accordingly amended to comply with the Constitutional Principles contained in Schedule 4 of the interim Constitution. It was signed into law on 10 December 1996.

The objective in this process was to ensure that the final Constitution is legitimate, credible and accepted by all South Africans.

To this extent, the process of drafting the Constitution involved many South Africans in the largest public participation programme ever carried out in South Africa. After nearly two years of intensive consultations, political parties represented in the Constitutional Assembly negotiated the formulations contained in this text, which are an integration of ideas from ordinary citizens, civil society and political parties represented in and outside of the Constitutional Assembly.

This Constitution therefore represents the collective wisdom of the South African people and has been arrived at by general agreement.

South Africa's National Constitution: Setting the Scene

Who ever thought peace could be so stressful? And exciting! And intensely interesting. When Nelson Mandela was released in 1992 after 27 years as a political prisoner, South Africans said we were 'over the waterfall' on a journey that could never reverse its tracks. Middle class whites had too much to lose to seriously consider taking up the weapon. The general public heaved a hesitant sigh of relief when one-by-one the major players agreed to embark on a negotiated process of transformation, rather than the otherwise inevitable route to armed conflict. This was closely followed by, and given content through the joint development of our National Constitution.

The first democratic elections in April 1994 resulted in a Government of National Unity, through which the major political parties co-ruled. The process of development of a National Constitution deliberately went broader than the

Government of National Unity to ensure a wider support base than the current ruling parties. The objective was to establish a Constitution that could outlive party politics and short-term governance changes.

Did this strategy work? Although it is early days, at least two anecdotes seem to confirm some measure of success.

Firstly, there were no major changes to the constitution when we entered the second five-year term of democratic governance in June 1999, even though the ruling ANC-party had won a two-third majority and thus one-party governance.

Secondly, and perhaps more significant, is that people across the spectrum, including the commercial farming sector and other major economic sectors in the country, clearly recognise the National Constitution as the highest institution in our legal system. In policy and legal reform processes, new institutions are frequently tested for their 'constitutionality'. If any sector is convinced that a particular provision is unconstitutional and they fail to achieve the desired changes through negotiation, petition or other alternatives, they can and will take the matter to the Constitutional Court.

The process of drafting the National Constitution and subsequent public consultation processes had some outcomes quite apart from the actual 'products'.

(a) The inclusive approach to drafting the Constitution (see Box 2.1 above) created a precedent for wide public consultation, expert input and international participation in policy and legal development processes. Previously, policy development was a very closed and official affair with limited, if any, public input, except through direct lobbying of politicians. In the words of a senior manager in the 'old' department in the early days of the new regime: 'we could never tolerate a situation where consultants develop policy on behalf of a department'. A few months later, consultants were employed in the formulation of almost all policy processes, although of course the final decisions remain the departments' prerogative.

(b) The very limited professional and social contact between races during the apartheid years created deep-seated stereotypes in the minds of people across the spectrum. The Constitutional development process brought individuals from across the deeply divided political and social spectrum together, where they had to interact very closely to come to an understanding of each other's points of departure and negotiate an acceptable middle road. Through this process, many professional relationships and even personal friendships were forged which would have been unthinkable under the previous regime.

(c) The regular media coverage of the issues debated in drafting the Constitution brought the debate and new perceptions right into the homes of ordinary citizens. This, together with coverage of other processes, such as the hearings of the Truth and Reconciliation Commission, helped to establish a whole new social consciousness and even conscience in the country. Stereotypes were challenged and norms re-evaluated. Ordinary middle class people became more aware of the widespread poverty and inequity, and became critical of the policy environment that had created these imbalances.

At the one end of the spectrum, access to resources to satisfy basic human needs is protected in the Constitution, including the right to clean water and sanitation (see Box 2.2, section 27). This was of particular relevance to those 12 million people who did not have access to the basic minimum standard of 25 litres of clean water per person per day, within 200m from the homestead. Recently, people's right to water for basic economic activity has been raised in the department (DWAF, unpublished minutes of meeting, March 2000).[4] This is important for strengthening the position of black smallholders in negotiations for water allocations.

Box 2.2 Extracts from the Constitution: clauses of particular significance for water legislation

---

(*italics* text indicate the author's emphasis)

From Chapter 3: Founding Provisions

CITIZENSHIP

3. (1) There is a common South African citizenship.
   (2) All citizens are -
   a. equally entitled to the rights, privileges and benefits of citizenship; and
   b. equally subject to the duties and responsibilities of citizenship.

From Chapter 2: Bill of Rights

ENVIRONMENT

24. Everyone has the right -
    - to an environment that is not harmful to their health or well-being; and
    - to *have the environment protected*, for the benefit of present and future generations, through reasonable legislative and other measures that -
        - prevent pollution and ecological degradation;
        - promote conservation; and
        - secure ecologically sustainable development and use of natural resources while promoting justifiable economic and social development.

PROPERTY

25. (1) No one may be *deprived of property* except in terms of law of general application, and no law may permit arbitrary deprivation of property.
    (2) Property may be expropriated only in terms of law of general application -
    - for a public purpose or in the public interest; and

> - *subject to compensation*, the amount of which and the time and manner of payment of which have either been agreed to by those affected or decided or approved by a court.
>
> (3) *The amount of the compensation and the time and manner of payment must be just and equitable, reflecting an equitable balance between the public interest and the interests of those affected...*
>
> HEALTH CARE, FOOD, WATER AND SOCIAL SECURITY
> 26. (1) Everyone has the right to have access to -
> - health care services, including reproductive health care;
> - *sufficient food and water*; and
> - social security, including, if they are unable to support themselves and their dependants, appropriate social assistance.
>
> (2) The state must take reasonable legislative and other measures, within its available resources, to achieve the progressive realisation of each of these rights.
>
> ACCESS TO INFORMATION
> 27. (1) Everyone has the right of access to -
> - *any information held by the state*; and
> - any information that is held by another person and that is required for the exercise or protection of any rights.
>
> (2) National legislation must be enacted to give effect to this right, and may provide for reasonable measures to alleviate the administrative and financial burden on the state.
>
> JUST ADMINISTRATIVE ACTION
> 28. (1) Everyone has the right to *administrative action that is lawful, reasonable and procedurally fair*.
>
> (2) Everyone whose rights have been adversely affected by administrative action has the right to be given *written reasons*.
>
> (3) National legislation must be enacted to give effect to these rights...
>
> INTERPRETATION OF BILL OF RIGHTS
> 29. (1) When interpreting the Bill of Rights, a court, tribunal or forum -
> - *must promote the values that underlie an open and democratic society based on human dignity, equality and freedom;* ...

Turning to content, the Constitutional principles of *equity* (see Box 2.2, section 3: Citizenship and section 39) and *environmental sustainability* (see Box 2.2, section 24, Environment) are of particular relevance to water management. This is succinctly captured in the slogan of the Department of Water Affairs and Forestry: 'some, for all, forever'. South Africa is a water scarce country with little potential for further water resource development and the Minister challenged his officials to 'vindicate' and 'give content to' (his words) the principles of the National Constitution through the review of the water law.

At the other end of the spectrum, the economic backbone of the country is dependent on continued access to water for its economic activities. Virtually all mining and industrial activities depend on water in their processes. Commercial irrigation farming, which accounts for more than half the national water use in South Africa, felt particularly vulnerable. To these sectors, the Constitution and the National Water Act ensures compensation (see Box 2.2, section 25, Property) if the economic viability of their enterprises is seriously jeopardised by resource reallocation. However, small general reductions across a sector to enable water transfer to more beneficial use in the public interest will not attract compensation.[5]

Major Driving Forces in the Water Policy Reform

The forces at play in the water law review can be summarised as political drive, visionary leadership, official irritation (with the old Water Act), scientific joy and intensive public participation.

The new political order saw a tremendous shift in the power base and an opportunity for major reform in South Africa. The successful exploitation of this opportunity in different sectors depended heavily on the vision, drive and political acceptability of the new Minister, as well as the support he or she was able to elicit from the responsible department. One has to bear in mind that the departmental staff at the start of the reform process, were the same staff that previously served under the Apartheid government. Among these staff there were some who were delighted about the new direction, and some who were bitterly opposed to it. The majority was unsure about their future and the security of their careers.

Comprehensive reform in the water sector was made possible by the broader context of reform and by the individuals who were responsible for the process.

- The new *Minister*, Professor Kader Asmal, was a visionary leader, with all the drive and energy necessary for major transformation of a very large organisation, the Department of Water Affairs and Forestry. The new Minister had a legal background and was a respected academic. In the beginning, the general public wondered somewhat about his political position, being an Indian and not an African, but these speculations soon dissipated when his reform initiatives started showing results. His achievements in the water sector enjoyed wide international recognition, including his appointment as Chairperson of the World Commission on Dams. Nationally, he was re-appointed as cabinet minister in 1999 for a second term of office. He was given the most troublesome portfolio to 'sort out', namely Education. He is a hard taskmaster whose focus remained on the process, as was evident in his ability to deal with disrupting forces at both ends of the political spectrum.
- In the Department of Water Affairs and Forestry, *senior officials* had been discussing the need for review of the Water Act (Act 56 of 1956) for many years, partly because it was difficult and expensive to administer. Water

legislation was reviewed in 1956 to reflect the shift in water use patterns since the start of the century, when the Irrigation Act of 1918 was still relevant in an agriculture-dominated economy. Since 1956 a further shift had occurred towards industrial uses. While irrigation had grown steadily, agriculture's share of the national Gross Domestic Product had shrunk to about 5% by 1995. These shifts, together with increasing water scarcity and a new political impetus for more equitable water distribution now again required a review of water related legislation. This common focus, albeit respectively from administrative and political perspectives, made for a workable marriage between the department and its new Minister.

- Another significant force, both during the reform and current implementation of water policy, is the *Director General* or chief executive officer of the department, Mike Muller, who was a powerful 'new broom' at the time. He came into the department in a relatively junior position early in 1994 to establish a new sub-division responsible for community water supply and sanitation, which was previously not a mandate of the department. This sub-division was quickly expanded into one of the major divisions in the department and Muller was soon appointed as Director-General of the department. This was part of a comprehensive transformation of the department, from an almost purely white male engineering concern to a much broader variety of disciplines, cultures and gender. Today the department has several women in senior and middle management positions and employs environmentalists, social scientists and community workers to address its broader mandate of integrated water management. Muller's greatest strength probably lies in his ability to bring pragmatism into the translation of political objectives into practical implementation. He commands the respect of his officials with vision, leadership and managerial ability; he has the intellectual capacity to understand the requirements, strengths and weaknesses of the various water sectors; and interacts successfully on the political interface.

- The review process was characterised by an explosion of creative energy, as *scientists and professionals* from a wide range of disciplines and institutions participated in the debates. Individuals were stimulated by the intellectual satisfaction stemming from a real opportunity to influence significant implementation of their convictions. For instance, after years of closed-loop arguments:
  - Groundwater specialists were now taken seriously when they argued that groundwater should legally be treated exactly the same as surface water;
  - Environmentalists were listened to when they argued that the integrity of the resource depended on the protection of the biota and their habitat in riverine ecosystems;
  - Development professionals now hardly had to point out that droughts tend to hit the poor hardest, with little legal recourse; and
  - Commercial farmers and organisational experts found a receptive environment for ideas of decentralised water management through catchment agencies.

- *The public*, in accordance with the needs of each water use sector, also sought to influence the review. The water law review must have been one of the most emotive processes in the new South Africa. The stakes were especially high for the communities without proper access to domestic water, and for the commercial irrigation sector that believed they stood to lose their livelihoods. The political transition shifted the historical balance of forces, and those who traditionally had access to power suddenly found their lobbying power almost completely eroded. On the other hand, traditionally excluded communities, such as emerging black farmers, suddenly found themselves able to give voice to their concerns and aspirations.
    - To commercial irrigation and livestock farmers, there was a direct threat to their livelihoods, in a totally changed political environment where they expected the worst, but were unsure of the shape the bad news would take. In contrast to the regular media coverage of instances of bad blood between commercial farmers and their labourers, the water law review saw farm labourers join their employers in a groundswell of fear and lobbying for sustained livelihoods.
    - The economic sectors, including mining, industry, commercial agriculture, the banking sector, the electricity supplier and Business South Africa were nervous about the potential implications for water pricing and security of investments.
    - Historically disadvantaged communities watched with bated breath to see if this new government would take their plight for daily drinking water seriously. For them it was difficult to accept that 'one can't drink a right',[6] yet that the policy and law review was important to create an environment in which their needs could be addressed.

By the time the National Water Bill was presented in Parliament, after nearly three years of consultation and debate, most sectors expressed support for the broad framework and several aspects of the National Water Bill. The following extracts from the public submissions reflect the outstanding issues according to the various sectors:

COSATU (Council of South African Trade Unions – the federated labour union in South Africa) and SAMWU (South African Municipal Workers' Union) 'supported the fact that the legislation provided for:

- A framework in which it will be possible for everyone to have access to a basic amount of water;
- A framework for the pursuit of social and economic development goals;
- Mechanisms to redress past discriminations and inequalities;
- The promotion of environmental sustainability; and
- The meeting of South Africa's international obligations, particularly our human rights obligations.

COSATU and SAMWU argued for wider requirements for consultation in the

Bill and an effective oversight role for Parliament, because the enabling nature of the legislation implied that much of the detail would be determined during implementation. They stressed the importance of ensuring adequate representation of disempowered sectors in Catchment Management Agencies and sufficient links to institutions under the Water Services Act. They opposed the tradability of water allocations, since they felt it would result in profiteering at the expense of disadvantaged communities.

The *Rural Development Network Services* requested clarification on the calculation of the reserve for basic needs, tariffs and pricing policy, ensuring adequate representation and limiting the powers of Catchment Management Agencies. They argued that inadequate provision had been made to redress gender inequalities and cautioned that public-private partnerships could be to the detriment of disadvantaged communities in the long term.

COSAB (Council of South African Banks) cautioned that provisions with intensive administrative requirements could exceed the resources and capabilities of the government and argued for an extended period for public comment. They strongly recommended that the concept of fixed term water use licences be dropped, and that licences in perpetuity be granted, because of the expected impact of this Bill on property values, in particular relating to land values pledged as security for loans. They also argued for compensation for any financial losses suffered and for free market trade in water allocations.

Business South Africa said that their comments were submitted 'not in an attempt to entrench past privilege' and accepted the time limit on licences, but insisted that the tradability of entitlements as foreseen by the Water Law Principles were not adequately provided for. Senior counsel had advised them that the constitutional court might find the failure to provide compensation for the taking of rights unconstitutional. They also argued for increased requirements for consultation with stakeholders.

Rand Water, the water supplier to the economic heartland of the country around Johannesburg, said that many important aspects of Rand Water's previous misgivings had been addressed in that the Minister's powers had been tempered by processes of consideration of comments by stakeholders and participation of persons affected by decisions. They wanted the principles addressed which underlie expropriation without compensation, the determination of existing lawful resource use, water allocation, storage, abstraction and licensing and the role and representation of a National Water Utility and Catchment Management Agencies.

The *South African Association of Water Boards* said provisions in the Bill should 'cut red tape by deregulation as far as possible; put the customer first by the establishment of participatory structures and offering the necessary choices; empower employees by means of appropriate tools and adequate decision making powers; and return to the basic of business i.e. minimise wastage, streamline workflow and to produce more with less'. They questioned the role of a National Water Utility, and saw a process of centralisation versus the commitment to subsidiarity in the Water Law Principles. They were concerned about the proposed notice period in terms of water use authorisations.

The *Civil Engineering Institute of South Africa* was concerned that the Bill, through lack of funds and personnel, may not be implemented properly and stressed the need for general authorisations for limited water use, without the need for licensing. Furthermore, the Department should ensure that the structures are in place to deal with applications promptly and efficiently. They argued for consultation before a national and catchment strategies were finalised and recommended that the Bill should be reconciled with other legislation, such as the Environmental Act. Funds generated from water use charges should only be used for the purpose for which they were levied. The CMA Board should be elected by, and accountable to, their constituency. They also wanted clarity on how CMAs would interact with each other, as well as other agencies/bodies.

Cape Water Programme at the Council for Scientific and Industrial Research argued that associated ecosystems should be protected in the ecological reserve. For instance, the groundwater requirements of vegetation in semi-arid areas, would have to be accounted for before issuing abstraction licenses, otherwise these critical ecosystems may be placed under stress and ultimately lost.

UCT Environmental Law Unit questioned several definitions in the National Water Bill.

Forest Industries Association said that the fundamental concerns that they had had with the drafts to date were 'the divorcing of any rights to use water from property ownership without any form of compensation being payable should water use rights be withdrawn, curtailed or in any way affected; the issuing of water licences for a limited period, up to a maximum of 40 years; the imposition of a "resource conservation charge" or a "run-off reduction levy" (i.e. a rainfall tax)'.

The *National Forestry Advisory Council* was concerned about the limited time period of licences and the lack of a minimum notice period. Forestry was classified as a 'stream flow reduction activity' (SFRA) and therefore a water use that could be charged for in a future pricing policy for water use.

South African Sugar Association was also concerned about the provision for 'stream flow reduction activities' as sugar plantation could be declared as such a water user in future, with the same implications for water charges. They wanted compensation and tradable water allocations and wanted to ensure that proceeds from water use charges remained within the relevant CMA. They wanted the pricing policy to include a charge for 'capacity building' with respect to the CMAs

Other comments from forestry and agriculture included the following.

- Greytown and District Chamber of Commerce saying that 'to restrict forestry to increase the run off of water into the sea is ludicrous';
- Kwa-Mbonambi Farmers and Timber Growers Association who wished 'to record our strongest objection to the New Bill', the
- South African Wood Preservers Association saying that 'term licences for the growing of timber will adversely affect our economic sector'
- Ugu Regional Council stating that 'the proposed Bill is restrictive to the agricultural sector upon which a substantial portion of our regional economy is dependent'; and

- Umvoti Agricultural Society said that they 'strongly object to the provisions made in the draft water bill' because 'Water use licences and the restrictive duration thereof will have the effect of de-basing assets and will lead to a lack of confidence in agriculture. The unlimited powers conferred on the Minister who need not even refer to parliament. A rainfall tax in the guise of a stream flow reduction levy goes totally against internationally accepted norms and standards. That rights to use water will be divorced from property ownership without payment of compensation is a fundamental concern. Those rights were paid for with the purchase of the land. Forestry is being singled out as a stream flow reduction activity'.

The submission of the South African Agricultural Union is not available on the parliamentary website and there was no submission by the small-scale irrigation sector (see the next section).

The Irrigation Sector and its Main Concerns about the Water Law Review

The agricultural sector has at least two very distinct sub-sectors, the 'commercial sector' and the 'small-scale or emerging farmer sector'. About 25,000 commercial irrigation farmers produce most of the agricultural export and generate about 25% of South Africa's agricultural output on 10% of the land under cultivation. They have access to 95% of the irrigation water used in the country. The small-scale irrigation sector has a much larger, but unknown number of farmers. This sub-sector uses only about 5% of the irrigation water and consists mainly of government-initiated irrigation schemes that are plagued by low levels of productivity, unsuitable organisational designs and under-utilisation. The present government views this under-utilised infrastructure and human capital as a major asset for poverty eradication and rural economic development.

The structure of these two sub-sectors and their respective major concerns regarding the water law is discussed in more detail below.

Structure and Concerns of the Commercial Irrigation Sector

The *commercial sector*, also known as 'organised agriculture', represents the predominantly white medium and large-scale farmers. They are very well organised geographically in about 1400 farmers associations, which are federated in district agricultural unions, under the umbrella of the national South African Agricultural Union (SAAU).[7] SAAU represents about 60,000 commercial farmers, including irrigation, dry land and livestock farmers.

The commercial irrigation farmers are represented nationally through SAAU's Irrigation and Water Affairs Committee, with membership from across the country. Further, several of the affiliated commodity organisations represent an irrigation interest, including the Potato Producers Organisation, Cane Growers Association, the various fruit growers' organisations, etc.

Commercial irrigation farmers can be roughly divided in terms of the nature of their access to water as follows.

- About 30% of the irrigated area in the country is supplied from government water schemes, which were mainly developed as poverty relief schemes after the depression and the World Wars in the first half of the 20th century.
- Another 30% of the irrigated area is supplied from Irrigation Board water supply schemes. These are essentially Water User Associations with farmer-owned and managed canals, pipelines and pumping schemes. There are around 400 of these private Irrigation Boards, which also played a significant role in voicing the concerns of commercial irrigation farmers in the review.
- For the remaining 40% of the irrigated area, water is abstracted individually from rivers and groundwater.

Water legislation in South Africa was previously based on Roman-Dutch principles, including secure water rights. Water rights were attached to land and were valid in perpetuity. Riparian farms had an automatic claim on water in the river. Many of these rights were never exercised, but were so strongly protected that they could not be reallocated for use by others. A distinction between 'private' and 'public' water and 'normal flow' and 'storm flow' created grey areas that often ended up in costly legal wrangles in the Water Court.

Commercial irrigation farmers and their organisations, as well as those supplying services to the irrigation sector, were concerned that the viability of the farms would be adversely affected by the proposed radical changes to the water law. The sector used different opportunities to voice their concerns: at workshops, in meetings with the Minister, through media releases. Some contributions were spontaneous individual reactions, while some were carefully planned and agreed, especially prior to major events. Two irrigation regions commissioned independent studies on the role of irrigation in the local economy, including job creation and backward and forward linkages to agriculture based industries. Some farmers provided transport to enable their labourers to attend workshops and raise concerns about the security of their employment. In some provinces the district agricultural union staged walkouts from public meetings to express their dissatisfaction.

The tactic and position of the commercial irrigation sector shifted significantly in the course of the water law review process. In the beginning there was no real debate, but reactions were rather in the form of an outcry and refusal to contemplate the possibility of such radical changes. There were even arguments against people's right to access for basic human needs. Later the debates became more sophisticated and there were some concrete proposals for more acceptable alternatives to proposed changes. However, when the SAAU presented their public submission to the Agriculture, Water Affairs and Forestry Portfolio Committee (AWFPC) in Parliament, near the end of the process, they seem to have reverted to earlier approaches. Parliamentarians asked them how they could take a position of 'demanding' and 'no compromise' even so late in the process and did not come with proposals for solutions. On the other hand, the South African Sugar Association, in their submission, complemented the Department on the 'progress which had clearly come about as a result of the consultation process...' (AWFPC minutes, May 1998).[8]

In addition to the compensation issue discussed before, the following proposed changes were of most concern to the commercial irrigation sector.[9]

- There would no longer be private rights to water and water allocations would be divorced from land ownership. Instead, water would belong to the nation and be held in custody by the Minister of Water Affairs and Forestry. Access to water would be regulated through time-bound authorisations to legal persona to use water for a specific purpose on a specific property. These allocations would be reviewed at least every five years. Conditions could be attached to these water allocations, requiring users to supply information or comply with specific restrictions.
    - In a water scarce country like South Africa, the value of a farm is closely related to its secure and legal access to water. Commercial farmers feared that the value of their property would plummet, and that they would no longer be able to negotiate production loans from banking institutions on the normal basis of land as collateral. Commercial farmers and banking institutions argued that a time-bound allocation provided insufficient security for investment, especially for long-term high-value enterprises, like export table grape production, which carries very high start-up costs. The department pointed out that Australia had licences with a 15-year term, while the proposal in South Africa was to allow up to 40 years (Minister's speech, Parliament, 9 Feb 1998).[10] While this topic was hotly debated throughout the process, the Director-General was convinced that, once implemented, water users and financial institutions would get used to the concept and that systems would adapt accordingly.
    - More directly, commercial farmers feared that they would no longer have access to the water they depended on to earn a living. Painfully aware of their weakened political bargaining power, they were particularly concerned about who would decide on these allocations. They had little reason to believe that the new regime would be sympathetic to their case and feared that incompetence would hamper the administrative procedures required to assess, allocate and review allocations. Another major concern was the potential that an administrative allocation system could create for corruptive practices.
- Decentralised and integrated water resource management in natural watersheds, river basins or catchments would replace the previous system of centralised water resource management by the Department of Water Affairs and Forestry head-office in Pretoria. New catchment management institutions, incorporating public participation, would be established to progressively take over the functions of the department, including the allocation of water use authorisations.

Box 2.3 Some differences between the first Draft Bill (May 1997) and the final National Water Act (Act 36 of 1998) of relevance to the commercial irrigation sector

Some of the provisions and conditions in the early drafts of the Bill were totally unacceptable to the commercial farming sector and were removed as a result of the consultation process.

Firstly, a water allocation would expire at the death of the person in whose name it was issued. This was untenable, since it would directly jeopardise the continuity of the farming practice, the estate of the deceased and the ability of his or her family to continue to make a living. The case was won, partly by the argument that while commercial farmers may be sophisticated enough to counter the effect of this provision by applying for allocations through a business entity or other ongoing legal persona, this option was inaccessible to most small-scale farmers.

Secondly, two of the conditions that could be attached to a water allocation, stipulated that government could both *prohibit* the growing of certain crops and *prescribe* the cultivation of specific crops. Farmers argued that these measures were draconian and that market forces and not government should dictate these decisions.

Thirdly, according to the early drafts, the Minister would unilaterally appoint members to the CMA Board from different user sectors. This was replaced by a system of nomination by the different sectors of their own candidates, from which the Minister would appoint Board members, or ask for alternative nominations.

- There was almost unanimous support by all sectors for the concept of catchment management. Managing water according to its natural flow and occurrence made eminent sense to environmentalists. It provided a mechanism for the department to iron out the major fragmentation in water management that developed during the apartheid period, when every homeland and province had its own sets of policies and approaches. Commercial farmers saw the decentralisation of water management, and particularly the decisions regarding water allocations, as an important mechanism to exert more influence over decision-making that would impact on their livelihoods so directly.
- Commercial farmers wanted to ensure that they had adequate influence in the proposed catchment management agencies (CMAs), while the department tried to prevent that any one sector would dominate these agencies. The agricultural sector are naturally the largest water user in almost every river basin, and generally very well organised and competent in certain elements of water management. Some of the larger Irrigation Boards could see themselves virtually becoming the CMA, but this was actively resisted by the department, who wanted to ensure an inclusive

decision-making body, representative of all concerns in a particular catchment (see National Water Act, section 81).

Box 2.4 Some differences between the first Draft Bill (May 1997) and the final National Water Act (Act 36 of 1998) of particular relevance to the small-scale irrigation sector

> Two issues related to their access to water and the legal position of small-scale farmers were raised by development professionals and later taken up by NAFU.
>
> *Firstly,* in the early drafts, water allocations would be issued to the landowner. It was pointed out that this would compound the difficulties faced by women farmers in tribal areas, where land allocations were mainly issued to men, although the women are in fact the farmers. The National Water Act now provides that water is allocated to the water user, not the landowner (National Water Act, section 28).
>
> *Secondly,* researchers had suggested that the development of strong independent small-scale irrigation was hampered by the lack of access to a legal institution through which small-scale irrigators could manage their own affairs, like the white Irrigation Boards (De Lange, 1994). Again because of the requirement for title deed in the previous legislation, black farmers could not become members of Irrigation Boards. The necessary changes were introduced through several steps:
>
> The Minister was initially of the opinion that the establishment of CMAs would provide adequate institutional capacity for decentralised management, and wanted to remove the Irrigation Boards completely from the statute. However, he was convinced by his officials and others of the value of an institution which would enable day-to-day management-by-the-users at a much more localised level than the catchments, some of which are more than 50,000 km^2 in extent (DWAF, Minister's speech, National Agricultural Workshop on Water Law, September 1997).
>
> It was decided to replace the Irrigation Board legislation with a new institution, Water User Associations, which would remove all discriminatory provisions. Existing Irrigation Boards would be transformed to become more inclusive and smallholders would also be able to establish their own WUAs (National Water Act, Part 8 and Schedule 5).
>
> In this context, the provision that water would be allocated to the water user means that women can now also become the members of, or form their own Water User Associations.

- The new water policy argued for greater emphasis on water demand management as an alternative to a singular strategy of supply-driven water resource development. Further, it argued that the private sector, rather than government, should increasingly fund water resource development. Those users with the greatest water demands, whose demand effectively results in the need for resource development, should bear the marginal cost of increasing supplies. In addition, environmentalists became vociferous about the evaporation losses and impact on riverine ecosystems of both large dams and the numerous small farm dams. Relocation and social disruption associated with large dam projects also came under scrutiny.

- The agricultural sector argued that a water scarce country with such variable and unpredictable climate could hardly afford that its water runs down to the sea unutilised and that water resource development should not be 'put on the back-burner'. The environmentalists maintained that the evaporation and seepage losses from dams were not being accounted for.
- The benefits accruing from irrigation apply not only to commercial irrigation farmers, labourers and their families directly, but also supports a supply and marketing industry with backward and forward linkages that contributes significantly to the national economy and export. Moreover, many rural areas depended on irrigation farming for their social and economic survival.

The water resource was now recognised as an indivisible national asset, encompassing all water in the hydrological cycle. Thus, the same legal provisions that applied to surface water would regulate underground water resources and rainfall. Consequently, the concept of stream flow reduction activities (SFRA) was introduced and the regulation of groundwater extraction brought into the fold.

- Together with the Forestry sector, the South African Sugar Association and the Cane Growers were concerned that they could now effectively be charged for rainfall, just as for the use of surface water abstraction. The sugar sector was able to demonstrate that they had supported and encompassed 38,000 small black cane growers in their membership, who would be at risk if the economic viability of the sector was negatively affected.
- Commercial farmers were sceptical about the practical implications of attempts to regulate groundwater extraction. Officials calculated that it would take the department with current staffing levels fifty years to register all the boreholes in the country, and the commercial farmers asked whether the department planned to place a policeman on each of these millions of extraction points?
- The Dendron irrigation area was able to relate how they had saved the economic future of their region from gradual ruin through over-exploitation of their groundwater resources, by implementing a system of self-monitoring and user management.

Structure and Concerns of the Small-Scale Irrigation Sector

The black *small-scale farmers* are not well organised. Although the National African Farmers Union has a very large potential membership, the large majority are resource poor, difficult to reach and can hardly spare a membership fee that would enable NAFU to provide meaningful services to its members. Three broad sub-sets of small-scale irrigation can be identified as follows.

- The largest area under small-scale irrigation farming in South Africa is found on government irrigation schemes in the former Homelands or *Bantustans*.[11] While exact statistics are not available, an estimated 40,000 small-scale farmers have plots of one to ten hectares each on some 50,000 hectares developed as government operated smallholder irrigation schemes. Many of these farmers are inactive for a variety of reasons and a large percentage produce mainly for home consumption. Most of the plots are registered to men, while the majority of the farm decision-makers are women.
- The largest, but unknown number of smallholder irrigators are poor rural women with holdings of 100-600 m^2 on communal food gardens of 1-10 hectares in size. These food plots are an important additional source of income and nutrition to resource poor rural families, where the majority of able-bodied men are migratory workers who may or may not send remittances to support their families.
- Finally, an unknown, but relatively small number of entrepreneurial black farmers have succeeded in establishing successful irrigation enterprises in a very hostile policy and support environment. Water rights and membership of Irrigation Boards was subject to land ownership and thus effectively excluded blacks, who could not legally own land. Since they had no title deed to the land they occupied, they also had no collateral to negotiate production loans or loan capital from the commercial banking sector.

The concerns of the small-scale irrigation sector were largely voiced by NAFU, through participation in public workshops, but more significantly in meetings with the Minister. Individual small-scale farmers attending public workshops, and NGOs and other development professionals participating in various aspects of the process, made several interventions. Since the department expected that the inputs from this sector would naturally be less co-ordinated than that of the commercial irrigation and other sectors, great efforts were made to identify potential participants and enable them to attend the workshops by organising and covering the costs of transport and accommodation. The following issues were of main concern to this sector.

- The provisions for financial support were the major concern of the black farming sector as voiced by NAFU, which represented the most sophisticated farmers from the small-scale or emerging farmer sector. The wording of the National Water Act places a duty on the Minister to consider the effects of past racial and gender discrimination in the allocation of financial assistance (National Water Act, section 61 and 62).
- NAFU was lobbied by the SAAU to support them in the argument for secure water rights, since this would affect farmers across the spectrum. These

discussions were friendly and mutually satisfactory, and the Minister met with SAAU and NAFU separately and independently on several occasions. The Director-General once said that DWAF could see the potential for strong collaboration between SAAU and NAFU and that this could result in a strong lobby that could make life difficult for DWAF.

Institutional Framework for the Water Law Review Process

The Minister of Water Affairs and Forestry is responsible for the water law in South Africa. The reform of policies and legislation has to be approved by the National Parliament, on recommendation of a standing parliamentary committee. In the case of water, this was the Agriculture, Water Affairs and Forestry Portfolio Committee (AWFPC). Meetings of the Portfolio Committees are open to the public and public hearings are also arranged, during which any individual or organisation or sector can make a public submission to the Portfolio Committee. Following approval by the AWFPC, the National Water Bill is referred to the second chamber of Parliament, the National Council of Provinces, for comment.

The relationship between the Ministry and its executive body, the Department of Water Affairs and Forestry is largely embodied in the interaction between the Minister and the Director-General of DWAF. This interaction was broadened and strengthened for the development of new policy and legislation through the establishment of a Policy and Strategy Team of 7-10 members, including the DG and some key departmental staff, together with ministerial advisors. Through the Policy and Strategy Team, the policy and legal drafting teams were given direction.

In diagram 2.1 only the Policy and Strategy Team, the Water Law Drafting Team (and the Task Teams mentioned below) are not a permanent feature of the administrative organisation of water matters in South Africa.[12] These committees were especially created to play their respective roles in the water law review process and then dismantled at the conclusion of that process. Officials and specialists were members of these committees, but no individuals who were directly representing any stakeholder group. It was believed that stakeholder representation in the Drafting Team would make it near impossible to make progress, since concepts would constantly be debated, hampering focussed attention on the technical process of drafting.

The Policy and Strategy Team was crucial in steering the reform process, and it is useful to reflect on its development over time. Since the objective of the Policy and Strategy Team was to ensure implementation of an entirely new political paradigm, membership initially included only those individuals with a clear connection to the new dispensation. This had specific implications for the irrigation policy process, because the appointment of consultants and approval of the irrigation policy process were managed by managers who were initially excluded from this forum of discussion. This contributed to the problems in the irrigation policy workshops, as described below.

Diagram 2.1 Main components of the institutional framework for water law review in South Africa

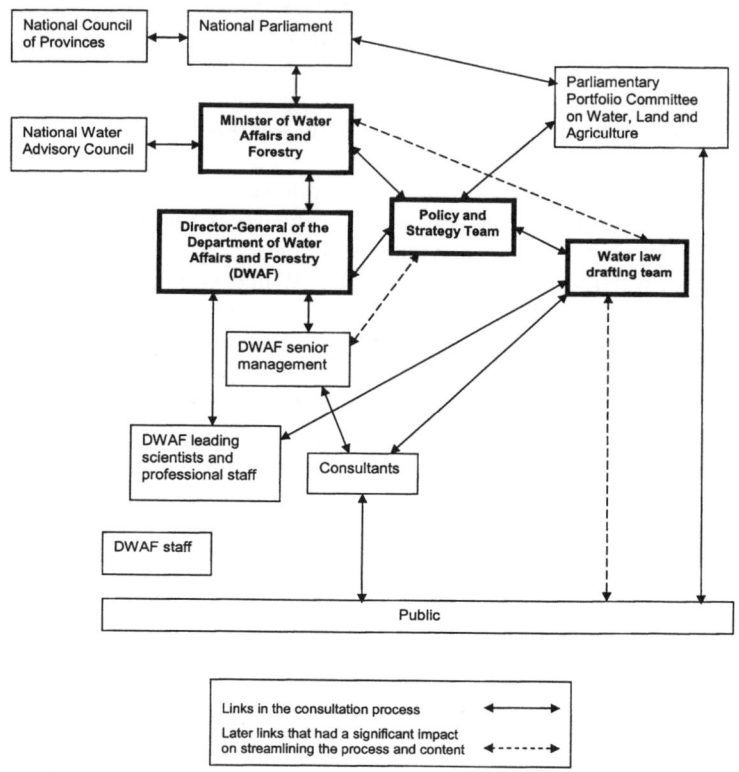

Another interesting feature of the framework was that it provided for direct interaction between the legal drafting team and selected officials and even consultants. This interaction helped ensure that the concepts and text in the Bill were scientifically sound and that a balance could be struck between new ideas and decades of experience with water use management and administrative processes.

Very late in the process, direct interaction between the Minister, the legal drafting team and departmental staff was instrumental in clarifying burning issues and streamlining finalisation of the National Water Bill. This interaction was in the form of a two-day meeting between the Minister and some 30 lawyers and officials, discussing the draft bill almost line-by-line, with reference to a compilation of public and expert comments received during the consultation process.

Towards the end of the drafting process, the Director-General initiated implementation task teams to critically assess the implementability of the provisions of the National Water Bill. There were institutional, water quality, water allocation, and water use and conservation task teams. This was the beginning of a process to restructure the department according to the implementation requirements of the National Water Act.

Quite separate from these committees with specific product responsibilities, the Minister also appointed a National Water Advisory Council of about 25 members. These members are representative of all water sectors, and meet monthly to advise the Minister on a wide range of water related issues. They assumed a watchdog role on the representative nature and inclusiveness of the water law review process.

Overview of the Elements of Water Policy Review and Implementation

The water policy reform was an open and consultative process that stretched over several years. This consultative approach has been carried forward in the legislation itself, which requires consultation with stakeholders on a wide range of issues.

The National Water Act itself does not attempt to anticipate and regulate through prescriptions for every potential future circumstance, but sets out the principles and an enabling framework within which elements can be implemented as they become relevant and necessary. An example is the provision for the regulation of groundwater extraction, which can be utilised in a specific geographic area if and when over-exploitation becomes a problem, but can otherwise be left unregulated.

Key Products of the Water Law Review

The water law review started in early 1995 with a document called 'You and Your Water Rights' which was aimed at raising awareness and public understanding of the current water rights. This was widely distributed and followed by the drafting of, and a round of public consultation on the 'Water Law Principles' in April to October 1996.

Following Cabinet approval of the Principles in November 1996, the 'White Paper on a National Water Policy for South Africa' (see Note 1) was launched in May 1997. Two new pieces of legislation were developed.

- The National Water Act (Act 36 of 1998), which deals with management of the water resources of South Africa, and which replaced the previous Water Act (Act 56 of 1956).
- The Water Services Act (Act 108 of 1997), which regulates water supply and sanitation service provision. This was drafted first because of the urgency to address the needs of the new Government's main constituency, the poor black communities, of which 12 million had no or inadequate access to these services.

Consultation Processes on the National Water Bill

The water law review process followed a natural progression that created almost a steamroller effect (see Diagram 2.2 for an overview of elements of the water policy and law review process). In the words of the Minister (see Note 10), the purpose of the document 'You and Your Water Rights', was to 'stimulate our people's emotions and thoughts on water legal issues, thereby sparking debate about water rights'.

Diagram 2.2 Overview of the elements of the water policy and law review in South Africa

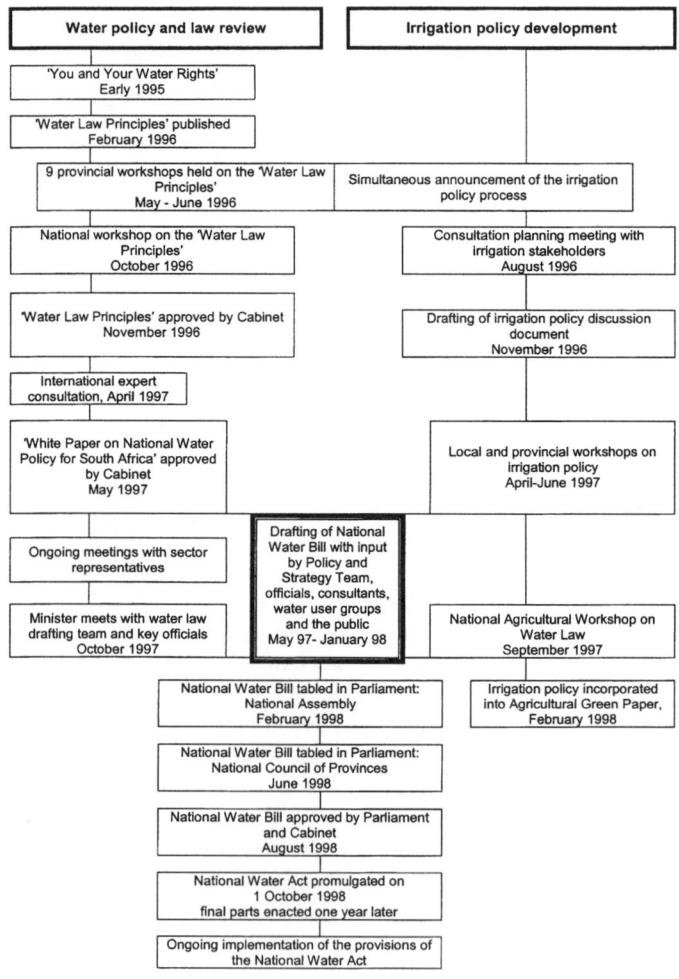

When published almost a year later, the 'Water Law Principles' reflected the spectrum of needs in South Africa and held the moral high ground in the new culture of human rights and environmental sustainability. The key principles defined the water resource as an indivisible national asset that included the entire hydrological cycle; it gave priority to basic human needs for survival and to protection of the minimum water required to ensure the integrity of the very resource itself; but it also argued for protection of investments people had made in harnessing the resource and being economically active and productive.

The Principles were debated for almost a year. First it was widely published in several South African languages (February 1996). Then it was debated in provincial workshops in May to June 1996 (see below). This was followed by a national workshop in October 1996 before the final 'Water Law Principles' were tabled for Cabinet approval in November 1996.

The 'White Paper on a National Water Policy for South Africa', which was approved by Cabinet six months later (May 1997), tried to reflect these principles as closely as possible. Key members of the Agriculture, Water Affairs and Forestry Portfolio Committee (AWFPC) in the National Assembly of Parliament participated in an international expert consultation, and were thus already familiar with the content of the White Paper when it was tabled. This strategy was effective, because when the White Paper was presented to Parliament, it was accepted with minimal changes. This White Paper then became the basis and yardstick against which the National Water Bill was drafted (June 1997 to January 1998).

The directive to the water law drafting team was clear: the law had to be an instrument to implement the policies articulated by the White Paper. The drafts produced by the team were repeatedly measured against the White Paper, and thus effectively against how well they translated the Water Law Principles into practice. (The drafting process is discussed in more detail below.)

In this period, the public had several opportunities to comment on the various drafts, some of which were posted on the department's website. Comments were received in writing, through meetings with the Minister or senior staff and in public forums. A series of consultation workshops with the irrigation sector were taking place throughout the country, culminating in a National Agricultural Workshop on Water Law in September 1997 (see below for more detail).

At the same time, the Director-General appointed several task teams within his department to ensure that the provisions of the draft bill would be practical to implement. With this, he set in motion a process of restructuring the department to prepare for the implementation of the new law when it was enacted.

The National Water Bill, that is, the finalised draft bill, was tabled in Parliament in February 1998, almost three years after the process had started with 'You and Your Water Rights'. The consultations, public submissions, debates and revisions in the National Assembly and the National Council of Provinces of Parliament went on for another six months before the President signed the National Water Bill into law on 26 August 1998.

The National Water Act was promulgated in two phases, the first on 1 October 1998 and the second on 1 October 1999. This allowed time for the department to develop the necessary instruments and capacity for the second, more challenging phase of water use registration, licensing and the establishment of a Reserve for basic human needs and environmental sustainability.

The implementation process is characterised by the ongoing development of supporting processes, policies, regulations and guidelines. Without trying to cover the detail and options within each, the following elements can be listed, with yet another note that the detailed content is being developed through consultative processes as required by the National Water Act.[13]

- A National Water Resources Strategy, including sector water conservation and demand management strategies.
- A water pricing policy.
- Registration of existing lawful water use by October 2000.
- Establishment of a licensing procedure for water use authorisation.
- Ongoing development of manuals for the determination of Resource Directed Measures, including the environmental Reserve.
- Development of guidelines for the establishment of Catchment Management Agencies (CMA) and Water User Associations (WUA), including guidelines on adequate public participation.
- Consultation and technical studies towards the establishment of the first of a total of 19 CMAs and catchment management strategies.
- Transformation of commercial farmer Irrigation Boards into Water Users Associations and the establishment of new Water Users Associations.

National Irrigation Policy and the Water Law Review

In parallel to the drafting of a new water law, national Irrigation Policy was being developed, essentially to explore how the policy objective of 'beneficial water use in the public interest' could be achieved in the agricultural sector. This phrase reflects the need to pay adequate attention to all three nodes of a conceptual triangle, namely (i) economic development and productivity, (ii) social equity and stability and (iii) environmental sustainability and integrity of the resource base. In this context, two main issues had to be addressed in the irrigation sector.

- Irrigation is by far the largest water user in the country and is generally viewed as the source for sector water transfers once other options had been exhausted. Debate was needed on potential water savings and the value of irrigation in an increasingly industrialised economy. Particular emphasis was given to sharing experience on proven improved management practices at farm and irrigation scheme level, and the potential for increased value of production through a switch to higher value crops and value-adding. National food self-sufficiency was no longer a policy objective, but rather household food security, which aims to ensure that individuals have access to enough and affordable nutrition. Sometimes basic foodstuffs can be imported more economically than they can be grown locally and the levels of protection for

local production have been reduced dramatically.
- There were unrealistic expectations among officials in several departments and among historically disadvantaged communities of the role that irrigation could play in eradicating poverty and bringing these communities into the mainstream economy. Seemingly prospering commercial irrigation is much more visible than the ailing and expensive experiments with smallholder irrigation schemes implemented by government in remote areas in the homelands. Debate and increased awareness was required to define and address the obstacles faced by the small-scale irrigation sector, including lack of access to loan finance, support services, markets, managerial capacity and appropriate technology. Realistic opportunities for sustainable smallholder irrigation development had to be explored.

One of the peculiarities of the South African administration is that irrigation matters have for decades been passed backwards and forwards between the department of agriculture and the department of water – there is no department of irrigation, as in many other countries.[14] There is ongoing negotiation between these departments about their relative responsibilities. Under the previous regime, an effective liaison structure was developed between the two departments concerned with white commercial irrigation, but this never had a mandate to deal with small-scale irrigation, which was the responsibility of the separate homelands. This reflects an important structural disadvantage, since there were inadequate networks within the existing bureaucracy to focus on small-scale irrigation.

The entire government structure was reviewed and integrated across these black/white boundaries after 1994, and irrigation was more than ever without a proper home. In the new dispensation, water is a national competency, and so there is only the national Department of Water Affairs and Forestry, with regional offices but no provincial departments. Agriculture, on the other hand, is a concurrent function, so in addition to the National Department of Agriculture, there are nine Provincial Departments of Agriculture.

Irrigation had a low priority in the new National Department of Agriculture, since only a relatively small number of small-scale farmers are involved in irrigation compared to the total small-scale agriculture sector, which includes dry land cropping and livestock farming. Further, in the new dispensation, the white commercial farming sector would no longer be the main client of government services, so there was little interest in commercial irrigation.

The Department of Water Affairs and Forestry's main interest in irrigation stemmed from the fact that this sector accounted for more than 50% of the country's water use and many officials believed that the sector dealt with the water irresponsibly and wastefully. The department was also under pressure to ensure that the new water legislation would do the utmost to support previously disadvantaged communities. Access to irrigation was the best way the department could see to make water contribute to poverty eradication. So it was DWAF that initiated the irrigation policy process, but following the consultation-planning meeting, the National Department of Agriculture was invited to co-host and co-fund the process, to ensure co-ordinated policy development and implementation.

The interaction between the water law review and irrigation policy development was initially very weak, but improved systematically, especially during the drafting of the National Water Bill, at which time the irrigation sector was being consulted in more than sixty local and provincial workshops. Following a directive of the Director-General, the consultant responsible for the consultation process on irrigation policy development regularly participated in the drafting meetings of the legal team, thus creating a direct link between the two processes.

The interaction between the water law review and irrigation policy processes culminated in a National Agricultural Workshop on Water Law, just before the final draft of the National Water Bill was developed. Although the workshop was co-hosted by the National Department of Agriculture and the Department of Water Affairs and Forestry, there was little interest and input by the Minister of Land and Agriculture. There was, however, significant participation by the Director-General, Dr Bongiwe Njobe-Mbuli and senior officials from the national and provincial departments of agriculture.

At this national workshop the Minister of Water Affairs and Forestry, senior officials of his department and the legal drafting team came in direct contact with some 200 leading representatives of both sub-sectors of the agricultural sector, as well as agricultural officials and the agricultural research community. It was a frustrating workshop and tensions ran high. However, this direct and public contact appeared to have had impact on the ultimate content of the National Water Act, because soon thereafter the Minister met with his officials and drafting team and directed some pertinent changes to the Draft Bill (see above).

Interaction between the irrigation policy and water law processes waned again when irrigation policy was incorporated into the Green Paper on Agricultural Policy where it received little further attention by the National Department of Agriculture at the time. Some of the provinces, especially those with significant government small-scale irrigation schemes, continued with their own processes to resolve existing issues. Later, the Director-General of the National Department of Agriculture asked the Northern Province Department of Agriculture and Environment to take the lead in implementing agriculture water policy as a pilot learning area for the country as a whole.

At the end of 1999, the interaction on irrigation matters between the National Department of Agriculture and the Department of Water Affairs and Forestry picked up again. Two developments shaped this train of events.

Firstly, in June 1999, the second democratic elections took place and Thabo Mbeki took over the Presidency from Nelson Mandela. A new Minister of Land and Agriculture was appointed when the new President selected his cabinet. Ms Thoko Didiza was the Deputy Minister of Land and Agriculture in the previous term of office, and now, as the new Minister, she announced that one of the main programmes in her term of office would be to pay major attention to the rehabilitation of smallholder irrigation schemes. This was set in a new focus on integrated and sustainable rural development dictated by the new Cabinet.

Secondly, the new Chief Director Water Use and Conservation of the Department of Water Affairs and Forestry, Ms Barbara Schreiner, took new initiatives to improve co-operation in the implementation of the National Water

Act in the agricultural sector. She initiated the following two processes in this regard.

- An interdepartmental committee to co-ordinate government support and responsibilities for smallholder irrigation; and
- The development of a water conservation and demand management strategy for the agricultural sector, with its main focus on commercial irrigation. One of the major initiatives of this interdepartmental committee is to develop co-ordinated policy and implementation programmes for irrigation management transfer and rehabilitation of government smallholder irrigation schemes.

Description of Some Elements of the Consultation Process

Some of the elements of the consultation processes are described below to give a sense of how these were approached. This is not an exhaustive account of the number or types of consultations engaged in during the water law review and irrigation policy development.

Provincial Workshops on the Water Law Principles

The drafting and public consultation of the Water Law Principles was the first major public step in the review of water policy and legislation in South Africa. These principles were drafted by a drafting team appointed by the Department of Water Affairs and Forestry, under the guidance of a consulting lawyer, Geoff Budlender, who was later appointed Director-General of the Department of Land Affairs.

A one-day public workshop was held in each of the nine provinces to discuss the Water Law Principles. Participation was drawn from all the water use sectors. These workshops were well attended, which is indicative of the importance the subject held in the minds of people. Also, 'workshop fatigue' had not yet set in, as it has in the meantime after five years of public consultation on virtually every subject imaginable.

An important feature of these workshops was their design to facilitate capacity building. Each of the provincial workshops was preceded by a preparatory workshop for the historically disadvantaged sectors, the day before the main workshop. During the preparatory workshop the content of the Principles were dealt with more thoroughly. This gave illiterate people an opportunity to discuss the implications of the principles for their own circumstances and prepare presentations for the main workshop the next day.

It is a natural and recurring phenomenon in public workshops with such abstract content (indeed in any discussion on water), that community representatives raise the specifics of their community water supply problems. The facilitator was particularly effective in explaining why the detail of each of these specific situations could not be discussed in this forum. He explained that the purpose of this particular process was to ensure people's right to adequate water.

However, he said, 'one can't drink a right' and therefore further processes would be implemented to create the physical supply systems. The department nowadays makes a point of noting and following up on these specific requests for assistance voiced at consultation meetings, or supplying participants with the contact detail of the relevant office for follow-up.

In those early days of political uncertainty other sectors sometimes insisted on attending these preparatory workshops too, but interest waned when it became clear that there were no hidden agendas. Also, the pace of these preparatory workshops were painfully slow, because of the need for translation, and the economically stronger sectors soon realised they had better ways to spend their time. Increasingly, such capacity building and preparatory workshops are being conducted in the vernacular, which greatly enhances participation and the richness of the discussion.

The main workshops were structured to elicit comments on the Principles on a sector basis. A brief introduction of the objectives of water law review and the content of the Principles was followed by sectoral group discussion of each of the Principles to indicate whether the group agreed or disagreed with respect to a specific Principle and what improvements they wanted to suggest. Each group reported back in a plenary session, followed by a session in which anyone could present a 10-minute submission to the workshop. While several submissions were made, one would have expected more organisations and individuals to use this opportunity. This can possibly attributed to the fact that the public was new to the concept of public consultations and submissions, and possibly that most people were satisfied by the systematic discussion of each Principle.

The approach of limiting group discussions to agreement, disagreement or proposed amendment of each Principle simplified reporting on the outcome of the process and the outcome could easily be analysed geographically across the provinces and across sectors. However, it narrowed the discussion and an opportunity for an improved recognition and understanding of local circumstances may have been lost in the process. Yet, any individual or institution could also comment directly to the department outside these workshops.

Dividing the discussion groups by sector was a powerful way of giving an independent voice to disadvantaged communities. This was further supported by the preparatory workshop, through which a significant number of submissions were generated, since each small group from the pre-workshop presented its findings in the 'submissions' session. In contrast to the Constitutional development process, the sector design of these workshops did not facilitate integration and discussion among participants from different sectors.

The consultant acted as facilitator at the nine provincial workshops and drafted a consultation report on the outcome of these meetings, recommending adoption of the Principles as the basis for water policy in South Africa.

The invitations and venues are a crucial element in consultation processes, because it affects the accessibility of workshops, especially for disadvantaged groups. These aspects were organised by a separate organisation. Although the workshops were well attended, as previously mentioned, there was always criticism of the attendance list and sometimes of the accommodation, meals, transport and other matters. Through all these processes I have come to the conclusion that it is impossible to please everyone. Event organisers and the department running the consultative process need to evaluate each event to an agreed standard and beyond that, live with the criticism. This seems to be especially true in a politically charged period when people need to let off steam in one way or another.

Having said that, it remains critically important to analyse attendance lists for representativeness and inclusiveness and to be prepared to take corrective measures if necessary. One of the provincial workshops in this first round of consultation was repeated because some sectors were absent at the first meeting.

Public consultation processes have matured considerably over the past few years and DWAF is currently drafting guidelines and standard practices. 'One of the principles that is present in the draft guidelines, and which tends to underpin most public participation processes in the department now is the need to get previously disadvantaged communities on board, to involve trade unions and to strive for a gender balance. Environmental NGOs are another increasingly recognised stakeholder. These are not always achieved, but DWAF strongly recognises the need to involve these sectors if any sense of legitimacy of process is to be created' (personal communication, Schreiner, November 1999).

Irrigation Policy Workshops

One of the first steps in the irrigation policy process was to convince the responsible managers at DWAF that it was necessary to consult the public. Initially, some managers felt it would require only one small workshop with key role-players to arrive at irrigation policy. Others had been exposed to the increasing requirement for consultation associated with relocation of people from new dam basins and convinced their colleagues of the need for thorough consultation. The department decided to organise a consultation planning meeting with the major irrigation stakeholders. At this meeting, the public insisted that more localised consultation was needed compared to the just completed provincial workshops on the Water Law Principles. Participants felt that discussion at provincial level would be inadequate to address the differences between agricultural regions. Another outcome of the workshop was closer co-operation with the national and provincial departments of agriculture in the irrigation policy process.

Local workshops were arranged in all major irrigation areas in the country. The White Paper on a National Water Policy for South Africa formed the basis for discussion, together with a discussion document with background information on irrigation in South Africa.

Group and plenary discussions aimed to capture local information in the five categories addressed in the discussion document.

- Economic and social aspects
- Using irrigation water efficiently

- Water management
- Small-scale irrigation farmers
- Women in irrigated agriculture

These workshops took place at the height of the insecurity of the commercial irrigation sector, when the first draft of the National Water Bill contained frightening provisions for agriculture as discussed in Box 2.1 above.

- The government could both prohibit and prescribe crop choice
- Water rights in perpetuity would be replaced by licenses for use with a limited time period
- These licenses would lapse at the death of the licensee, thus eroding the long-term viability of family farms
- No compensation would be payable if a water allocation was revoked

In this climate of fear for their livelihoods, it is little wonder that the majority of the discussion at these workshops focussed on the new water law and implications for the sector. However, the facilitators had no mandate to discuss these burning issues and no access to current political decision-makers to guide them in handling these public fears. Their mandate was to discuss irrigation policy. Some of the commercial farmers' unions accused the department of attempting to side-track attention away from the real burning issues through this irrigation policy process. In some provinces the unions staged dramatic walkouts from these workshops to protest against the water law review.

This was clearly an untenable situation, which was only partially resolved through the increased interaction between the water law review and irrigation policy processes. Participation in the legal drafting team meetings served to feed information and reaction from both the agricultural sub-sectors into the drafting process, but probably contributed little to improving the interactions in the remaining irrigation policy consultation workshops.

In retrospect, the timing for these consultations on irrigation policy was probably unfortunate from an irrigation policy development perspective. The workshops did however play a major role in creating awareness and understanding of the provisions of the White Paper on a National Water Policy for South Africa, the water law review process and its implications for the irrigation sector. It might have been advantageous to recognise this as the overriding need of the period and to have boldly reoriented the remaining consultations halfway through the process to serve this purpose more directly.

Drafting of the National Water Bill

One of the interesting aspects of the drafting of new water legislation was that so many people were involved who had no legal background. Many had to learn that a 'bill' was the draft legislation that would become an 'act' the day it was finally approved, accepted and enacted or promulgated.

The Minister was adamant that this piece of legislation should be written in simple, understandable language rather than legalese, that it should be 'enabling' rather than prescriptive and prohibitive, and that it should be practical to implement and administer.

The members of the water law drafting team were a relatively balanced mix of lawyers from DWAF and the private sector, and included officials with decades of experience with the old Water Act of 1956, as well as new blood with socially advanced ideas bordering on the extreme. A senior judge chaired the team and was directly answerable to the Policy and Strategy Team.

The drafting process was peculiar. Once a preliminary structure for the bill was agreed, small groups of 2-3 members were assigned to draft certain portions of the Bill, which were then distributed to all the other members of the team, and later others, to study. The drafting team would then meet to discuss the proposed text clause-by-clause. Any participant in the drafting team meetings could comment on the proposed text, but to limit drafting time, these comments had to be in the form of a proposal for alternative wording. In this process, the meaning and implication of a specific provision could be changed completely if the drafters could be convinced that it made sense. The resultant new draft would then be scrutinised and commented on by the Policy and Strategy Team and directives given on aspects that were departing from the provisions and objectives of the White Paper. This process of to-and-fro went on for weeks and months, with drafters and other people attending the drafting team meetings trying to influence the concepts in this roundabout and rather stressful way. Later in the process the concepts became better defined and there was more clarity on the options. At this stage, clear directives from the Policy and Strategy Team resolved many of the earlier tensions.

While there was near outrage in reaction to Draft 1, the Bill gradually matured to a somewhat more acceptable Draft 4, following which it was significantly changed. The resulting Draft 5 was further refined and after Draft 7, the National Water Bill was ready for submission to Parliament. In this maturation process, challenges in terms of the constitutionality of some provisions were taken seriously and analysed and adapted in consultation with constitutional lawyers. Again, the Minister was clear: he did not want major amendments by Parliament and he did not want the Act to be challenged in the Constitutional Court. This had an impact on the provisions for compensation in case of a major reduction in a water allocation, as discussed above.

The drafting team worked incessantly over many months under extremely stressful conditions and time constraints. It is a tribute to their tenacity and capacity that the National Water Act is a well-balanced piece of legislation. It reflects many divergent needs across the South African social and economic spectrum, and contains scientifically advanced concepts, such as a Reserve for environmental and basic human needs, integrated water resource management on a catchment basis, inclusion of all water in the hydrological cycle into the same legal framework, etc.

As mentioned earlier, final clarity on outstanding issues was achieved through direct interaction between the Minister, the drafting team and key officials near the end of the drafting process. During these meetings, which took place after the

National Agricultural Workshop on Water Law, some of the major concerns of the agricultural sector were resolved (see Box 2.1 above). However, some issues were never conceded, notably the time limit to which future water allocations would be subjected.

The National Agricultural Workshop on Water Law

The Director-General organised a National Agricultural Workshop on Water Law when the feedback from the irrigation policy workshops highlighted the tension throughout the commercial agricultural sector about the proposed changes to the water law. The significance of this national workshop was that it brought together representatives of the most threatened water use sector – irrigation – with the Minister, senior DWAF officials and members of the water law drafting team. In particular, it was the only public meeting where the drafting team could be called on to explain their thinking directly to the agricultural sector. Also, this meeting was not limited to the handful of senior officials of the SAAU and NAFU, who had had formal meetings with the Minister and his staff. The workshop gave a much larger spectrum of people a chance to interact. Approximately 220 commercial and small-scale farmers, government officials from various departments, researchers, consultants and the media attended the two-day workshop.

The first day of the meeting was designed as group discussions focussing on the various concerns of the agricultural sector. The question session following the presentations in the morning session was so charged, that a decision was taken to have plenary discussion instead, with the Directors-General of the Department of Water Affairs and Forestry and the National Department of Agriculture in a panel. Questions were taken from the floor and answered by either DG, or a key staff member, or a member of the legal drafting team, all of whom were in the audience. This was an extremely difficult day. Some participants were very uncomfortable with the unexpected change in programme. Some senior members of SAAU felt that these matters are better discussed in small meetings between the leaders. Most participants did not like the answers to their questions, because the department seemed to be standing firm on some of the most contentious issues.

However, the workshop had some impact. A heart-felt appeal of nearly ten minutes, eloquently addressed to the Minister in that forum of over two hundred people, may have had a significant impact on subsequent changes to the Draft Bill. A respected senior white farmer appealed to the Minister, explaining how his irrigation farm constituted his life's work, his only source of pension and his children's only inheritance. An arbitrary cancellation of his water allocation would, in one blow, erode all of that. Soon after this workshop, the provision that water allocations would expire at the death of the licence holder was removed.

The Minister had expressed his satisfaction at the racial and gender mix at the workshop. In fact, he always insisted on an adequate mix of participants, and he never wanted to see only white faces in a panel facing the audience. However, as at almost every workshop, there was a question from the floor about the representativeness. In this case someone asked why there were not more coloured farmers from the Northern Cape region. The organisers explained that invitations in the Northern Cape province had been made through the provincial small-scale farmers' union. Further, this is an unusually convenient channel, since in most other provinces, such an organisation does not exist, making organisation of events of this nature extremely difficult. NAFU has branches in most provinces, but is not sufficiently resourced to focus on more than the most commercialised of the small-scale farming sector. In Western Cape alone there are 42 unaffiliated farmers' unions. This makes it virtually impossible to give a representative group of small-scale farmers notice of these meetings of national import. In reality, the attendance of so many small-scale farmers at this meeting was made possible because (i) many of them were personally known to the organisers, (ii) all their travel and accommodation was organised and paid for by the department. The small-scale irrigation sector had to realise the importance of being well organised and should make every effort to achieve this. The need for organisation of the small-scale irrigation sector is further discussed below.

Similar to the approach at the provincial workshops on the Water Law Principles, an additional event was organised for the small-scale irrigation sector. In an evening meeting before the main event, some 50 small-scale farmers had a pre-workshop discussion with the Minister of Water Affairs and Forestry. He joked that he was being called the 'Minister of Rural Development' (there is no such portfolio in South Africa), because of his department's efforts to supply water to so many people in the rural areas. He was not prepared at the time to seriously consider the 'one-stop shop' that small-scale farmers were requesting. The farmers argued that they had no clarity which department to approach on which aspects related to irrigation. Isolated and impoverished as they were (and still are), it becomes an impossible task to go from one department to the other to get permission and assistance for water, land, finance, agricultural advice and other matters related to irrigation farming. The Minister pointed out that many of their needs should be addressed by the Minister of Land and Agriculture. As explained before, irrigation was not a priority of that Minister at the time, who was primarily focussing on land reform and transformation of the internal and international marketing arrangements in agriculture. This was a friendly and enjoyable meeting, but largely inconclusive due to a lack of clear purpose and strategy for the meeting.

International Consultations

The South African water law review benefited from international consultation through ongoing input by individual foreign consultants and two expert consultations, spaced several months apart. The wide variety of experience in a range of countries was considered and assessed in relation to South Africa's own unique circumstances. These countries included both first world and developing economies, to reflect our own mixed situation, and included Mexico, Chile, USA, Australia and others.

The approach was to bring consultants and senior government officials from these countries to South Africa, expose them to our situation and explain the past problems and new objectives with our water law. Following this, workshops were held where the foreign experts presented their experience and made recommendations according to their understanding of our requirements. Local senior officials, consultants, the water law drafting team and members of the Parliamentary Portfolio Committee then debated these concepts.

The Finland government financially supported the water law review process, but international donors did not play a significant role in influencing the content of the water law review.

Interaction with the international community is ongoing and broadening with increasing interest particularly from international researchers and water managers in a wide range of aspects related to the National Water Act.

There is ongoing interaction with neighbouring countries in the SADC region on their respective water reforms, on developing and revisiting existing treaties. Virtually every significant river in the region is shared between two or more countries. The SADC protocol on shared river systems is an important instrument guiding water policy and legislation in the region.

Questions Often Raised About the Water Law Review Process in South Africa

South Africa's water law review has been a very rich experience and raises many interesting questions, most of which fall outside the scope of this paper. One could, for instance, analyse the role of political skill and management in the success of the review process; the role of the media; the transformation of the department, its old and new staff members and the changes they all have had to undergo to adapt in the new environment; the psychological impact on water users brought about by this and other concurrent processes, and so on.

This paper does not pretend to provide a rigorous analysis of any of the questions below.

Why Was It Possible For Such A Comprehensive Review To Take Place?

This water law review could not have taken place without the *fundamental power shifts* that took place in the South African political landscape at the end of the *apartheid* era. One cannot enter here into an analysis of how the broader process of democratisation created this new landscape, but the strong economic sectors, who previously also held the political power, in a sense became 'politically disadvantaged' in the new dispensation. The new regime had a clear commitment to address the imbalances of the past, to protect the rights of the historically disadvantaged, economically poor sectors of society.

Even the eloquence of the more affluent sectors became less of an advantage, because as they speak, they tend to be heard with a certain measure of suspicion that they act only in their own self-interest, even when they voice concern for the disadvantaged sectors. At the same time, it became fashion to let the poor speak and a deliberate and concerted effort was made to create the opportunity for them to speak.

The power shift was not confined to the upper echelons of politics and governance either: blacks were systematically filling decision-making positions in local, provincial and national structures, also in the water sector. In this process some of the basic tools of power were redistributed.

Ironically, the physical separation between races created by *apartheid* may have contributed to the context within which such a complete shift could take place. The direct interdependence between the rich and the poor does not seem to be nearly as comprehensive as for instance that of landlords and tenants in rural India. For instance, in South Africa only a small percentage of black employees live near or with their employers and even these move in separate social circles.

In this dramatically changed landscape, the more appropriate question might be 'how was it possible to review the water law so comprehensively without serious detriment to the economy and stability of the country?' Despite the dramatic power shift, the temptation to dictate was resisted and a consultative approach adopted.

Professor Kader Asmal, the Minister of Water Affairs and Forestry acknowledged the challenge of this balancing act in his address to Parliament when he presented the National Water Bill on 9 February 1998.

> While policy is always relevant, this Bill is not driven primarily by ideology. It is driven simply by finding innovative and practical answers to the question: 'What should be done in water legislation to ensure stability and prosperity for this country in the short term as well as in the long term.' This has to take due account of the past, i.e. those who were detrimentally affected by laws of that era, those who made investments based on the then allocations, but it also has to look very far into the future. At all times in this government one has to keep in mind that while we have an opportunity to improve the quality of lives of all our people, depending on what choice of action we take our action could yet turn out to be the ruin this country. Thus one has to be both visionary and juggler – not easy!

The *vision, drive and political skill* of this 'visionary and juggler' were indeed major determinants of the scope of the undertaking and its successful completion. He also emphasised the need for equity, or 'fairness', to ensure broad-based support for the new legislation.

> It is important therefore to ensure that the most downtrodden as well as the most privileged of our society have some feeling of ownership of this Bill. If this Bill is not good for the poor, then it surely is not good for the rich. Whether you are black, or white or whether you are rich or poor, our destinies are intertwined. We all face a water emergency by 2020 unless we act in concert, and now.

The process was indeed given impetus by the memory of a series of serious droughts since the early nineties. The water law review responded to a *real issue*: water scarcity is an ongoing and increasing threat and the existing legislation did not provide the necessary tools to address this adequately.

The *political embeddedness* of the process was ensured by the Minister's personal interest in the matter and his success in placing it on the political agenda of the country. The less successful irrigation policy process, run in the same period by the same department under the same Minister clearly illustrates the importance of political commitment. The irrigation policy process was conceived outside the political arena by senior technical officials and subsequently changed hands between the Minister of Water Affairs and Forestry and the Minister for Land and Agriculture. Eventually it was largely ignored until two years later, when the implementation of various elements of the National Water Act and the interest of a new Minister of Agriculture and Land revived the official attention in the matter. It remains to be seen if the current focus on support for small-scale irrigation will succeed in establishing the sound political base required for completion and significant implementation of policy. In contrast, the water law review is completed, an entire department has been restructured according to its implementation requirements and is buzzing with activity aimed at achieving that goal.

The water law review was part of a comprehensive process of reform in the country and sought to comprehensively reform the entire water sector, not just problematic sub-sectors. The process was designed to move *from the general to the specific*, from principles applicable to all water users, to comprehensive national policy, to increasing specificity in the implementation in various water use sectors.

The *strategy* that the water law review process followed, whether deliberate or not, was very effective. In a step-wise process it created enough interest in the process to elicit participation by all relevant water use sectors. The media was used extensively throughout the process, both by the department and by other stakeholders, to stimulate public interest. The proposals were dramatic enough to spur the expenditure of huge amounts of energy and cost by very diverse groups and institutions over a period of almost three years.

Several aspects in the *management of the process* also played a role. The commitment to consultation was part of a new culture of commitment to the systematic democratisation of South African society. One of the interesting aspects is probably clarity on where *not* to involve public stakeholders directly. Three broad arenas of interaction come to mind.

- The public arena, where the department and its consultants and sometimes the Minister interacted in open forums with stakeholder groups to develop a better understanding of the potential impacts of proposals and to hear suggestions for alternatives;
- The official arena, where departmental staff and consultants interacted to generate the content of the various documents leading up to the final Act; and
- The political arena, where the Minister and senior staff interacted with Cabinet, Parliament and representatives of political parties.

There was no direct involvement of public stakeholders in the official processes of drafting documentation, especially the various drafts of the Bill, except through written submissions or invited presentations in reaction to proposals. The reason for this is that there would be very little progress if the drafting process were combined with a negotiating process to agree on concepts.

Several other processes within the official arena may have had an impact on the course of the process. The following observations are made from an outside perspective.

- The establishment of the Policy and Strategy Team consolidated the discussions from which the legal drafting team was given direction. The Policy and Strategy Team initially consisted exclusively of members of the new regime, which probably gave these individuals an opportunity to work as a team and reach agreement on the political objectives of the process before members from the old department were included. Thus the Policy and Strategy Team was an important instrument in the political transformation of the department.
- There was further a systematic transformation of the department from a virtually single discipline – engineering – to a variety of disciplines, race and gender. This does not only mean that broader perspectives are brought into water management discussions, but that new topics are discussed in the water management field.
- The involvement of old and new staff in debates and drafting of documentation created ownership of the new legislation and capacity for its implementation within the official arena. It was noticeable in that period that those staff who were not involved felt insecure and even some resentment about the changes.
- Near the end of the drafting process, there was yet another restructuring of the department – this time towards implementation. The Director-General did not hesitate to overhaul the structure of the department for this purpose and with this signalled that he was serious about the implementation of the new legislation. It also gave staff clear direction on their roles and responsibilities towards the new goal.

The *time was right* for this process to take place. Quite apart from the political imperatives for change in water legislation, the water management requirements in South Africa had changed substantially over the four decades since the previous water law review in 1956. Nature contributed in her own way: the droughts of the nineties raised public awareness of the increasing scarcity of the resource. The importance of timing is also illustrated by the irrigation policy process, which was premature and tried to share the public attention with something as contentious as the water law review.

Did Consultation Really Have an Impact on the Final Content of the National Water Act?

During the water law review process the department was often accused of 'window-dressing' – organising the consultation process only to pacify stakeholders, but without any real intention of taking submissions into account, especially those by the strong economic sectors. However, by the time the National Water Bill was submitted to Parliament, several sectors expressed appreciation for the manner in which the process has lead to substantial amendment of some of the original ideas.

Many were still very unhappy with some of the provisions in the National Water Bill, but none of them now accused the department of lack of consultation. The Kwa-Mbonambi Farmers and Timber Growers Association wrote that they 'wish to record our strongest objection to the New Bill' and all the other submissions highlighted the outstanding issues from their perspective. It is open to anyone's judgement which of the following statements were strategic and which sincere, but it was common then, as it still is, to question the representivity and inclusivity of processes. Instead, most stakeholders acknowledged the process, as is evident from these extracts from the public submissions to Parliament.

- The South African Sugar Association said in its memorandum to the Portfolio Committee: 'There has been a significant improvement to the Bill as a product of consultation and review, and the Association wishes to express its appreciation for the manner in which this process has been conducted.'
- 'Business South Africa which represents the overwhelming majority of formal business in South Africa welcomes the further opportunity to comment on the National Water Bill. The constructive engagement with the Department on previous drafts is acknowledged, particularly the agreement to review areas of concern.'
- 'COSATU (COSATU is the SA labour union, MdL) and SA Municipal Workers Union (SAMWU) welcome this opportunity to make a submission on the National Water Bill before the Portfolio Committee on Water Affairs and Forestry. Through the NEDLAC [National Economic Development and Labour Council] process, organised business and labour have had the opportunity to contribute to policy formulation underlying this important piece of legislation, nonetheless – as a number of issues were not resolved in NEDLAC – we welcome this additional opportunity to input into the process.'
- The UCT (University of Cape Town) Law Unit said: 'The Bill is an impressive document, which appears to meet the challenge of blending South Africa's environmental socio-economic and new democratic dispensation with remarkable skill.'
- The South African Institute of Civil Engineering said: 'The Bill, as a whole, represents the thoughts and inputs from many people, and the Minister of Water Affairs and Forestry and his Department are to be congratulated on the completion of an immense task in a relatively short period of time. The Bill will certainly lead to some significant changes in water resource management.'
- The South African Association of Water Boards: 'This Association welcomes the many fundamental changes in the management of South Africa's scarce water resources which are embodied in the Bill. The drafting team should be

commended on their endeavours and the Department of Water Affairs and Forestry and the Parliamentary Portfolio Committee for Agriculture, Water Affairs and Forestry thanked for providing an opportunity to comment on the document.'
- Rand Water, the water supplier to the economic heartland of South Africa in and around Johannesburg, and the largest Water Board in the country 'evaluated the above Bill and established that a few remaining concerns that Rand Water has with respect to the National Water Bill and its effect on Rand Water's rights and its operations, still need to be addressed. Many important aspects of Rand Water's previous misgivings have, however, been addressed in that the Minister's powers are tempered by processes of consideration of comments by relevant organs of state, organisations representing water users and the public in general, attempting to regulate water resource management on a consensual basis, with participation of persons affected by decisions.'
- The Forest Industries Association emphasised their 'overriding support for the principles of the new water bill' but said that 'in providing this submission on the National Water Bill, as approved by Cabinet in January 1998, the Portfolio Committee should be aware of the fact that the Forest Industries Association has been involved in an extensive period of discussion and negotiation with the Department of Water Affairs and Forestry on all the previous drafts of the Bill. Whilst these discussions have undoubtedly resulted in a better understanding of the standpoints of the parties involved and have lead to changes being made to the Bill as it now stands these are not considered to have been substantive, and have done little to address the more fundamental concerns that the Forestry Industry has registered.'
- The National Forestry Advisory Council, were 'pleased to note that progress has been made in reaching a greater level of consensus on the National Water Bill. We continue to believe that wherever possible, efforts should be made to foster agreement between the Department and various stakeholders, including the forestry industry.'

The National Forestry Advisory Council's wish for continued consultation was realised, because the department prepared a substantive comment to all the issues raised in the public submissions to Parliament, and this was considered in detail in a series of meetings of the Portfolio Committee. But in fact, the process was coming to a closure. Little new evidence or ideas came to light at this stage of the process. Critical comparison of the National Water Bill and the final National Water Act, and studying the list of amendments approved by the Portfolio Committee, shows that the Parliamentary process did not result in substantive changes to principle items in the National Water Bill.

This could be seen as a testimony to the thoroughness of the preceding consultation, on the one hand, but also as evidence of the department's insistence on some provisions even though some stakeholders contested these throughout the process.

Not only did the development of the National Constitution set the precedent for such a comprehensive consultation process as enjoyed by the water law review process, it also contributed in a substantive way to the outcome of the water law review. Probably one of the most significant changes from the strong economic sectors' point of view, was the provision for compensation, while the Minister initially did not want to compensate for loss of access to water at all. This change also came about before the parliamentary process. Various water user sectors raised the issue in the consultation process and several of them sought legal council on this matter. Finally, in response, the Minister arranged for senior council to the department[15] and the necessary changes were made to the draft Bill.

The response to the question 'did consultation really have an impact on the outcome of the water law review?' is an indisputable yes, but it is also true that some of the fundamental changes suggested in the consultation process were not conceded.

Who Had Most Impact in The Various Stages of The Water Law Review?

It would be difficult to reach a categorical answer to this question without much more rigorous analysis of the various drafts of the Water Law Principles, the White Paper, the Bill and the list of parliamentary amendments, together with the events in the public, official and political arenas that helped shape each of these. Without such analysis, one can only raise a tentative opinion.

In general, first drafts were developed by groups of specialists (both staff and consultants) from a broad spectrum of disciplines and background, acting within what I call the official arena. Their debates of the technical and scientific requirements for future water management, shaped by the political imperatives for social equity, produced the first ideas and set the stage for debate. Outsiders to this group, whether staff, public stakeholder groups or any individual with a valid point and the ability to communicate it, could have a significant impact on these original ideas, as long as their suggestions were not seen to transgress the substantive boundaries set by the political agenda.

The legal drafting process in the first few drafts tended to refine the wording of concepts. Substantive changes to the concepts themselves came about mainly through directives of the Policy and Strategy Team as a result of contact with stakeholder groups and advisors, but also through direct contact of the drafters with scientists and consultants selected by the Policy and Strategy Team.

The Minister was thorough in his consultation, thus ensuring that there were no surprises by the time the National Water Bill went to Parliament. He not only met with every stakeholder group, but wrote to every major political party several months before the Bill went to Parliament, inviting comment and participation in debate on the fundamental policy imperatives.

In both committees of Parliament, namely the Agriculture, Water Affairs and Forestry Portfolio Committee in the National Assembly, and the National Council of Provinces, the ruling African National Congress party had a quorum and a majority. Thus, although there was a thorough process of several months of debate, public submissions and further input by the department, the vote on amendments to the National Water Bill was carried by the ANC majority. Thus, the final issues that could not be resolved through three-and-a-half years of consensus-seeking, were decided politically.

Has There Been Any Real Shift in Access to The Resource and Water Management Decision-Making?

The first part of the question focuses on access. The enabling nature of the National Water Act implies that implementation will be a gradual process. The Minister promised that there would be no 'water grabbing' of the poor from the rich, of blacks from whites.

The plight of the millions of South Africans without access to primary household water basic has received priority. The test case for increased access of the historically disadvantaged sectors to water for economic purposes still lies ahead (also see below). The current focus is on the establishment of the first Catchment Management Agencies and the next phase would involve each CMA's development of their catchment management strategy, including an allocation plan for a five year period.

In the meantime, the department is engaging in several initiatives to create access for historically disadvantaged sectors to water for economic purposes, through direct allocations, financial assistance for the development of water supply infrastructure and share equity schemes. For instance, the department facilitated loan capital for the (white) Blyde River Irrigation Board to upgrade their bulk water supply infrastructure, on condition that they forfeit a percentage of the saved water for 800 hectares of small-scale irrigation development.

The second part of the question, on water management decision-making, requires analysis at various levels. Decision-making on water management takes place within the headquarters, regional and field offices of the Department of Water Affairs and Forestry, but also in Water Boards, municipalities and local government structures on domestic and industrial water supply, and in Irrigation Boards and schemes for irrigation water supply. Integrated water management in river basins will be implemented through Catchment Management Agencies.

In general terms, there is commitment to the political objectives of the water law reform at senior management level within the department and a good level of understanding of the tools and provisions of the National Water Act. At the lower management and staff levels, there has been less commitment and understanding, and this is being addressed through staff training courses on the implications of the National Water Act.

There has been a varying degree of transformation in the Water Boards, municipalities and local government structures responsible for domestic and industrial water supply. The capacity building and training of the new local government structures has been a major challenge. The department established the Community Water Supply and Sanitation Training Institute to train local government staff in the basics of water management and administration and 'how to be a good client', including the management of contractors and suppliers.

In the commercial irrigation sector, all Irrigation Boards are being transformed into Water User Associations under the National Water Act. The

Water Act of 1956 limited membership of Irrigation Boards to those with title deed to the land, thus effectively excluding historically disadvantaged sectors. The transformation aims to broaden membership to include all water users who are influenced by the decision-making of such a Water User Association. The impact of these changes on water management decision-making is currently the subject of a study funded by the Water Research Commission in South Africa. The National Water Act also enables groups of small-scale irrigators to establish their own Water User Associations and several such proposals are currently under development.

The functions of the department will gradually be delegated to nineteen Catchment Management Agencies to be established in the country, including the water allocation function. Current debates in the pilot catchments revolve around representation in the CMA Board, which will appoint a Chief Executive Officer to carry out the day-to-day catchment management functions under the guidance of the Board. The success in balancing interests in the CMA Board will largely determine the quality of water management decision-making at this level in future.

Where is Small-Scale Irrigation in All This and What About the Future?

It is notable that the small-scale irrigation sector made no submission on the National Water Bill during the public hearings in Parliament. Why would this be, if they were arguably the sector that stood to gain most by the change in legislation? Were they already satisfied with the provisions of the Bill? Were they comfortable that the Minister and the department would ensure that their case was protected? Or were they simply not well enough organised to put together their arguments and submit these into the process? If the latter is the case, what are the implications for the road ahead? Is government doing enough to equip this sector for meaningful participation in the new decision-making environment it has created?

The small-scale irrigation sector indeed enjoyed a measure of preferential treatment in the water law review. Their case was heard more through the political channel, on the initiative of the Minister who invited NAFU to meet with him, than through the public consultation channels, although the Minister also mobilised resources to enable people from this sector to attend workshops. However, this preferential treatment may have lulled the sector into a sense of security that militated against bottom-up organisation to protect their rights. There was no attempt by either the Department of Water Affairs and Forestry or the National Department of Agriculture to organise the sector from the bottom up, to build their capacity and to enable them to debate amongst each other the issues in the context of their own sector. One could argue whether this should be a role of government, but the effect of the adopted mode of consultation was that no real capacity has been built within the small-scale irrigation sector to prepare them for the next processes in the implementation of the National Water Act.

Is this a significant problem, or could DWAF continue to handle these issues on behalf of small-scale irrigators? Two provisions of the National Water Act work together to make it particularly important that the small-scale irrigation sector develop its own capacity. These are the hierarchy of access to water and the decentralisation of decision-making on water allocation.

The hierarchy of access to water set out in the National Water Act places the small-scale irrigation sector in direct competition with the strong economic sectors for this scarce resource. The water law review naturally focussed on safeguarding access to primary household water to address the desperate needs of the majority of water users in the country. Indeed this requirement, together with water to ensure the ecological integrity of the resource, enjoys the highest protection as the Reserve, which may not be allocated for any other purpose. Next in the hierarchy is a volume determined nationally by DWAF for international requirements and national inter-basin transfers. After DWAF has also reserved an additional volume from each catchment for 'contingency requirements', the balance becomes available for allocation by the CMA within the catchment. With primary water requirements safeguarded in the Reserve, the focus in that sector has shifted from allocation to the development of the necessary infrastructure to ensure physical access. This is not a function of the CMA, but is dealt with in a separate institutional framework involving DWAF, the Water Boards and local government. This train of events leaves the small-scale irrigation sector as the most vulnerable grouping within the CMA, certainly in terms of water allocation.

A simple comparison of the participation in the water law review shows how even the least affluent of the large economic sectors, namely commercial irrigation, is significantly better equipped to state their case. Despite the considerable expenditure on travel and accommodation this required of individual farmers, they were able to attend workshops and voice their concerns. Some irrigation regions were able to commission their own reports and one group even invited the Director-General to visit their area in a private aeroplane, to increase his understanding of their situation. None of these are even remotely possible for the small-scale irrigation sector. Their case was stated by NAFU, which represents only a very small percentage of small-scale farmers and has severe difficulties in communicating with its members, as discussed before.

In the current consultations and departmental activities towards the establishment of the pilot CMAs, this pattern is repeating itself. There is no evidence of spontaneous participation by the small-scale irrigation sector and a persisting tendency to refer to the 'agricultural sector' as if it were a single coherent entity. This seems to imply that it would be inadequate to rely on DWAF to continue acting on behalf of small-scale irrigators, as it did during the water law review. The establishment of a reserve for basic economic activity could go a long way towards safeguarding some access for small-scale production. However, it would not contribute to the ability of the sector to influence other water management decisions, such as the timing of releases from dams, decisions on resource development, etc.

Has small-scale irrigation already lost its first water battle against the mining sector? In fact, there was no battle, just a simple transfer of the allocation from a small-scale irrigation scheme to a group of mines, without the knowledge of the irrigators. Admittedly, this is a temporary transfer, to be returned to the scheme in the next few years when the capacity of the dam will be increased. Due mainly to the withdrawal of government support to the scheme over the past few years, the scheme is hugely under-utilised. Since it is estimated that it would take a couple of years to reactivate the scheme, it is a neat water management solution to use this allocation in the meantime for mining development that contributes to the sorely needed economic development of the area. However, it does raise a very fundamental question about the relative value of different water uses, a question that will be posed more and more frequently as water becomes scarcer.

Indeed, where is small-scale irrigation in all this? Is it in the national interest to spend resources on such an embattled sector, which will increase the use of a very scarce resource for relatively low value production? This question has to be faced squarely, especially in its consequences for poverty eradication, redress of gender imbalances and community stability.

Once government has taken this very basic decision, it needs to adapt its actions accordingly. If the answer is 'no', then it needs to be clear about it so that the sector knows where it stands, but if 'yes', then a serious rethink is necessary about the capacity building and support requirements of the sector.

How could the capacity of the small-scale irrigation sector be developed to take their stand in the water management debates that has become a permanent feature of the water sector in South Africa? The pre-workshop actions in the water law review process (described above) were effective in ensuring that the historically disadvantaged sectors who were present at the workshops had an opportunity to formulate arguments and nominate a spokesperson to state their case in the main event. Participation in these events also increased those participants' knowledge about the current water management issues.

However, this approach has some shortcomings. These pre-workshops remained isolated events with limited participation. Limited capacity building was achieved and only of those who happened to be present. These consultations did not draw on, nor in any way contributed to organized representation of the sector. This is partly due to the belief that achieving organised representation is a daunting task that would take too much time and thus slow down the progress of the overall process. Today this trend continues and professionals in the field try to get around this obstacle by seeking the 'perspective' rather than a 'representative' of the sector in question. The capacity building of a small number of individuals is perpetuated, because somewhere in some process their names happened to be captured on an attendance list.

There is no real substitute for bottom-up, federated organisation through which small-scale irrigators could debate water management and other issues and develop and put forward a growing corps of leaders from their midst to speak on their behalf. DWAF has recognised this and is currently experimenting with the establishment of a small-scale irrigation forum in one of the pilot CMA areas. The preliminary idea is that the resources to enable meetings and communication among the members of the forum would form part of the CMA's budget. New questions are raised in the process. Some resistance is being experienced from a political leader in one of the sub-catchments, who argues that there is already a forum representing disadvantaged communities' water issues in his area. This is not true of the rest of the catchment, but raises the question if small-scale irrigation should be represented separately or as part of an overall organising effort around water management issues in rural communities. Or would such a step mean that the needs of small-scale irrigation would again be subsumed?

The answers to these questions are not clear yet, but requests are increasing for training in the National Water Act, so that small-scale irrigators would know their rights and obligations. Recently, infrastructure was demolished of small-scale farmers who were unaware whether they were acting within their rights when they built these, or not. Small-scale cattle farmers report that commercial farmers across the river tell them that they are not allowed to water their cattle in the river without specific permission. This may have been true under the old Water Act, but is no longer the case under the National Water Act. In practice, the scales are still out of balance and small-scale farmers, maybe more than any other sector, need to know their rights.

Notes

1   The following websites contain information relevant to the water law review process and this paper: The South African Constitution (Act 108 of 1996), the National Water Act (Act 36 of 1998) at http://www.polity.org.za/govdocs; the White Paper on a National Water Policy for South Africa at http://www.polity.org.za/govdocs/white_papers/water.html; Parliamentary proceedings, minutes of the meetings of the Agriculture, Water Affairs and Forestry Portfolio Committee and the National Council of Provinces, public submissions at http://www.pmg.org.za.
2   The Southern African Development Community (SADC) consists of 14 Southern African countries, including Angola, Botswana, Democratic Republic of the Congo, Lesotho, Malawi, Mauritius, Mozambique, Namibia, South Africa, Seychelles, Swaziland, Tanzania, Zambia and Zimbabwe.
3   The International Water Management Institute's world map indicating expected water scarcity by 2025 can be found on the following website: http://www.iwmi.org.
4   Minutes of the Coordinating Committee on Support for Small-scale Irrigation, 27 March 2000.
5   Initially, the Minister was opposed to any form of compensation for reduced or lost water allocations. However, in meetings with the Minister and through the press, the agricultural sector pointed out that this would be unconstitutional, a position that was confirmed by constitutional lawyers advising the Minister.
6   In the policy workshops, community representatives invariably raised the specifics of their plight for domestic water. Len Abrams, a consultant who facilitated the workshops on the Water Law Principles, used this phrase to explain that it was impossible to address all the infrastructure requirements of communities through the water law review process. The immediate focus was to ensure that people had a legal right to the water. In later consultation processes, DWAF noted and followed up these specific requests for assistance through the on-going infrastructure development programmes of the department.
7   SAAU recently changed its name to Agri-SA and established a national Agriculture Business Chamber, partly to deal with the specific challenges of globalisation and national restructuring. In addition to the water law review, many major changes, both locally and internationally, has affected the agricultural sector, including the dismantling of the national marketing boards, changes to labour legislation, land

reform and other policies and increasing criminal attacks on farm families.

8   See Note 1 for the website where the minutes of the parliamentary committees, including the Agriculture, Water Affairs and Forestry Portfolio Committee (AWFPC) can be found.
9   This is not an exhaustive list of the changes introduced by the water law review, but only tries to reflect some of the concerns of the commercial agricultural sector, how they were voiced and how they were dealt with.
10 A copy of the speech in Parliament of Minister of Water Affairs and Forestry (1994-99), Professor Kader Asmal, is available at http://www.pmg.org.za.
11 Under the policy of 'separate development' or *apartheid*, several nominally independent homelands were established within the boundaries of the country, according to the tribal origins of the various peoples of South Africa. These homelands were characterised by overcrowded concentrations of resource poor people with little or no economic opportunities.
12 Consultants are a regular feature of the department's resources.
13 See the Department of Water Affairs and Forestry's website for more detailed information on current processes: http://www.gov.za/dwaf.html.
14 This is primarily because irrigation plays a relatively minor role in the industrialising South African economy, at less than 5% of GDP.
15 For a lay person it remains a mystery that the results of this council were never made public, despite the provision in the Constitution that 'everyone has the right of access to any information held by the state'. A motion by one of the political parties in the final meeting of Parliament's National Council of Provinces that this information should be made available, was voted down by the African National Council majority on the grounds that it would delay finalisation of the water law review process.

References

De Lange, Marna (1994), *Small-Scale Irrigation in South Africa*, WRC Report 578/1/94, Water Research Commission, Pretoria, South Africa.

DWAF (Department of Water Affairs and Forestry) (1997), *Proceedings of the National Agricultural Workshop on Water Law*, September.

DWAF and NDA (Department of Water Affairs and Forestry and National Department of Agriculture) (1997), *Towards an Irrigation Policy for South Africa*, Pretoria, South Africa.

Chapter 3

The Politics of Creating Commitment: Irrigation Reforms and the Reconstitution of the Hydraulic Bureaucracy in Mexico

Edwin Rap, Philippus Wester and Luz Nereida Pérez-Prado

Introduction

In the past ten years water management in Mexico has been radically reformed under the influence of neo-liberal government policies. The creation of the *Comisión Nacional del Agua* (CNA; National Water Commission)[1] in 1989, the transfer of government irrigation districts to water users' associations (1989-present) and the promulgation of a new water law (1992) exemplify this. Internationally, Mexico's Irrigation Management Transfer (IMT) program has been heralded as a success and has drawn widespread attention for its rapid implementation.[2] Consequently, it has been propagated as a model for other countries seeking to improve the performance of their public irrigation systems and cut burgeoning public expenditures (Groenfeldt, 1998).

Surprisingly for a reform of this magnitude no attempt has been made to explain how is was articulated. Policy documents single out the strong commitment of the political leadership and policy managers to the IMT program and the creation of appropriate legal and institutional frameworks as explanations for the origin and success of IMT (Gorriz et al., 1995; Groenfeldt, 1998). However, how and in which arenas this commitment was created and which actors were fundamental to this process remains unexplained. The literature presents the occurrence of IMT in Mexico as a logical and unavoidable outcome of the economic crisis of the 1980s. The argument goes that this crisis led to a large decrease in government funding for irrigation and a reduction in the payment of water fees by water users, resulting in a poor performance of the publicly managed irrigation districts and a widespread deterioration of the irrigation infrastructure. The irrigation reforms were thus an inevitable response of the Mexican government to this state of affairs[3] (Gorriz et al., 1995; Johnson, 1997a, 1997b; Kloezen et al., 1997; Palacios, 1997, 1998). We posit that this line of argument does not suffice when it comes to understanding the emergence of the IMT policy and the strong commitment it enjoyed from senior politicians and bureaucrats.[4]

To understand why IMT became a reality in the 1990s it is necessary to broaden the analysis to include the historical, political and bureaucratic[5] processes that engendered and sustained Mexico's water reforms. Such an analysis, which takes policy actors as the unit of analysis and the articulation[6] of reforms as the focus of attention, clarifies why and when water reforms are effectuated and how alliances are negotiated through which reforms gather momentum. By focusing on their political and bureaucratic embeddedness it becomes possible to understand the trajectories and variable content of water reforms, the conflictive dynamic of their articulation and their ramifications. In this regard it is important to acknowledge that policy-making in Mexico '(...) does not result from pressures exerted by mass publics, nor does it derive from party platforms or ideology, nor from legislative consultation and compromise. Rather, it is an end product of elite bureaucratic and political interaction (...)' (Grindle, 1977:7). As a consequence we need to focus on the political and bureaucratic actors, arenas and conditions that play a role in engendering policy ideas and bureaucratic transformations.[7]

The water reforms in Mexico were part of a larger set of neo-liberal reforms enacted during the presidency of Salinas de Gortari (1989-1994) with the support of international funding agencies. Decentralisation policies such as the transfer of irrigation districts are assumed to weaken the position of the bureaucracy by devolving authority and control over resources to organisations external to the government, reducing government subsidies and downsizing bureaucratic staff. This has led many people to believe that the hydraulic bureaucracy[8] in Mexico had no part in defining or supporting the water reforms and that either the Mexican president or international funding agencies must have imposed it. This perception of reform processes also partly explains why policy documents place such emphasis on the need for the strong commitment of policy managers to reforms.

In contrast to this reading of the water reforms, this article argues that the composition of and the commitment to the water reforms emerged from a complex and protracted process of interaction and enrolment between policy actors such as senior hydrocrats,[9] the Mexican presidential candidate and World Bank officials. Segments of the hydraulic bureaucracy played a crucial role in the generation of policy ideas and the articulation of the water reform package in the 1980s, as part of an ongoing struggle within the Mexican bureaucracy. We argue that IMT became feasible because it was embedded in a broader reform package that resulted in greater autonomy for the hydraulic bureaucracy through the creation of a single water authority and the reordering of control over bureaucratic domains and resource flows. Thus, these neo-liberal water reforms were only possible as they combined decentralisation with the concentration of bureaucratic powers and generated important benefits for segments of the hydraulic bureaucracy.

The aim of this article is to stimulate debate on the role of hydraulic bureaucracies in water reforms. The management of irrigation by bureaucracies became a target of criticism in the late 1970s and 1980s rural development literature (Bottrall, 1981a, 1981b; Chambers, 1988; Moore, 1981; Repetto, 1986; Wade, 1978, 1982; Wade and Seckler, 1990). This literature stressed the need to study the hydraulic bureaucracy, but only focused on the discrepancies between policy objectives and implementation, highlighting the problematic role of low level field staff. By sustaining the divide between policy formulation and implementation, this literature viewed the bureaucracy solely as an instrument for attaining policy objectives, thus disregarding the role of senior hydrocrats in policy-making activities both before and after policy legislation (Clay and

Schaffer, 1984; Long and van der Ploeg, 1989; Yanow, 1988). To overcome this, we view bureaucracies as 'the accumulated product of a social history of past policies [that] become congealed in institutional form and develop a network of interests around them, both inside and outside the bureaucracy' (Beetham, 1987:51). These networks of interests and concerns are rooted in the history and culture of particular bureaucracies, their relationship to larger socio-political constellations as well as in actors' education and professional experiences. By focusing on the historical and cultural embeddedness of these overriding concerns we do not attribute them to isolated individuals and prevent that too much emphasis is placed on the intentional behaviour and strategic action of individuals. Hence, no criticism of individuals and their motives, values and beliefs is implied. Although our analysis of the water reforms in Mexico is by no means a comprehensive account of 'what really happened' and is at times speculative,[10] it demonstrates the benefits of bringing the bureaucracy back in when analysing water reforms.

The following section explores how the Mexican hydraulic bureaucracy developed a set of overriding concerns between 1926 and 1976 related to the control over irrigation infrastructure and bureaucratic and financial autonomy. These overriding concerns help us understand, in Section 3, why the merger between the hydraulic and agricultural bureaucracies in 1976 was so traumatic and led to an energetic struggle for renewed autonomy. Against this backdrop, the consolidation of the water reform package is analysed in Section 4 while Section 5 reviews how this consolidation resulted in IMT and the reconstitution of the hydraulic bureaucracy through the formation of the CNA. Lastly, Section 6 discusses the conclusions and implications of our analysis.

The Construction Era (1926-1976) and the Formation of the Hydraulic Bureaucracy

To understand policy articulation processes in Mexico it is necessary to place the struggles between policy actors in the broader frame of historical, political and bureaucratic transformations. The answer to the question of what is distinctive about the link between the Mexican political system, its bureaucracies and the policy process can be found in several key developments in the country's political history. The political authorities that have ruled Mexico since the 1920s managed to establish a relatively stable political regime compared to other parts of Latin America. The Mexican revolution (1910-1917) was appropriated by the triumphant political elites gathered in the *Partido Revolucionario Institucional* (PRI; Institutional Revolutionary Party). Since its foundation in 1929, the PRI has ruled the country uninterruptedly, and, until a decade ago, without much politically organised challenge. According to some authors, the PRI owes its 'success' to its early establishment of political and economic mechanisms for solving conflicts within the elite and for ensuring mass support and political control. Through clientelism and corporative representation and control the party successfully incorporated workers, farmers and the middle class (Camp, 1999; Grindle, 1996).

In agriculture the expansion of irrigation[11] and the centralisation of decision-making was used by the Mexican regime to keep in check sharp socio-economic differences and increase political control (Vargas, 1996). This centralisation[12] entailed that the federal government strengthened its role in water management and consolidated the state's control over water through the formation and expansion of a hydraulic bureaucracy that enjoyed a large degree of bureaucratic and financial autonomy.[13] This section reviews this process from 1926 to 1976 to understand how the network of interests and overriding concerns of the hydraulic bureaucracy were formed and points out those aspects of bureaucratic transitions, culture and politics important for elucidating the creation of the CNA and the articulation of the IMT policy. We view these concerns not as historical or cultural constants, but as accumulated products of a history of legal, bureaucratic and policy transformations, embedded in a larger process of centralisation.

In his insightful work *El Agua de la Nación* Aboites (1998) shows that the process of centralisation by which the federal government of Mexico increased its control over water already started at the end of the 19th century during the Porfirian regime. This trend was consolidated in Article 27 of the constitution drawn up after the Mexican Revolution in 1917, in which water in Mexico was defined as belonging to the nation and falling under federal jurisdiction. Article 27 also established that the only legal way to gain access to the nation's water is through a concession granted by the federal government. Based on Article 27 the centralisation of water management began in earnest in the 1920s, when President Calles launched a program for the construction of large-scale irrigation systems (termed irrigation districts in Mexico). This program found its legal expression in the Irrigation Law issued in 1926, which also created the *Comisión Nacional de Irrigación* (CNI; National Irrigation Commission), the first government agency solely devoted to the design and construction of irrigation districts and their subsequent management. Although exact data on the area irrigated before the creation of the CNI are not available, Orive-Alba (1970) estimates it to have been some 800,000 ha. In twenty years time (1926-1946) the CNI doubled this figure through the construction of 816,200 ha of large-scale irrigation systems and 21,343 ha of small-scale systems, mainly in Northern Mexico (SRH, 1975). Dam storage capacity in Mexico, which amounted to 88 million cubic meters (MCM) in 1900 increased to 22,717 MCM by 1946 (SRH, 1976).

Control over Irrigation Districts

The construction, settlement and management of medium and large-scale irrigation districts entailed control over large sums of money as well as political control over the selection of beneficiaries of government programs and their access to irrigated agriculture (Martínez, 1988). As a result, the control over the irrigation districts and the resources allocated to this function was the subject of much bureaucratic competition between 'functional rivals'. Greenberg (1970) argues that the functional rivals of the hydraulic bureaucracy consisted of those agencies whose activities were similar enough that their personnel felt them to be competitive with each other. The Ministry of Agriculture in particular fit this bill.

Between the 1920s and 1940s the content of irrigation policy was subject to both inter-bureaucratic struggles as well as the ups and downs in the relationship between the state and the peasantry. This revolved around the long standing tension between policies targeting private capital as a means of increasing agricultural production and those directed at peasants to retain political support in rural areas (Fox, 1992; Gates, 1988; Stanford, 1993). Under President Calles in the 1920s, the construction of irrigation districts served to replace large landholdings (*latifundia*) with medium-sized family farms. The CNI was instrumental in the creation of this new class of farmers through the settlement and land distribution efforts it oversaw in the irrigation districts. In the mid-1930s President Cárdenas dealt with this challenge in quite a different manner by proceeding to make true the revolutionary promise of giving the 'land to the tiller', especially in regions where large landowners were amongst his political opponents. During his term, nearly half of the irrigated surface was incorporated into the *ejidos*[14] (Wionczek, 1982). As the CNI was too closely linked with Calles' policies, Cárdenas transferred the administration and colonisation of the irrigation districts to his trustees at the Bank of Agricultural Credit in 1934. The CNI was vehemently opposed to this move and in 1943 won the inter-bureaucratic struggle and regained control over the administration and colonisation of most of the irrigation districts (Orive-Alba 1970).

During this period responsibilities for water development were spread over several federal ministries and agencies: irrigation with CNI, flood control with the Ministry of Communications and Public Works, potable water with the Ministry of Health and hydroelectricity with the Federal Electricity Commission. During the presidential election campaign of 1946 the then director of CNI convinced Alemán, the presidential candidate, of the need to correct this dispersion of administrative efforts. After Alemán became president this came to pass, and in December 1946 the *Secretaría de Recursos Hidráulicos* (SRH; Ministry of Water Resources) was created. This was a pioneering move in many respects and was the first time water was elevated to the level of a ministry in the Western Hemisphere. The SRH integrated the use of water resources and concentrated the government's efforts in this field in a single organisation. However, when the ministry was still being formed, the *Secretaría de Agricultura y Ganaderia* (SAG; Ministry of Agriculture and Livestock) persuaded Alemán that it should have authority over the irrigation districts, thus continuing its struggle with the hydraulic bureaucracy (Orive-Alba 1970). Once again this met with severe resistance from the hydraulic bureaucracy and Agriculture's control over the irrigation districts was to last only a short period, with the SRH regaining control over the irrigation districts in 1951.

The above shows how legal and policy transformations concerning irrigation management are subject to bureaucratic struggles between functional rivals, such as the SRH and the SAG. These transformations revolve around two crucial political phenomena in Mexico: presidentialism and the presidential term of six years (the *sexenio*). Presidentialism refers to the dominant role that the president plays in reordering bureaucratic domains and in the materialisation of political and economic reforms during his *sexenio*. The *sexenio* amounts to a calendar of political time, as described by Grindle (1977). A relative rupture with the preceding administration typically characterises the beginning of each *sexenio*, through changes in the leadership at all levels of the federal administration. At the end of the *sexenio* bureaucratic groups align themselves with and offer their support to close allies of the presidential candidate that will probably be appointed to top posts in the upcoming administration (Greenberg, 1970). These prospective senior government officials influence the presidential candidate's views on institutional and policy reforms as well as on the division of bureaucratic functions and allocation of resources. Policy changes and bureaucratic transitions are thus most frequently shaped and defined at the end of a *sexenio*, to be initiated at the beginning of a new *sexenio*. When viewed in a longer time frame, it becomes apparent that the radical ruptures in the control over the irrigation districts are structured by the political calendar of the *sexenio* and in that regard turn out to be an element of continuity.

During the following decades until the creation of the *Secretaría de Agricultura y Recursos Hidráulicos* (SARH; Ministry of Agriculture and Water Resources) in 1976, the SRH consolidated its control over the irrigation districts and managed to keep the Ministry of Agriculture out. This cumulated in 1972, with the promulgation of the *Ley Federal de Aguas* (Federal Water Law). Article 46 of this law establishes that the SRH is completely responsible for the irrigation districts, from construction to management, effectively forbidding user management of the districts (Diario Oficial, 1972). However, the various water laws promulgated between 1926 and 1947 contained provisions for the creation of water users' associations or water boards to manage irrigation districts. The 1929 water law already mentions water users' associations and confers legal status to them (SAyF, 1929).[15] Subsequent water laws also allowed for the operation of irrigation districts by user associations or water boards to only become illegal under the water law of 1972. For example, the irrigation law of 1947 mentions that:

> Until the (...) the users are organised and sufficiently trained as to take charge [of the operation of the irrigation districts and irrigation units], the Ministry of Agriculture and Livestock will be in charge of the maintenance of the structures, water distribution and the overall operation of irrigation districts. For this reason it will make efforts to organise water boards and water user associations, to whom in the end the operation of the irrigation districts will be handed over. When it is judged opportune, the districts or units will be turned over to their respective users, so they can operate them directly, albeit under the vigilance and supervision of the Ministry of Agriculture and Livestock (Diario Oficial, 1946:16; our translation).

In line with the law's provisions water boards were charged with the operation of several districts in the north, such as Rio Yaqui, Rio Mayo, Rio Colorado and Delicias. But water monopolisation by commercial farmers combined with the lack of strong government support was in most cases detrimental to the water boards enduring. The control of most of these districts was returned to the SRH from SAG in 1951, effectively ending the experiments with water boards[16] (Palacios, 1993; Vargas, 1996).

Financial Autonomy

The monopoly of the CNI and the SRH in the construction of irrigation systems secured it a large and steady income flow between the 1930s and the 1970s. These resources represented an important element of continuity for the hydraulic bureaucracy and largely accounts for the financially wealthy and autonomous bureaucracy that the SRH became during this period. The SRH's budget was one of the largest among the federal agencies with 61 to 100% of pubic investments in the agriculture sector going to the construction of irrigation works between 1926 and 1976. Further, it managed its own funds and had relative budgetary freedom from other bureaucratic entities, although subject to presidential and party priorities (Durán, 1988; Greenberg, 1970; Grindle, 1977; Wionczek, 1982).

It was only in the early 1960s that foreign loans started to become important for SRH (Durán, 1988; Greenberg, 1970; World Bank, 1983b). Because of an international reputation of being an efficient and technically competent ministry the SRH was very successful in acquiring international loans for irrigation construction purposes, thereby generating urgently needed foreign currency for the government. From 1966 to 1975 foreign loans constituted more than 15% of SRH's irrigation investments on average (Durán, 1988; World Bank, 1983b). Mexico and more in particular its hydraulic bureaucracy became favoured clients of the World Bank. As a major recipient of external funding SRH was granted privileges not given to other ministries, such as a large degree of autonomy in making technical decisions and a significant budget to hire a cadre of well-trained professional engineers (Greenberg, 1970).

Another source of income for the hydraulic bureaucracy was the water charges that it levied on irrigators. However, this source of income was much less stable and controllable. Apart from the fact that the water charges collected in the districts were never sufficient to fully cover Operation and Maintenance (O&M) costs, the fees were not paid directly to SRH but to the Ministry of Finance (van der Zaag, 1992). The initial policy intention under Calles was that those who benefited from state-built irrigation works would reimburse the state for its investment as well as fully cover the O&M costs of the irrigation systems (Wionczek, 1982). This objective was reiterated in the 1947 irrigation law. Nonetheless, fee payment generally covered only a fraction of irrigation investments and O&M costs (Aboites, 1998). Between 1950 and 1964 cost recovery averaged 60% (Orive-Alba, 1970). From 1965 to 1976 this average slipped slightly to around 56%, but between 1977 and 1982, it dropped drastically to around 20% (Johnson, 1997a).

Although it is unclear how fees were actually established by the different agencies responsible for the irrigation districts throughout the years, political criteria were often more relevant than technical and financial ones (Ascher, 1999; Wionczek, 1982). The argument often used to justify low water charges was that the poorer farmers in the districts would not be able to pay the fees (Wionczek, 1982). However, this does not explain why the fees were not adjusted in accordance with the increased value of irrigated land. Policies favouring low fee levels and the government's priority to invest in large construction works rather than in optimal use of the available infrastructure led to a sub-optimal use and deterioration of the irrigation infrastructure.[17] However, this deterioration presented itself selectively, especially harming small farmers. Middle and large producers were financially able to solve maintenance problems for themselves. Although supported in the name of social equity, low water service fees were mainly beneficial to larger farmers.

The above shows how financial autonomy became an important concern of the hydraulic bureaucracy. This concept refers to the degree that a bureaucracy can generate and control its income flows, set its own budgets and decide on expenditure and investment independently from other bureaucratic entities. The more affluent a bureaucracy is and the more budgetary freedom it has the larger its degree of financial autonomy.

Bureaucratic Autonomy

At the centre of the Mexican political system stand the president and the party. Officially, they decide on the programs that the bureaucracies should undertake, in line with the political projects that they intend to carry out in the coming *sexenio*. The party has the power to appoint and remove officials at all levels of the bureaucracy (Greenberg, 1970). Hence, officials must maintain good relations with the party and party activity is an important prerequisite for high level appointments in the bureaucracy (Grindle, 1977).

The co-ordination of water management at the ministry level during the SRH era implied that senior hydrocrats stood in direct contact with the president. The minister was the central figure in the SRH, appointed by the president and directly responsible to him for all the actions of the Ministry. The president allowed his minister, often a friend or political confidant to appoint his own team of trusted collaborators (*equipo*) and left the internal operations and the management of funds to the discretion of the minister. The minister thus enjoyed a considerable degree of operative and budgetary autonomy, within the broad policy lines negotiated with the president. Historically, well-qualified men, all trained as civil engineers and with experience in the Ministry led the SRH. The minister's technical qualifications were considered important to impress donor agencies during loan negotiations and to convince them that their money would be well spent. At times the SRH minister played an important role in national politics and at least once in the presidential succession (Castañeda, 1999; Greenberg, 1970; Orive-Alba, 1970).

During its 30 years of uninterrupted reign the SRH was not constrained by any superior bureaucratic entity, such as for instance the Ministry of Agriculture. An important concern of senior hydrocrats thus became their bureaucratic autonomy, i.e. the degree of operative freedom and internal control that a bureaucracy has and the extent to which it can prevent external influence on decision-making. In part this was possible because of the close relations with several Mexican presidents and the PRI and the important role that the hydraulic bureaucracy played in their political projects.

To illustrate how these overriding concerns are embedded in the history and culture of the hydraulic bureaucracy it is necessary to review the composition of its staff and the consequences this had for the internal relationships, bureaucratic culture and network of interests that developed. The expansion of the hydraulic bureaucracy and the professional and bureaucratic formation of its cadre occurred in a complex of education and research institutions, professional associations, private construction companies and international organisations with which the hydraulic bureaucracy maintained close relations.

A distinct attribute of the hydraulic bureaucracy was the relatively homogeneous composition of its staff, with similar academic and bureaucratic careers, which contributed to the closed and hierarchical culture of the SRH and the strong sense of identity of hydraulic bureaucrats. The professional staff of the CNI and SRH consisted of civil and irrigation engineers that had similar educational backgrounds and bureaucratic careers. The majority of these engineers were trained in two major Mexican engineering schools: the Faculty of Engineering of the National Autonomous University of Mexico (civil engineers) and the Chapingo National School of Agriculture (irrigation engineers). Apart from professional skills, students developed strong friendship bonds and clientele networks in these schools, which benefited them in their bureaucratic careers. In addition, the relation between the bureaucracy and these education and research institutions was actively maintained, with SRH officials returning to their universities as lecturers and vice versa.

The similar training and bureaucratic trajectories of most of the engineers had an important impact on the bureaucratic culture of the SRH. The SRH was a ministry with a small staff of well-paid engineers that worked in a structured and organised manner. There was little room for disobedience and both orders from superiors and the periodic transfers of officials ordered by headquarters were followed and accepted. In comparison with the Ministry of Agriculture, the SRH was known for its closed, conservative and authoritarian culture and the strong discipline of its staff. During the first decades of the hydraulic bureaucracy its engineering staff developed a keen sense of its hydraulic mission and possessed a strong esprit de corps. Officials of the hydraulic bureaucracy identified with the grand tradition of Mexican hydraulic engineering and its major accomplishments and defined themselves as a distinct group with its own bureaucratic history and culture. This permitted a sense of pride towards the profession and the institution, strengthened by closed networks (to the outside world) of friendship and mutual support.

These networks of the hydraulic bureaucracy extended to construction companies. In the early forties SRH decided to tender contracts for the construction of dams and large irrigation works. This resulted in the formation of several Mexican construction companies, which played a major role in the development of the SRH, since they served as contractors and consultants to the Ministry. In addition, former members of the hydraulic bureaucracy frequently staffed them and senior hydrocrats were advisors for these companies or had financial interests in them. The hydrocrat thus fulfilled different roles, namely that of bureaucrat, politician and businessman. Understandably, the close links with contractors resulted in pressures to give priority to construction projects and partly explains the strong construction bias of the hydraulic bureaucracy (Greenberg, 1970).

The relations with international organisations form another important field of institutional interactions. Due to the good reputation of the Mexican hydraulic bureaucracy, senior SRH engineers did consultancies for international organisations, such as the FAO, the Inter-American Development Bank and the World Bank. These contacts facilitated the negotiation of international loans in which the SRH became reasonably successful (Greenberg, 1970) and played an important role in obtaining the World Bank's support for the prestigious National Water Plan at the beginning of the 1970s, as detailed in the next paragraph.

To recapitulate, three main points emerge from this section. First, from the 1920s to the 1970s irrigation development and management were characterised by increasing federal government intervention. Second, this centralisation of water management, coupled with the government priority of development of large-scale irrigation led to the formation, expansion and specialisation of a hydraulic bureaucracy. Over time, this bureaucracy developed specific overriding concerns, namely control over irrigation districts and bureaucratic and financial autonomy, and cemented its relative autonomy through the relations it maintained with the president, the party and a broader set of state institutions and funding agencies. Third, it is clear that user management of irrigation districts is not new, although the government did not actively support the operation of irrigation districts by users, partly explaining why most of them did not endure.

Loss of Autonomy and the Emergence of Policy Ideas in the SARH Era (1976-1988)

The SARH Fusion

In 1976 President López-Portillo merged the smaller but financially affluent SRH with the larger but financially poorer SAG to create the *Secretaría de Agricultura y Recursos Hidráulicos* (SARH; Ministry of Agriculture and Hydraulic Resources). The creation of the SARH was linked to several political, administrative and economic considerations. Apparently senior SRH officials backed another presidential candidate in 1975, which persuaded López-Portillo to 'punish' SRH and remove certain bureaucratic groups from the political stage. In addition, as a Minister of Finance during the Echevarría administration (1970-1976) he had experienced severe problems with SRH's lack of budgetary discipline. Although economically the tide was against him in 1976 López-Portillo decided to expand the state's role in the economy.[18] This was made possible by the discovery of new oil deposits and the substantial increases in oil prices (Grindle, 1996). López-Portillo announced a substantial administrative reform to rationalise the wide array of bureaucracies created in previous decades (Martínez, 1988). This included the creation of the *Secretaría de Programación y Presupuesto* (SPP; Ministry of Programming and Budget), a ministry that unified financial planning functions that had previously been distributed among various ministries.[19] The creation of SARH was part of this administrative reform and served to unify all activities related to agriculture in one ministry (Arce, 1993).

With the creation of SARH, the SRH was dissolved and effectively downgraded to the level of an Under Ministry. The SARH was divided in three under ministries, each headed by a Deputy Minister: Agriculture and Operation, Planning and Hydraulic Infrastructure, to which most of the old SRH officials were assigned. As a result, senior hydrocrats were no longer in direct contact with the president. The Deputy Minister now had to submit his policy initiatives to the SARH minister, significantly curtailing his discretionary powers. We will illustrate below that senior hydrocrats lost control over crucial bureaucratic domains and resource flows to other bureaucratic groups, which increasingly started to dominate the SARH. The hydraulic bureaucracy thus lost its dominant position and bureaucratic autonomy and was subjected to the control of the agricultural bureaucracy.

The bureaucratic merger entailed a serious demotion for the hydrocrats and provoked 'institutional turmoil' (Mestre, 1997). The SRH top opposed the fusion, as they clearly understood that it would entail a significant loss of autonomy. For many hydrocrats the fusion was traumatic and they experienced the demise of the SRH as the end of the grand era of hydraulic engineering in Mexico. To make matters worse they were fused with an old-time functional rival.[20] In 1970, unaware of the fusion to come, Greenberg wrote that 'The sharing of power with the Agriculture Ministry is the result of a long history of struggle which saw first one, and then the other agency, in a position of dominance' (1970:87). The merger started a new phase in this historical struggle and engendered an energetic and politically expressed demand for renewed autonomy on the part of the hydrocrats.

Although the irrigation districts were left intact, they became the responsibility of the Under Ministry of Agriculture and Operation, implying that the hydraulic bureaucracy lost control over the irrigation districts as well as the income flows related to the districts (Palacios, 1994). In addition, after an initial phase in which the hydrocrats continued to dominate in the state delegations and the irrigation districts, many of them were replaced by agronomists (Arce, 1993). The subordinate position of the hydrocrats led to intensive conflicts between groups of ex-SRH and SAG officials. Not only were the academic and bureaucratic careers of the hydrocrats dissimilar, also the conservative and authoritarian culture of the SRH, the bureaucratic discipline, the professional identity they shared and the strong pride they felt towards the grand tradition of hydraulic engineering made them feel very distinct from the SAG-agronomists. These antagonistic cultures contributed to severe tensions between these two groups over the internal operation and control of the SARH. Ex-SRH officials at the time jokingly referred to the merger of SRH and SAG as the confusion instead of the fusion.

The displeasure of the hydrocrats over the loss of control was not only directed at the agronomists, but also at the growing influence of 'politicians' and 'administrators' in the ministry, i.e. non-engineers without experience and interest in hydraulic matters. The first SARH minister (1976-1982) was a politician without a professional degree and the second minister (1982-1988) was a lawyer, something which radically broke with the SRH tradition of being led by well-qualified men, trained as civil engineers and with a career in the SRH itself (Greenberg, 1970). In the SARH, ex-SRH officials were thus confronted with politically appointed administrators in positions that used to be occupied by engineers. The changed set of decision criteria and especially the ignorance or neglect of specific technical criteria was a recurrent source of frustration for them.

The hydrocrats in SARH were also severely constrained in their financial autonomy. In the SRH, the major source of income had been the funds for the construction of irrigation systems. In the SARH decisions over the construction funds were taken out of the hands of the hydrocrats and their Deputy Minister. Agronomists interfered with the decisions over construction funds and succeeded in diverting much of these funds to other purposes. Especially the construction companies, the traditional beneficiaries of the contracts tendered by the SRH, were affected by this shift in decision-making power.

At the end of López-Portillo's *sexenio* influential groups of civil engineers started to lobby for renewed bureaucratic autonomy and explicitly expressed their demand for an autonomous water authority during the election campaign of presidential candidate De la Madrid in 1982. During this campaign a working group was formed to define water policies for the upcoming *sexenio* and several campaign meetings were held on water, co-ordinated by Dr. González-Villareal in close collaboration with De la Madrid's campaign manager, Carlos Salinas, who later became Mexico's president (IEPES, 1982).

Several politically influential ex-SRH ministers and senior hydrocrats participated in the working group and identified policy ideas strikingly similar to the ones adopted six years later with the creation of the CNA. The working group agreed on the need to create a 'new water culture' among users and to increase their active participation. They also suggested the need to manage water at the level of river basins. Another recommendation was the proposal to organise a 'financial system for water' that would give the hydraulic bureaucracy large discretionary powers over funds destined for the water sector. Lastly, the working group repeatedly stated that the authority to manage water should be located in 'a single water agency' (González-Villareal, 1982a).

This attempt to re-establish autonomy did not succeed. After De la Madrid became president in 1982 the Under Ministry of Hydraulic Resources was maintained as a separate but dependent part of SARH. It is likely that a more autonomous authority was not feasible at this time because of the economic crisis that held the country in its grip. Thus, the ex-SRH engineers had to accept their subordinate position in the SARH for another six years, during which time their financial and bureaucratic autonomy was further curtailed. The working group of ex-SRH officials achieved only a partial success through the appointment of their co-ordinator as the new Deputy Minister for Hydraulic Resources.

The financial situation of the SARH severely worsened between 1982 and 1988, largely as a result of the unprecedented economic crisis that hit Mexico in 1982. To deal with the crisis, De la Madrid, the first economist after a row of lawyers as president, adopted a neo-liberal approach that strongly departed from the populist and interventionist economic policies followed by previous presidents. This resulted in a restructuring of the SARH, the reduction of barriers to commerce, the abolishment of subsidies for agricultural inputs and a liberalisation of agricultural prices (Vargas, 1996). According to some observers De la Madrid gave much less priority to agriculture than previous administrations not only because of the economic problems but also because it had lost its strategic importance for the country. This is understandable as agriculture's share of the Gross Domestic Product (GDP) dropped to 6% at the beginning of the 1980s, down from 11% in the 1960s (Palacios, 1994).

The Emergence of a Planning Team, New Experiences and Policy Ideas

Partly in response to the difficult SARH era new ideas concerning irrigation policy and the restoration of autonomy for the hydraulic bureaucracy were developed by senior hydrocrats. To understand their origins it is necessary to back up a little. The *Plan Nacional Hidráulico* (PNH; National Water Plan), a water master planning organisation created in 1973 to provide a frame of reference for future lending programs in the field of water resources, played an important role in this. This new bureaucratic body led to the formation of a team of water resource planners that departed from the traditional construction bias of the hydraulic bureaucracy by developing a broader vision on water resource planning and management. The planners working in the PNH developed policy ideas and accumulated particular experiences favouring participation of water users and handing over of government tasks in water management, which can be seen as precursors to the creation of the CNA and the IMT policy. Other groups within the bureaucracy, more related to construction or O&M departments, were much less receptive to these ideas.

In 1973 the Mexican Government, the World Bank and the UNDP signed a tripartite agreement to develop a National Water Plan by 1975 (Herrera-Toledo, 1996). For this purpose the World Bank created a special office in Mexico with four permanent staff members, to assist in the formulation and evaluation of policy ideas and to advise on policy decisions (Buras, 1983). The SRH created a special Plan Commission as a semi-independent body under the Under Ministry of Planning to organise a multidisciplinary study group that would produce the Plan. Dr. González-Villareal, a civil engineer of the Faculty of Engineering of the UNAM, was appointed as the General Co-ordinator of the study group. He composed a team of young, dedicated and specialised professionals for this purpose, who became his close collaborators.[21] The staff of the Plan Commission was divided over national and regional planning groups, in which foreign resident advisers of the World Bank also participated (Herrera-Toledo, 1996). They generated an impressive set of studies on land and water resources and their use at both the river basin and national level. These studies attempted to match estimates of future water demands by the domestic, industrial and agricultural sectors with estimated future supplies and specified alternative courses of actions for meeting the projected shortfalls (Cummings et al., 1989; Herrera-Toledo, 1996).

In an evaluation of the first National Water Plan, Buras,[22] (1983) points out that at the start the PNH study group had a 'definite engineering orientation' with relatively little input from economic, agronomic or social disciplines. He states that the first PNH started as a super-project in hydraulic engineering and that solutions to the discrepancies in water availability were initially sought in the construction of inter-basin transfers and the expansion of irrigation.[23] Cummings (1974) mentions that the Mexican government considered the *Sistema Hidráulico Interconectado del Noroeste* (SHINO; North-western Interconnected Hydraulic System), an ambitious plan to expand the irrigated area in the arid Northwest by a massive inter-basin transfer of water from the south. However, several studies showed the technical and financial unfeasibility of this plan (Buras, 1983).[24]

Although the first PNH had a clear construction bias and this bias did not fully disappear in subsequent revisions of the Plan, the perspective of the study group on water resources was significantly broadened amongst others due to the integration of the foreign advisers. The PNH opened up space for the formation of a team of water planners and enabled them to write plans and studies concerning water resource management. They developed policy ideas and gained experiences that departed from the traditional construction bias of the hydraulic bureaucracy, more specifically of groups within the Under Ministry of Construction and their network of beneficiaries. Ex-staff of the Plan confirm that the influence of World Bank advisors on the development of policy ideas was important. This is not surprising because it was an explicit intention of the Mexican government to incorporate such advice on policy decisions. The World Bank was thus intimately involved in putting certain policy issues on the agenda and further developing them into concrete options.

The first Plan was received well by both president Echevarría and the World Bank in 1975 and two of its major recommendations were immediately implemented (Herrera-Toledo, 1996). Firstly, the PNH was converted into a permanent planning agency falling under SARH in 1976, thereby institutionalising the planning process. Dr. González-Villareal continued to co-ordinate the PNH and basically kept on the same team. He reoriented the objectives of the PNH to support the implementation of policies and programs contained in the first Plan and to continue studying present and future water needs. During this time the Commission gained sufficient technical authority to play an important role in policy formulation and decision-making at the highest levels of government (Herrera-Toledo, 1996).

Secondly, the recognition that traditional large-scale irrigation development would not work in the humid tropical lowlands of the Gulf Coast resulted in the *Programa de Desarrollo Rural Integrado del Trópico Húmedo* (PRODERITH; Program for the Integrated Rural Development of the Humid Tropics). Interestingly, PRODERITH fell under the PNH Commission, which was directly charged with executing works. In 1976 several pilot projects were started, centring on drainage, small-scale supplemental irrigation and intensive agricultural development. PRODERITH aimed to stimulate the social and productive development of traditionally marginalized villages of these regions (Herrera-Toledo, 1996). The World Bank played a role in preparing the program and also partially financed it. By 1985 when the pilot projects had been successfully implemented, a second phase was entered, in which the efforts were concentrated on transferring the developmental process and decision-making to organisations of beneficiaries (Herrera-Toledo, 1996).

Some of these regions had a conflictive history of authoritarian government intervention in terms of land development and forced resettlement schemes. The explosive social situation lead to popular protests and military interventions. PRODERITH entered with an alternative approach, based on 'social participation', to the paternalist manner in which the government had intervened and tried to develop these regions in the past. The approach entailed negotiations with the communities in which people could participate in developing a local development plan on the basis of their problems and priorities. These plans served to involve and organise the communities and to determine the program support activities in the region. The program successfully established relations with the communities and carried out the development plans.

Different SARH officials we interviewed acknowledged that the experience with this model of community participation in decision-making was important in the development of policy ideas that also applied to the irrigation districts. Some observers state that it was a pilot or try out for what later became the transfer of irrigation districts. The group of planners accumulated valuable experiences with the organisation of producers, their participation in decision-making and the handing over of responsibilities that were traditionally exercised by the government. In circles of senior SARH officials, concrete policy ideas started to develop in favour of more substantive user participation and the handing over of government tasks in the irrigation districts.

Shifting Policy Agendas

The hydraulic bureaucracy especially felt the financial consequences of the economic crisis of 1982. During the *sexenio* of De la Madrid only 44.5 billion pesos were invested in the irrigation sector, compared to 89.8 billion pesos during the previous *sexenio* (in constant 1979 pesos) (Palacios, 1994). To make matter worse, the World Bank temporarily stopped lending to Mexico, as a response to the moratorium on payments of foreign debts that the government had declared.[25] For the irrigation sector this was a strong departure from the past twenty years during which the Bank loaned more than $800 million (World Bank, 1983b). In the World Bank the construction bias of the hydraulic bureaucracy also started to be a matter of debate around this time. Buras (1983) mentions that the World Bank wanted to see the orientation of the hydraulic bureaucracy change and was no longer willing to support the construction projects that many hydrocrats and construction companies had in mind. Although construction activity in these years did not stop, it was seriously reduced.

In line with international debates on irrigation management (Bottrall, 1981b) the 1983 World Bank review of the Mexican irrigation sector concluded that Mexico needed to shift attention away from the construction of new irrigation systems to improving the management and efficiency in existing systems. The report outlines a program of repair and upgrading of existing systems and an increase of the role of users in decision-making (World Bank, 1983a). The report goes on to recommend the bulk sale of water by the government to users' associations in the districts, who would then sell and distribute the water to its members, citing the successful use of this system in the Río Yaqui irrigation district. Although 'turn-over' or 'IMT' is not mentioned as such in the report it is striking to see how its recommendations strongly coincide with the core of the IMT program in the 1990s. The 1983 report signified an important shift in the Bank's agenda for the irrigation sector and its recommendations influenced both the emergence of the IMT policy idea and its content.

Although user management of irrigation districts has been a recurrent theme throughout the history of irrigation development in Mexico transfer as such was not on the policy agenda in the early 1980s, although discussions on user participation in irrigation increased during the 1970s and early 1980s. A key recommendation of the 1975 PNH was to reduce the subsidies to the irrigation districts and to increase user participation in the financing of irrigation development, operation, maintenance and rehabilitation (SRH, 1975). During the period of the oil boom (1977 to 1982) the government's subsidies to the irrigation districts sharply increased to 46 billion pesos, compared to 22 billion pesos between 1971 and 1976 (in constant 1979 pesos) (Palacios, 1994). However, the 1982 economic crisis led to a reconsideration of the amount of subsidies going to the irrigation districts. During De la Madrid's election campaign in 1982 several meetings on water issues were held, at which it was suggested to improve the 'participation' of users in the financing of the irrigation districts.

> The Mexicans that avail over (...) drainage and irrigation services are in a privileged situation. But those who benefit from these services also have a greater responsibility towards the nation; a responsibility which should imply an increased participation (...) in the financing of the administration of water and the works to utilise it (González-Villareal, 1982b:21; our translation).

This recommendation was mirrored in the conclusions of an important World Bank study on the irrigation sector in Mexico, which singled out an 'across-the-board increase in water charges' as the most important policy decision that needed to be taken by the Mexican government concerning the irrigation sector (World Bank, 1983a). This proposal materialised in an amendment of the *Ley Federal de Derechos* (Federal Rights Law) for 1983. It enacted that in 1984 the payment of the water charges would have to cover the costs of operation, maintenance and improvement of irrigation districts. In 1985 it would have to be sufficient to create a fund for improvements and expansion of the districts, and in 1986 to recover government investments in irrigation (World Bank, 1983b).

Actual cost recovery seriously lagged behind these ambitious objectives. Even though water charges were raised in many districts inflation was much higher due to the severe economic crisis, rendering the raises largely superfluous. In addition, the government severely reduced the subsidies for the irrigation districts. The dramatic failure to persuade farmers to pay much higher water charges made it clear to senior hydrocrats that more drastic measures were needed. An increased role of water users in the management of irrigation districts became more important as an option to improve cost recovery in 1983 and 1984.

> It is fundamental to look for new ways in the operation of the districts that would permit a more comprehensive participation of users in their administration and

financing, with the aim of rationalising possible subsidies and preventing excessive bureaucratic intervention (González-Villareal, 1982a:123; our translation).

Loss of Control over the Irrigation Districts and Further Developments

As one of the many reorganisations of the SARH during De la Madrid's *sexenio*, the irrigation districts were combined with the rainfed districts in 1985 to form rural development districts (Palacios, 1994). Ostensibly, the reason behind this policy was to reduce costs and improve the use of resources. However, it also entailed that the hydraulic bureaucracy's control over the irrigation districts was further reduced. The new rural development districts, which fell under the Under Ministry of Agriculture and Operation, became to an increasing extent the domain of agronomists. Moreover, due to the severe lack of funds the irrigation districts' infrastructure was deteriorating quickly. This situation was completely unacceptable to senior hydrocrats and the need to 'rescue' the irrigation districts was to play an important role in the definition of the IMT policy (Vargas, 1996).

In 1985 a working group was formed between SARH and the Ministry of Finance to review the tax law and to further define a fiscal policy for the use of water, with the aim of achieving financial self-sufficiency in the water sector. Besides industry and the commercial service sector, it was decided to include irrigated agriculture in the law, implying that farmers would have to pay an annual tax for the right to use water on top of water charges. The proposal was sent to Congress at the end of 1985, but due to the serious problems agriculture faced as a result of the economic crisis Congress established a zero tariff for the agricultural water tax. (Palacios, 1996) This was a severe blow to SARH, as it had been agreed in negotiations with the Ministry of Finance that the irrigation districts would no longer be subsidised and that the proposed water tax would be used to finance the irrigation sector.

According to several senior irrigation engineers we interviewed, the then Deputy Minister of Hydraulic Resources Dr. González-Villareal seriously started to consider the possibility of transferring the irrigation districts to water users in 1985. The ramifications of transfer and how to initiate it were discussed at a breakfast meeting in 1985 between the Minister of Agriculture, Dr. González-Villareal and other senior SARH officials. They saw no way to reverse the already serious deterioration of the districts and to resolve the financial problems without drastic changes in the way the districts were administrated. They also understood that to obtain desperately needed external funds they had to accommodate the Bank's new agenda. It was clear at this point that such elements had to be incorporated in the policy agenda for the next *sexenio* if it were to accomplish something for the hydraulic bureaucracy, i.e. the renewal of autonomy. To formally initiate the transfer of irrigation districts was politically not feasible at the middle of the *sexenio* and the 1985 earthquake in Mexico City, which disrupted the country and destroyed the central SARH offices, seriously slowed down concrete initiatives.

In 1986, the World Bank resumed talks with the SARH and started to suggest possible new irrigation loans for the upcoming *sexenio*. The negotiations were led by Dr. Gonzalez-Villareal, who proved to be a skilful negotiator and came up with a number of home-grown policy initiatives,[26] among which the loan proposal for transferring a limited number of irrigation districts. At the same time a number of experiments with user management of irrigation districts in the North-West and the formation of a river basin council and agency in the Lerma-Chapala Basin were initiated by the Deputy Ministry of Hydraulic Resources. From 1985 to 1987 transfer remained a policy idea with its modalities and characteristics still largely undefined, although its financial and institutional basis was seriously worked on.

The further loss of financial and bureaucratic autonomy and a weakening of the control over the irrigation districts by the hydraulic bureaucracy during the De la Madrid *sexenio* was unacceptable to most hydrocrats. For them the major issue in water management was the dispersion of responsibilities and resources over different bureaucratic agencies. Although SARH was legally responsible for the nation's waters, urban and industrial water use, water for hydropower and water quality fell under other ministries (IEPES, 1987). It was argued that the lack of inter-ministerial co-ordination made it very difficult to manage water adequately. To senior hydrocrats it was clear that radically different policy scenarios had to be explored to extract the hydraulic bureaucracy from its worst crisis. Ideally, this would entail the re-constitution of an autonomous water authority that would concentrate the responsibilities and financial resource flows related to water. The emergence of IMT as a policy idea was closely intertwined with the aim of the hydraulic bureaucracy to re-establish financial and bureaucratic autonomy and control over the irrigation districts. To achieve this aim, different groups of ex-SRH engineers started exerting political pressure towards the end of the sexenio of De la Madrid, setting the stage for the creation of an autonomous water authority.

Towards a Reform Package

The Policy Actors

In January 1989 Salinas created the CNA, less than six weeks after he became president and in June 1989 the National Development Plan, which endorsed IMT[27] and a wider water reform package, was released. These rapid developments indicate that in 1987 and 1988 disparate policy ideas, such as the transfer of irrigation districts, the creation of an autonomous water authority and water pricing were articulated further and combined in a single reform package. This occurred during the run-up to the presidential elections in 1988 when an influential segment of water resource planners within the hydraulic bureaucracy negotiated the water reform package with the presidential candidate. Although international lending agencies were not directly involved in these negotiations, their position on necessary water reforms and the prospect of new loans played a crucial role in defining the reform package. This section outlines the different positions and agendas of these three policy actors and shows how they agreed on a reform package that included both a concentration[28] and a decentralisation[29] of bureaucratic domains and resource flows related to water. In particular we highlight the role of the group of water resource planners in the articulation, packaging and consolidation of the reform package.

The team of water resource planners formed by the PNH and led by Dr. González-Villareal took the lead in proposing IMT to Salinas and convincing him of the need for an autonomous water authority during his election campaign.[30] In this they were supported by different groups of civil engineers working for the Under Ministry of Hydraulic Resources in SARH, construction companies or stationed at the Faculty of Engineering of the UNAM. Especially the construction companies played an important role in supporting the demands of these engineers in line with their interests. When Salinas became a presidential candidate and started galvanising support from the bureaucracy for his campaign, this coalition of engineers offered its support to him in return for the creation of an autonomous water agency (van der Zaag, 1992).

The three key policy actors involved in the articulation of the reform package were very familiar with each other's dispositions and agendas due to frequent interactions in the past. These can be summarised as follows:

1. Salinas espoused a neo-liberal agenda that aimed to modernise state-society relations and reduce government expenditures through decentralisation, shared responsibility between the public and private sector and social reconciliation.
2. Dr. González-Villareal and his team of water resource planners represented the hydraulic bureaucracy in its strong will to re-establish bureaucratic and financial autonomy, whilst supporting policy ideas concerning user participation and a financial system for water.
3. The World Bank supported policies to reduce government intervention and expenditure in irrigation, amongst others through decentralisation, water pricing reforms and increased users' participation in decision-making. As a favoured client of the Bank Mexico had become eligible again for new loans after a period of austerity implying that reform initiatives would be looked upon favourably.

During the *sexenio* of Salinas (1989-1994) the neo-liberal agenda of De la Madrid was continued with increased intensity. During his election campaign in 1988 Salinas launched the plan to modernise rural Mexico, through a set of economical, political and social reforms. He emphasised the need to break with the paternalist practices of the government in the rural areas through a strategy that would allow for more participation of the social and the private sector. Key concepts in Salinas' discourse were shared responsibility (*coresponsabilidad*) and social consensus building or consultation (*concertación social*). These ideas originated from his academic research on popular support for the political regime in rural communities and experiences with organising farmers (Cornelius et al., 1994; Gordillo de Anda, 1988).

In Salinas' vision shared responsibility would be reached through social reconciliation efforts, both in the rural and urban areas. He proposed a mode of governance termed 'social liberalism', which sought to avoid the excesses of both unfettered free market capitalism and heavy-handed state intervention, thereby leading to the reduction of absolute poverty and an increase in social well-being (Cornelius et al., 1994). Salinas' ambitious agenda aimed to modernise the relations between state, society and the market and strongly favoured decentralisation and participation of the social and private sector in water management. During his *sexenio* Salinas followed a policy of liberalising trade, deregulating the economy, privatising parastatals, reforming the financial sector and fractionalising the corporate structure of the PRI (Grindle, 1996).

The group that played a key role in mapping out the course of the hydraulic bureaucracy and the irrigation districts during the 1980s and 1990s was the team of water resource planners under the direction of Dr. González-Villareal. They were the intellectual authors of many of the policy ideas presented to Salinas during his election campaign. Through his different contacts and positions in the bureaucracy Dr. González-Villareal came to represent the broadly shared claim of the hydraulic bureaucracy for bureaucratic and financial autonomy.[31] He started his bureaucratic career during Echevarría's *sexenio* (1970-76), after completing an academic training in civil engineering at the UNAM and obtaining a Ph.D. at the University of California at Berkeley, specialising in water resource planning. He worked as General Co-ordinator of the PNH from 1972 to 1982, first under the SRH and later under the SARH. The Plan was important for the formation of his *equipo*, which developed a more encompassing vision on water resource planning and management. This group of planners developed policy ideas favouring water user participation, water pricing and institutional reform that can be seen as the precursors of CNA policies and IMT.

The co-ordinator of the PNH was widely respected for his vision and expertise concerning the planning and management of water resources. In addition, he knew irrigation from a practical perspective. His father was a producer in the irrigation district of Río Yaqui in the northern state of Sonora, a system that was user-managed for several decades. His respected position was acknowledged when he co-ordinated campaign meetings on water for De la Madrid in 1982. During these meetings he developed a set of ideas that were basically in embryo what was proposed at the end of the 1980s to Salinas.

As a SARH Deputy Ministry during De la Madrid's administration, Dr. González-Villareal led the hydraulic bureaucracy in a time of severe crisis. In this period he frequently interacted with Salinas, who had an important say in setting SARH's budgets as minister of Budget and Planning. Their discussion and interaction on water reforms in view of the state's difficulty to continue financing the irrigation districts and other forms of water use, date from this period. Their public discourse suggest that they shared certain opinions and discursive topics, for example on 'social participation', although their formation is very different. Both also showed interest in the shared responsibility of the social and private sector in the financing of water management (González-Villareal, 1982a; Poder Ejecutivo Federal, 1989). Finally, the IMT policy idea fitted in Salinas' plans to 'modernise the countryside', in which strategies of 'social reconciliation', deregulation and decentralisation played an important role (Poder Ejecutivo Federal, 1989).

In addition, Dr. González-Villareal and his team of water resource planners knew the world of international funding agencies well. They developed good relations with officials of the World Bank during the PNH and were well informed about international trends in loan policies and irrigation management reform. The World Bank stopped making loans available for irrigation projects after 1982 and indicated the need for a shift in irrigation policies. As the Under Minister of SARH González-Villareal stayed in touch with the World Bank and initiated discussions on possible loan packages.

Thus, in 1988, Dr. González-Villareal found himself centrally positioned to propose water reforms. He represented different groups of hydrocrats in the SARH and maintained good relations with academia and the influential construction sector. His respected vision and expertise on water resources and irrigation and his political participation in the PRI[32] were important for his capacity to convince a wide array of political, bureaucratic and societal actors of the transformations that were needed in the water sector. In him Salinas saw a person with a vision of how to reform the water sector, the skills and authority to deal with potential resistance in the bureaucracy or among water users and the necessary relations with influential interest groups and the World Bank. He thus became a central actor for Salinas in bringing about a reduction in public spending, to give a strong push to his policy agenda of modernising the relations in the countryside through decentralisation, to reorganise the bureaucracy and to acquire international funding to support the proposed transformations.

The Election Campaign

In December 1987 and January 1988 five national meetings on water were held as part of Salinas' election campaign. That Salinas made this effort indicates that he saw the political and electoral importance of problems in the water sector. During these meetings, that were co-ordinated by Villalobos-Guerrero, an influential civil engineer, the contours of the water reform package for the next *sexenio* became clear. In the first meeting in Acapulco Salinas asked Rovirosa Wade, who had been the SRH Minister at the beginning of the 1970s, what he thought of the SAG/SRH fusion. Rovirosa Wade responded:

> I was slightly concerned [as] the conception of the water resource could get lost. Giving all the power to an independent entity, I think, would be a solution that the whole country is demanding for the management of water as a vital resource. In concrete, I propose not that the SRH comes back, but that at least there would be an independent authority located in the Presidency of the Republic and that it is given all the power necessary for water management (IEPES, 1987:10; our translation).

It is remarkable that he rejects the possibility of recreating a ministry, something that certainly not everybody agreed with him at that point. During the same meeting Salinas asked Dr. González-Villareal his opinion on the risks of transferring irrigation districts to the users. His answer is illuminating:

> The transfer of irrigation districts to users already was an established policy of this administration [of De la Madrid], which has encountered some difficulties. (...) Those of the Northwest are prepared to adopt their own administration. As a matter of fact, in the Rio Yaqui irrigation district users already manage the maintenance, administration, collection of fees and the delivery of water at lower levels. (...). In a program that will be financed in the near future with international credit, called 'modernisation of irrigation districts', a subsequent phase after the original construction of the districts is proposed, consisting of the bulk delivery of water to the users and an administration directed by them (...). However, in the districts of the centre of the country (...) we believe that the process has to be more gradual. First, some rehabilitation and public investment will be needed, before a first phase of user organisation, if the process is to be effective (IEPES, 1987:7; our translation).

This indicates that the discussions with international lending agencies that were initiated in 1986 to negotiate loans for modernising and transferring a number of irrigation districts had reached such an advanced stage that Dr. González-Villareal felt confident enough to announce publicly to Salinas that international loans for IMT would be forthcoming.

Another element of the reform package proposed during these meetings was the 'financial system for water', which was already presented in the PNH (SRH, 1975). It was argued that a key aim of reforming the administration of water was to:

> Allocate sufficient economic resources that show the will of our society to solve the problems of water, with larger federal, state and municipal contributions, and with a larger participation of the users. The financial system for water, consisting of the investments, the water rights, the payment of differentiated tariffs for services, would be more sound if the fees and taxes collected are re-invested in the sector (PRI, 1988:41; our translation).

The financial system for water clearly reflects the concern of the hydraulic bureaucracy for financial autonomy. During the SARH period the income flows related to water were not controlled by the hydraulic bureaucracy, but were dispersed over different ministries. A concentration of the control over these income flows would significantly strengthen the financial autonomy of the authority.

During a meeting two months later it becomes clear that the various policy ideas had been discussed in more detail with Salinas, as he publicly accepts the need for institutional reform in the water sector and the transfer of the irrigation districts. He states that he takes Wade's remarks on an independent water authority seriously, but emphasises that he does not want to 'take a step back on this track' (IEPES, 1987:14), thereby countering the demand for a new ministry. He does call for a careful reflection on how the co-ordination of different water uses could be strengthened and agrees with the need to create 'one single water authority'. He attributes many capacities to this new authority, including the authority to decide over its own programs and budgets, something that the existing Under Ministry did not have. Lastly, he acknowledges that the creation of this new authority is a precondition for his proposed policies of decentralisation and social reconciliation and reaffirms that the irrigation districts will be transferred (PRI, 1988). The reform package as a whole containing both IMT and the creation of a new water authority came together at the end of the campaign meetings on water in a PRI document published in June 1988 outlining the party's election platform.

> The integrated management of water quantity and quality, by an authority which is constituted as a water authority creates the need to realise legal adjustments that leave no doubt about the national property of water, on the one hand, and that make the administrative reorientation possible on the other. (...) It is a priority to rehabilitate and modernise the irrigation and rain fed zones with the participation of the users and the state government, so that when they are in a condition to operate efficiently, they can be transferred gradually to the users (IEPES, 1988:33; our translation).

The same document considers four options for the legal structure of the new water authority.

1. Consolidation of the current administrative structure.
2. Integration of the water authority in the SARH by strengthening the Under Ministry of Hydraulic Resources or creating a deconcentrated authority[33] in the Ministry with full responsibility for all water related activities.
3. Creation of a new Ministry, which would entail high political and financial costs and would go against the policy of austerity and a slimmer government, according to the document.
4. Creation of a public decentralised authority (IEPES, 1988).

Although the creation of a new ministry was unlikely the group of water resource planners formulated its charter, regulations and organisational structure in full detail in the latter half of 1988. In November 1988, Salinas resolutely ruled out this possibility, as the PRI did not have a majority in parliament to approve the creation of a new ministry. Instead, it was decided that the new water agency would become a deconcentrated authority that would fall under the SARH. The hydraulic bureaucracy was thus forced to accept the 'second best' option and had to go back to the drawing board to define the legal and financial structure for the new water authority.[34]

The above shows how different policy ideas became an integral part of the water reform package and how as a consequence it became more articulated during the election campaign of Salinas. This transition occurred in a small circle of policy actors, consisting of the president, senior hydrocrats and World Bank officials, as part of the policy agenda setting for the upcoming *sexenio*. Combining these different policy ideas in a reform package increased their feasibility, as it had become clear during De la Madrid's administration that merely recreating a water ministry as an individual reform was politically and financially not feasible. To make the move of creating an autonomous authority feasible, it had to be accompanied by a set of apparently paradoxical reforms: a concentration of bureaucratic domains and resource flows, a decentralisation of the irrigation districts and an active water pricing policy. If successful, this composite strategy would reduce government expenditure in water management, secure higher and more stable income flows from water use in different sectors and attract international loans. In addition, it would enjoy the political support of the president and international lending agencies. When Salinas stated during his campaign that the new water authority was a precondition for his proposed policies of decentralisation and social reconciliation (PRI, 1988), he acknowledged the link between concentration and decentralisation measures. The composite strategy of concentration, decentralisation and water pricing made the reform package a financially, politically and bureaucratically viable option.

The reform package offered another advantage. Many observers state that Dr. González-Villareal faced groups of middle and senior hydrocrats that were opposed to the transfer of the irrigation districts. The advantage that he could project to them was that the IMT policy opened up the possibility of reconstituting an autonomous water authority and that it would return a certain level of control over the districts, which at that time fell under the agricultural bureaucracy.

The World Bank did not directly influence the creation of the CNA or the selection of its Director, but its financial and ideological support for the proposed reforms were crucial in making the creation of an autonomous water authority feasible. The role of the World Bank in promoting IMT was more direct, as it was clear that new irrigation loans would become available if IMT was implemented. However, it would be too simple to posit that the World Bank unilaterally imposed IMT and water reforms on the Mexican government and its hydraulic bureaucracy. This is not to deny the element of financial and ideological coercion, as every international loan is accompanied by some 'arm-twisting', as one CNA official expressed it. The policy ideas leading to IMT were a product of historical experiences and concerns of a particular group of water resource planners, admittedly developed in close interaction with World Bank officials. If IMT was a condition to loans 'it did not cost the World Bank much trouble to convince the Mexicans', according to a well-informed interviewee. Considering our description of the guild of hydraulic engineers it is difficult to see how they could accept a completely 'foreign' imposition of a policy that would affect them in such a drastic manner. The World Bank is limited in its influence when its policy agenda is not supported by and does not create benefits for the hydraulic bureaucracy.

From a definition of presidentialism, which attributes a dominant role to the president in policy and bureaucratic transformation, it could be argued that Salinas imposed the reforms on the hydraulic bureaucracy. However, we have shown that the hydraulic bureaucracy proposed concrete policy ideas to the presidential candidate and actually enrolled him in their effort to re-establish autonomy. As a presidential candidate Salinas needed the political support of the hydraulic engineers as well as their support for achieving his ambitious reforms in the rural sector. Reciprocally, the authorisation of Salinas was crucial for the senior hydrocrats to achieve their objectives. His full support was especially needed for overcoming potential resistance within the bureaucracy. Also it is clear that Salinas did not concede all of the proposals made by the hydraulic bureaucracy, exemplified by his refusal to create a new Ministry.

Without denying the political, financial and ideological coercion that is needed for a reform program of this magnitude, it is clear that for all three policy actors there was a limit to the realisation of their agendas. We have highlighted the active role that segments of the bureaucracy played in the definition of the reform package, in this case driven by a concern for bureaucratic and financial autonomy and control over the irrigation districts. This is something that is often underrated by approaches that analyse policy formulation and implementation by focussing on the required commitment of politicians and international funding agencies to a single, isolated and black-boxed policy. Our analysis shows that there was not one policy, well defined and isolated, to which the policy actors could choose to commit themselves to or which was imposed by one of these actors. The individual elements of the reform package meant different things to the policy actors involved. At the end of the 1980s the interaction between the different policy actors and their agendas had already gone through a trajectory which started with the PNH. During Salinas' election campaign a process of interplay and mutual enrolment occurred in which the exact shape of the individual reforms was defined and hardened and the reform package as a whole became more irreversible. The commitment of the political leadership and influential segments of the bureaucracy to an individual policy such as IMT was the outcome of this protracted process of interplay.

The Creation of the CNA and Irrigation Management Transfer

Bureaucratic Autonomy

In January 1989 Salinas created the *Comisión Nacional del Agua* as a deconcentrated agency of the SARH and designated Dr. González-Villareal as its Director. This appointment was long kept uncertain by Salinas, as the co-ordinator of his campaign meetings on water, Villalobos-Guerrero, was also considered an eligible candidate.[35] The CNA was the first 'modern institution' created by Salinas. However, its institutional set-up remained unclear for several months after its creation as all the effort in 1988 had gone into developing a charter and regulations for a ministry instead of a deconcentrated authority. A SARH official remembers that it basically took the whole of 1989 to establish the normal functioning of the CNA. This caused delays in implementing the ambitious reform package agreed in the campaign. The argument that was used for the creation of CNA was as follows.

> The 'sectorization' of the Federal Government's role in relation to water caused serious problems in co-ordinating water policies that were defined and implemented as a result of sectoral objectives, with very little intervention of SARH in its managerial role, thus aggravating the already critical problems of scarcity, conflicting uses and pollution in several of the major river basins in the country. Due to these existing institutional problems, (...) the Federal Government created the Comisión Nacional del Agua (National Water Commission) as an autonomous agency, (...), as the sole federal authority dealing with water problems and conflicts (CNA, 1990:5).

The sectoralization mentioned above refers to the dispersed government responsibilities and resource flows regarding water in the SARH period. During the *sexenio* of Salinas, these dispersed bureaucratic domains and resource flows were gradually concentrated in the CNA. The CNA integrated the quantitative aspect of water management, regarding for instance drinking water, industrial use, and irrigated agriculture in both irrigation districts and the smaller irrigation units, but it also became responsible for the qualitative aspects such as water pollution. The concentration of these different domains strengthened the CNA in its bureaucratic autonomy.

As a deconcentrated agency the CNA fell under the SARH, but could function with relative autonomy. The choice for a deconcentrated authority was partly motivated by the intention to confine the financial autonomy of the new agency. In matters such as negotiations with funding agencies over international loans, the CNA would depend on the SARH.[36] However, in spite of the chosen legal form, in practice the CNA gained a large degree of financial and bureaucratic autonomy.

When González-Villareal was appointed as director of the CNA he brought his *equipo* with which he had worked in the PNH and later in the Under Ministry of SARH with him. An important part of his group were water resource planners from the PNH, others were civil engineers with a more traditional construction background. González-Villareal also included experts from different disciplinary backgrounds and experiences, such as lawyers and social scientists. He appointed the members of his team in middle and high level posts, creating a group of subordinates that he could trust and with whom he had unambiguous hierarchical working relations. He knew what he could expect from them because of his intimate knowledge of their disciplinary background and bureaucratic trajectories. Several people have indicated that this was of crucial importance for the difficult bureaucratic reforms that the CNA was facing. During this period the CNA was an institution in which civil engineers (re)gained a certain dominance, at the expense of other bureaucratic groups, particularly agronomists.

It is generally observed that Dr. González-Villareal had a direct relation with the president. The co-ordination of the IMT policy, which was a central activity of the CNA during its early years was done between the director and the president, in which the SARH minister interfered very little.[37] The federal status of the Commission was strengthened with the creation of its technical council or governing body in 1991, which consisted of all the Ministers of the Ministries involved with water policy. Institutional and sector co-ordination is set at the federal level within the technical council (Herrera-Toledo, 1996:41).

Control over the Irrigation Districts

To regain control over the irrigation districts and to increase the levels of water charges paid by farmers, the CNA followed a seemingly paradoxical strategy of decentralising the management of the irrigation districts to the users. Although the involvement of user organisations in irrigation management sharply increased as a result of IMT, at the same time the hydraulic bureaucracy's control over the irrigation districts was reasserted through IMT. To unravel this paradox, this section indicates how the transfer of the irrigation districts strengthened both the financial and bureaucratic autonomy of CNA.[38]

The main objective of the Mexican IMT program was to reduce public expenditure on irrigation through creating financially self-sufficient WUAs that would recover the full O&M costs of the irrigation systems (CNA, 1994a; Espinosa de León and Trava, 1992; Gorriz et al., 1996; Johnson, 1997a, 1997b; Trava, 1994). Although the IMT program had the full backing of the Mexican president and its objectives were clear, it got off to a slow start in 1989, for several reasons. Firstly, the irrigation districts were still part of the rural development districts, over which CNA had no control. Secondly, the transfer of irrigation districts to water users' organisations was illegal under the 1972 water law. Thirdly, the large majority of the irrigation districts were not financially self-sufficient, with the government subsidising O&M costs for 85% on average. Achieving 100% cost recovery seemed difficult, especially as efforts to convince farmers to pay significantly higher water charges had consistently failed in the previous forty years. Lastly, most of 1989 was dedicated to establishing CNA as an autonomous organisation and defining its structure.

A particularly vexing problem facing IMT in 1989 was that the legal framework was inadequate as the water law of 1972 prohibited the transfer of the irrigation districts. To resolve this issue, a group of CNA lawyers at the national level devised the legal trick to artificially divide the districts into irrigation units (termed *unidades de riego* in Mexico). Under Article 77 of the 1972 water law irrigation units are defined as farmer-managed irrigation systems with users' associations fully responsible for the financing and execution of O&M. Additionally, Article 78 of the law indicated that two or more irrigation units could be joined to form an irrigation district (Diario Oficial, 1972). Based on these clauses the lawyers argued that an irrigation district could be considered to exist of various irrigation units, which were called *módulos* to prevent confusion (Palacios, 1994). An additional spin given to this legal construction by the CNA lawyers was to constitute the Water Users' Associations (WUAs) as civil associations. This was necessary to ensure that the WUAs would fall under the control of CNA, as 'normal' WUAs for irrigation units as provided for by the 1972 water law would fall under the responsibility of SARH (Espinosa de León, 1994).

The resolution of the legal problems in 1989 removed one of the hurdles to really start with IMT in the field. Equally important, CNA regained control over the irrigation districts from SARH during the second half of 1989 and started establishing itself as the only water authority in the country (Palacios, 1996). Based on these achievements the National Program for the Decentralisation of the

Irrigation Districts was drawn up towards the end of 1989. This program indicated that 20 districts, covering 1,963,230 ha, were to be transferred between 1990 and 1994 and that efforts would be made in the other districts to raise water charges to fully cover O&M costs. The 20 districts, mainly located in the north of Mexico, were carefully selected based on a social, economic and institutional assessment by the CNA of the likelihood of success of transfer in these districts (Espinosa de León and Trava, 1992).

The above schematically shows how CNA regained control over the irrigation districts through IMT. The state, represented by the CNA, retains ownership of the nation's waters as well as the irrigation infrastructure. The CNA remains ultimately responsible for the management of the irrigation districts, while the responsibility for operation, administration and maintenance is transferred to the WUAs (CNA, 1994). IMT did not imply a complete transfer of bureaucratic authority and control over the irrigation districts, because the CNA retains control over the head works and the main canal system and important oversight functions. Rather, the hydraulic bureaucracy regained and reordered its control over the irrigation districts, which contributed to its bureaucratic autonomy.

Financial Autonomy

Another important benefit of the water reforms was that it strengthened CNA's financial autonomy. A crucial component of the creation of the CNA was the establishment of a 'financial system for water'. Under this system, CNA gained direct control over a range of financial resource flows as it became responsible for 'the collection and administration of the resources originating from the payment of rights for the use of the nation's waters. According to the law all revenues generated by these rights are specifically allocated to the CNA (CNA, 1994b:50; our translation). Through this legal provision CNA succeeded in concentrating the income flows related to water use in different sectors without interference by other ministries, thereby strengthening its financial autonomy. These water rights represented a significant and growing source of income for the CNA: water rights for the exploitation of the nation's waters (industry and service sector), bulk water delivery to the urban sector (drinking water levies) and water charges paid by the WUAs.

Analysing the income flows from the collection of water rights during the first six years of the CNA reveals a rapid increase: from 498.6 million pesos in 1989 to 2,341.3 million pesos in 1994 (see Table 3.1). The shares of these different income flows are significant, about 63% originates from water rights taxes, 18% from bulk water delivery to cities and a mere 5% from water charges paid by irrigators. This indicates the major importance of the water rights and urban water for CNA's financial autonomy. Although the share of irrigation charges in the total income of the CNA between 1988 and 1994 was relatively small this needs to be seen in the context of the radically reduced government subsidies to the irrigation districts after their transfer and their increased self-sufficiency. In 1988 the level of cost recovery in the irrigation districts was 18%. In 1994, the water charges paid by water users in transferred districts fully covered O&M costs, while at the national level cost recovery significantly increased to 80%. In addition, the WUAs were paying the CNA for the operation of the head works and main system in the transferred systems (CNA, 1994b).

As all funds collected by CNA remain within the bureaucracy, its degree of financial self-sufficiency sharply increased, from 51% in 1989 to 92% in 1993 (see Table 3.2). Although the hydraulic bureaucracy increased its financial autonomy significantly through the establishment of the 'financial system for water', it could not directly negotiate international loans with funding agencies and was dependent on the SARH in this respect. However, new international loans for the irrigation and drinking water sector were initiated after a sexenio of financial drought, further strengthening the financial position of the CNA.

Table 3.1 Development of CNA's income from water rights and fees (millions of constant 1994 pesos)

	1989	1990	1991	1992	1993	1994
Bulk water delivery to cities	263.0	219.2	291.0	338.2	437.9	413.7
Water rights taxes	188.0	682.1	964.0	1,376.1	1,639.6	1,460.7
Irrigation service fees	45.0	85.5	129.3	144.8	127.1	112.6
Other income sources	2.6	13.6	25.8	118.1	307.0	354.3
Total	498.6	1,000.4	1,410.1	1,977.2	2,511.6	2,341.3

Source: CNA, 1994b.

Table 3.2 Financial self-sufficiency levels of the CNA (millions of constant 1994 pesos)

	1989	1990	1991	1992	1993	1994
External credit (1)	152.4	262.1	996.9	1,527.7	881.1	726.0
Fiscal budget (2)	818.2	1,531.0	1,797.2	1,781.4	1,871.8	2,184.0
Total Budget (3=1+2)	970.6	1,793.1	2,794.1	3,309.1	2,732.9	2,910.0
Income (4)	498.6	1,000.4	1,410.1	1,977.2	2,511.6	2,341.3
Financial Self-sufficiency (=4/3)	51%	56%	50%	60%	92%	80%

Source: CNA, 1994b.

It can be concluded that with the creation of the CNA as a deconcentrated authority, the hydraulic bureaucracy achieved its objective of re-establishing a relatively large degree of bureaucratic autonomy. The concentration of resource flows related to the exploitation of nation's waters and the international loans secured a growing income source, thereby accomplishing to a large extent the aim of financial autonomy. Through a carefully thought out composite strategy, consisting of concentration, decentralisation and water pricing the CNA thus succeeded during the *sexenio* of Salinas in establishing a large degree of autonomy.

Conclusions

In a sense, user management of irrigation districts has come full circle in Mexico if a longer historical view is taken. By the early 1990s user management of irrigation districts was once again enshrined in policy and law, as it had been in the 1930s and 1940s. However, there was nothing logical or unavoidable about this happening. This article shows that to understand the articulation of reform policies it is necessary to centre on the interactions between policy actors and the short and long-term circumstances that shape the ways in which they try to advance particular policy ideas. As stated in the introduction, the aim of this article is to stimulate debate on the role of hydraulic bureaucracies in water reforms, thereby complementing the manner in which IMT and bureaucratic reform in Mexico have been discussed in the literature to date. We have shown how the creation of the CNA and the launching of IMT were strongly linked with the engagement of the Mexican hydraulic bureaucracy in policy articulation. It is clear that to understand water reforms, or the lack thereof, it is necessary to bring the hydraulic bureaucracy back in to the analysis.

Many observers assume that IMT was imposed on the hydraulic bureaucracy by the Mexican president and the World Bank, as they perceive an inherent contradiction in senior hydrocrats supporting a policy that apparently weakens their bureaucracy. This article highlights the inadequacies of this line of argument by widening the frame of analysis to include political and bureaucratic processes, indicating that the reforms were part of an ongoing struggle within the Mexican bureaucracy. This shows how the reform package did not entail a complete devolution or reduction of bureaucratic powers, but that it served to reorder bureaucratic control over essential domains and resources. The reform package enabled senior hydrocrats to express and materialise some of their pressing concerns, such as regaining bureaucratic and financial autonomy. Contrary to popular belief, the water reforms in Mexico generated important benefits for segments of the hydraulic bureaucracy, indicating why they were committed to a broad reform program of this nature.

The commitment to the water reform package was a temporal, situational and content specific outcome of a complex interplay between the presidential candidate, the World Bank and a group of senior hydrocrats. By focusing on these policy actors and their agendas this article elucidates how the reform package emerged as an outcome of mutual persuasion, compromise and coercion between these policy actors. It gained momentum when the concentration of bureaucratic domains and resources in a single water authority, the decentralisation of the irrigation districts and active water pricing policies became part of the reform package. This packaging strategy made the reform viable financially, politically and institutionally as it would attract international funding, reduce government subsidies, secure a steady income flow for the hydraulic bureaucracy, receive authorisation from the president and find sufficient support among the upper reaches of the hydraulic bureaucracy.

We have argued that it is necessary to rethink the relationship between water policy-making and senior levels of the hydraulic bureaucracy. That commitment and political will are outcomes of policy articulation rather than prerequisites for reform has several implications for researching water reforms. Firstly, it suggests that it is necessary to conceive of contemporary water reforms as effects of specific political and bureaucratic policy practices and experiences. Secondly, it entails analysing how officials of international funding agencies, researchers, consultants, politicians and hydrocrats engage in their institutional reproduction through articulating reforms that reorder the control over contested bureaucratic domains, redirect essential resource flows and redefine themselves and their clientele.

Acknowledgements

We wish to thank all the senior hydrocrats and other policy actors, who unfortunately must remain anonymous, for openly sharing their insights and recollections with us during interviews. The constructive and insightful comments by Roberto Melville and Sergio Vargas on earlier versions of this paper and the invaluable support of Sergio Ramos vastly improved the contents of this paper, as did the interesting discussions with Dr. González-Villareal, who gracefully agreed to referee our paper at the Hyderabad conference. The International Water Management Institute (IWMI) gratefully acknowledges the financial support provided by the German Government's Bundesministerium für Wirtschaftliche Zusammenarbeit und Entwicklung (BMZ) and the Deutsche Gesellschaft für Technische Zusammenarbeit (GTZ) GmbH for this study under the Research Program on Institutional Support Systems for Sustainable Local Management of Irrigation in Water-Short Basins. The staffing support provided by the Dutch Ministry of Foreign Affairs under its Associate Expert program and the Netherlands Foundation for the Advancement of Tropical Research (WOTRO) is also gratefully acknowledged.

Notes

1. Throughout the text Spanish words are italicised and Mexican acronyms and abbreviations are used.
2. The speed with which irrigation districts in Mexico were transferred to users' associations between 1990 and 1994 surprised donors, water experts, consultants and researchers alike. This is understandable, as transferring nearly 2.5 million ha in five years is no mean feat (CNA, 1994a).
3. An eloquent summary of this policy narrative is given by Groenfeldt when he states that 'This process [IMT] was initiated as a result of mounting budgetary pressures during the financial crisis that Mexico experienced during the 1980s. Investments in the irrigation sector fell dramatically, resulting in deterioration of the schemes, poorly maintained irrigation and drainage canals, roads and infrastructure. This period of structural adjustment *forced* drastic changes in Mexico's agricultural and irrigation

policies. The program to transfer management of the irrigation districts to water users was adopted *out of necessity*' (Groenfeldt, 1998:55-56; emphasis added).

4   This becomes apparent if one takes into account that many of the conditions that are said to have led to IMT were often in place throughout the history of irrigation development in Mexico, without this leading to transfer. For example, Mexico suffered various serious economic crises between 1930 and 1980, with drastic consequences for irrigation, without this resulting in IMT. In addition, nearly all the works on IMT in Mexico single out the declining levels of water fee payments during the 1980s as the most important reason for transfer. Nevertheless, the concern for cost recovery is not new in Mexican irrigation policy circles and dates back to the Irrigation Law of 1926, which decreed that irrigation districts were to be financially self-sufficient. At several points in time, the Mexican government considered the level of water fee payment too low and made attempts to increase them, however in most cases without a lasting effect. Likewise, the poor maintenance of irrigation districts has been a recurrent theme in Mexico and the need for extensive rehabilitation was already identified in the 1960s. Lastly, ideas of increased user involvement in irrigation management were present in irrigation policy circles since at least the 1930s and several districts were actually managed by users' associations from this time onwards. Although the legal conditions for such involvement were in place for several decades, this never resulted in a substantial number of irrigation districts being managed by its users.

5   The use of the term bureaucratic to qualify actors or processes is not intended to be derogatory, but simply refers to the actors and interactions within and between government agencies that influence the articulation of policies and reforms.

6   Defined here as the process by which policy actors support, modify, displace and translate different interests concerning a policy idea with as outcome that a policy or reform package becomes irreversible, or not. Seen in this way commitment to policies is the outcome of struggles and negotiations between different policy actors. A 'successful' policy follows an unstable trajectory from a policy idea to a policy likelihood, and finally to a policy reality, i.e. it becomes more articulated, through which its content and composition is redefined and transformed. By focusing on policy articulation the artificial divide between policy formulation and implementation disintegrates.

7   This poses the methodological challenge of 'studying up', where participant observation and other research methods 'may not be readily portable to elite contexts' (Gusterson, 1997:116). In our research 'studying up' meant interviewing senior members of the hydraulic bureaucracy and other key figures in the Salinas government as well as staff of international organisations. We also interviewed Mexican scholars knowledgeable on irrigation development and policies and extensively read officials documents and newspaper articles.

8   Defined here as the various government agencies that were responsible for the allocation, distribution and use of the nation's waters and the construction and administration of hydraulic infrastructure.

9   This term is a contraction of hydraulic bureaucrats and is used here to refer to engineers working in water bureaucracies. We thank Alex Bolding for suggesting this term to us. Others have pointed out that it may be better to designate senior hydrocrats as hydro-politicians, as this more accurately reflects their activities.

10 Due to the limitations of our study, such as the lack of public information on the processes underlying reforms in Mexico and the methodological difficulties in accessing this type of information, we can only present a schematic overview of the

events, actors, arenas and processes that were fundamental in defining and constituting the water reforms.
11 Irrigation has been practised in Mexico since pre-Hispanic times, and by 1919 some 800,000 ha were irrigated (CNA, 1994b). At present, some 6.1 million ha are irrigable, of which 3.3 million ha are contained in 81 irrigation districts, constructed and until recently managed by the state, while 2.8 million ha are either in private or farmer-managed irrigation systems. Irrigated land contributes about 50% of the total value of agricultural production and accounts for nearly 70% of agricultural exports. Although agriculture only accounts for around 6.5% of Mexico's GDP it employs 22% of the economically active population (INEGI, 1998).
12 See Aboites (1998) for a legally precise and Mexico specific definition of centralization/federalization and an excellent historical overview of this process from 1888 to 1946.
13 The term autonomy is not used to suggest absolute independence or isolation, but is used in a relative sense to express the position of the hydraulic bureaucracy in comparison with other bureaucracies.
14 Land reform communities created after the Mexican Revolution of 1910. Before the revision of Article 27 of the Constitution in 1992, *ejido* land belonged to the state, with a combination of community (*ejido*) and private (*ejidatorio*) usufruct. *Ejido* members are called *ejidatarios*.
15 Instances of user management of irrigation districts in the 1930s based on this law were mentioned to us in interviews. However, we have been unable to document this and it remains unclear why and how many user associations were formed and what their attributes were.
16 Water boards continued to function in several irrigation districts until the transfer program started in the 1990s, such as in the Ciudad Juarez and the Tula districts. The most frequently mentioned and only documented case is that of Rio Yaqui (IMTA n.d.), which was managed by water boards since 1947.
17 In 1960 the SRH estimated that more than 200,000 ha of the irrigation districts, or 10% of the irrigable area at that time, could not be used due to deteriorated or incomplete infrastructure. That same year, a proposal to increase water charges led to nothing. It was opposed by a group of 10,000 large producers with a strong commercial and productive weight at the national level (Wionczek, 1982).
18 The number of federal government employees grew from 0.3 to 1.3 million between 1969 and 1976, the public sector deficit rose from 2.8% of GDP in 1972 to 4.6% in 1976 and inflation and foreign debt were also on the rise. This forced the government to devalue the peso in 1976 and to sign a stabilisation agreement with the IMF, in which it pledged to reduce government spending (Grindle, 1996).
19 In the following two *sexenios* this Ministry functioned as a stepping stone for its two subsequent ministers Miguel de la Madrid and Carlos Salinas to become president (Castañeda, 1999).
20 A telling joke that recalls the fusion of these long-time rivals narrates that it was like merging *America* and *Chivas*, the two major football clubs of Mexico with a long tradition of mutual rivalry.
21 According to Grindle (1977) senior bureaucrats use their appointive powers to build loyal and efficient teams, which are called *equipos*.
22 Buras was appointed to the Advisory Council of the PNH as a foreign expert on water resource planning and thus intimately involved.
23 It was calculated that it was necessary to double the area under irrigation, to 10 million ha, by 2000 (SRH, 1975; Herrera-Toledo, 1996).

24 It is remarkable that in spite of this, SHINO was continued and given a strong push forward during the SARH era (CNA, 1994b). This illustrates the continued construction bias of the hydraulic bureaucracy.

25 This was only the case for new loans. Loans that had already become effective before 1982, for example those to the irrigation sector of in total $335 million, continued lthroughout the 1980s (World Bank, 1983b).

26 Insiders stress that during the negotiations the concrete policy proposals came from the Mexican side.

27 The Plan mentions that 'the formation of organisations (...), which will become responsible for the operation and maintenance of the hydraulic infrastructure, is considered expedient. It is expected that the irrigation districts will be financially autonomous and administratively independent' (Poder Ejecutivo Federal, 1989:77; our translation). This phrase was used repeatedly in subsequent CNA policy documents to justify IMT.

28 Concentration is used to refer to the integration of existing bureaucratic domains and resource flows regarding the exploitation of the nation's waters, formerly dispersed over different government agencies, into a 'single water authority'.

29 Decentralisation is used here as the delegation of authority and financial resources concerning water management from the federal government to user or other lower level organisations, generally with the aim to reduce government expenditure through the creation of self-sufficient water management organisations. Our choice for this definition is pragmatic. The disadvantage of the term is that it suggests a delegation of political power to lower levels in a territorial hierarchy, implying a political redefinition of the territory (Smith, 1985). It is questionable if this is so in the case of IMT. We choose not to enter this debate here, but to warn the reader that this is not implied by our use of the term.

30 In October 1987 De la Madrid designated Salinas as the presidential candidate for the PRI. The Mexican tradition of presidentialism and the dominance of the PRI for more than 60 years created a system of political transitions according to which the old president 'uncovered' the presidential candidate. This has been an important element of stability for the political system (Camp, 1999; Castañeda, 1999; Grindle, 1977).

31 Dr. González-Villareal had good relationships with leaders of the guild of hydraulic engineers (ex-ministers and deputy ministers) and with important political actors. Through his position as president of the College of Civil Engineers he also maintained good relationships with civil engineers in the government, the academic world and in construction companies.

32 Dr. González-Villareal was PRI-candidate for governor of his home state of Sonora more than once.

33 In Mexico, a deconcentrated authority is a semi-autonomous federal agency with the power to set its own policies, levy taxes and fines, issue permits and carry out acts of authority. This contrasts with decentralised public agencies, which are also semi-autonomous, but depend on their mother ministry for overall policy guidelines and direction.

34 A senior SARH official mentioned this course of events to us at the time.

35 Villalobos-Guerrero became the CNA Director during the next *sexenio* of president Zedillo (1994-2000).

36 Based on interviews with senior SARH officials.

37 Based on interviews with ex-CNA and senior SARH officials.

38 This presentation is necessarily schematic and does not pay attention to the actors involved in transfer, how the CNA overcame resistance from the users and parts of the hydraulic bureaucracy, nor how the IMT policy was further articulated in the field.

References

Aboites, L. (1998), *El Agua de la Nación. Una Historia Política de México (1888-1946)*, CIESAS, Mexico City.
Arce, A. (1993), *Negotiating Agricultural Development: Entanglements of Bureaucrats and Rural Producers in Western Mexico*, Wageningen Studies in Sociology no. 34, Wageningen, the Netherlands: Agricultural University Wageningen.
Ascher, W. (1999), *Why Governments Waste Natural Resources. Policy Failures in Developing Countries*, The Johns Hopkins University Press, Baltimore and London.
Beetham, D. (1987), *Bureaucracy: Concepts in Social Thought*, University of Minnesota Press, Minneapolis.
Bottrall, A. (1981a), 'Improving Canal Management', *Water Supply and Management*, 5: 67-79.
Bottrall, A. (1981b), *Comparative Study of the Management and Organization of Irrigation Projects*, World Bank Staff Working Paper no. 458, World Bank, Washington, D.C.
Buras, N. (1983), *Water Resources Planning in Mexico: The First National Water Plan*, Paper presented at the Second U.S.-Mexico Conference on the Regional Impacts of U.S.-Mexico Economic Relations: Challenges and Opportunities. Tuscon, Arizona, May 25-27, 1983.
Camp, R.A. (1999), *Politics in Mexico. The Decline of Authoritarianism*, Oxford University Press, New York and Oxford.
Castañeda, J.G. (1999), *La Herencia. Arqueología de la Sucesión Presidencial en México*, Extra Alfaguara, Mexico City.
Chambers, R. (1988), *Managing Canal Irrigation. Practical Analysis from South Asia*, Oxford and IBH, New Delhi.
Clay, E.J. and B.B. Schaffer (eds) (1984), *Room for Manoeuvre: An Exploration of Public Policy in Agriculture and Rural Development*, Heinemann Educational Books, London.
CNA (1990), *Water Policies and Strategies*, CNA, Mexico City.
CNA (1994a), *Transferencia de los Distritos de Riego en México*, CNA, Mexico City.
CNA (1994b), *Informe 1989-1994*, CNA, Mexico City.
Cornelius, W., A. Craig and J. Fox (1994), *Transforming State-Society Relations in Mexico: The National Solidarity Strategy*, Center for US-Mexico Studies, University of California, San Diego.
Cummings, R.G. (1974), *Interbasin Water Transfers: A Case Study in Mexico*, Johns Hopkins Press, Baltimore.
Cummings, R.G., V. Brajer, J.W. McFarland, J. Trava and M.T. El-Ashry (1989), *Waterworks: Improving Irrigation Management in Mexican Agriculture*, World Resources Institute, Washington, D.C.
Diario Oficial (1946), *Ley de Riegos*, Diario Oficial, Mexico City.
Diario Oficial (1972), *Ley Federal de Aguas*,: Diario Oficial, Mexico City.
Durán, J.M. (1988), *Hacia una Agricultura Industrial? México 1940-1980*, Universidad de Guadalajara, Guadalajara.
Espinosa de León, E. (1994), *Transferencia de los Distritos de Riego. Libro Blanco General*, CNA, Mexico City.

Espinosa de León, E. and J.L. Trava (1992), *Transferencia de los Distritos de Riego a los Usuarios*, Tercera Conferencia Regional Panamericana, Noviembre de 1992, ICID, Mazatlán, Sinaloa.

Fox, J. (1992), *The Politics of Food in Mexico*, Cornell University Press, Ithaca, N.Y.

Gates, M. (1988), 'Codifying Marginality: The Evolution of Mexican Agricultural Policy and its Impact on the Peasantry', *Journal of Latin American Studies*, 20: 277-311.

González-Villareal, F. (1982a), *Agua y Desarrollo. Documento de Profundidad*, IEPES, Mexico City.

González-Villareal, F. (1982b), 'La Infraestructura Hidráulica de México', in *Consulta Popular en las Reuniones Nacionales, Agua y Desarrollo*, pp. 20-21. IEPES, Mexico City.

Gordillo de Anda, G. (1988), *Campesinos al Asalto del Cielo, de la Expropiación Estatal a la Apropiación Campesina*, Siglo XXI, Mexico City.

Gorriz, C.M., A. Subramanian and J. Simas (1995), *Irrigation Management Transfer in Mexico. Process and Progress*, World Bank Technical Paper no. 292, World Bank, Washington, D.C.

Greenberg, M.H. (1970), *Bureaucracy and Development: A Mexican Case Study*, Heath Lexington Books, Lexington, Massachusetts.

Grindle, M.S. (1977), *Bureaucrats, Politicians, and Peasants in Mexico. A Case Study in Public Policy*, University of California Press, Berkeley and Los Angeles.

Grindle, M.S. (1996), *Challenging the State. Crisis and Innovation in Latin America and Africa*, Cambridge University Press, Cambridge.

Groenfeldt, D. (1998), *Handbook on Participatory Irrigation Management*, The Economic Development Institute of the World Bank, Washington, D.C.

Gusterson, H. (1997), 'Studying Up Revisited', *Political and Legal Anthropology Review*, 20(1): 114-19.

Herrera-Toledo, C. (1997), 'National Water Master Planning in Mexico', in A.K. Biswas (ed), *National Water Master Plans for Developing Countries*, pp. 8-53, Oxford University Press, New Delhi.

IEPES (1982), *Tema Agua. Documento de Referencia. Grupos de Trabajo para la Elaboración del Plan de Gobierno 1982-1988*, IEPES, Mexico City.

IEPES (1987), *Dialogo Nacional. El Agua: Recurso Vital*, IEPES, Mexico City.

IEPES (1988), *Perfiles del Programa de Gobierno 1988-1994. Agua: Recurso Vital*, IEPES, Mexico City.

IMTA (n.d.), *Organizacion y Administracion*. IMTA.

INEGI (1998), *Estadísticas del Medio Ambiente, Mexico, 1997*, INEGI, Aguascalientes.

Johnson III, S.H. (1997a), *Irrigation Management Transfer in Mexico: A Strategy to Achieve Irrigation District Sustainability*, IIMI Research Report no. 16, IIMI, Colombo, Sri Lanka.

Johnson III, S.H. (1997b), 'Irrigation Management Transfer: Decentralizing Public Irrigation in Mexico', *Water International*, 22(3): 159-67.

Kloezen, W.H., C. Garcés-Restrepo and S.H. Johnson III (1997), *Impact Assessment of Irrigation Management Transfer in the Alto Rio Lerma Irrigation District, Mexico*, IIMI Research Report no. 15, IIMI, Colombo, Sri Lanka.

Long, N. and J.D. van der Ploeg (1989), 'Demythologizing Planned Intervention: An Actor Perspective', *Sociologia Ruralis*, XXIX(3/4): 226-49.

Martínez, T. (1988), *Los Campesinos y el Estado en México*, Colegio de Postgraduados, Chapingo.

Mestre, E. (1997), 'Integrated Approach to River Basin Management: Lerma-Chapala Case Study – Attributions and Experiences in Water Management in Mexico', *Water*

International, 22(3): 140-52.
Moore, M.P. (1981), 'The Sociology of Irrigation Management in Sri Lanka', *Water Supply and Management*, 5: 117-133.
Orive Alba, A. (1970), *La Irrigación en México*, Editorial Grijalbo, Mexico City.
Palacios, E. (1993), *Diagnostico sobre la Administracion de los Modulos Operados por las Asociaciones de Usuarios*, Colegio de Postgraduados, Montecillo.
Palacios, E. (1994), *La Agricultura de Riego en México*, FAO/CNA, Mexico City.
Palacios, E. (1996), *El Agua: Recurso para el Futuro de México. Marco Legal e Institucional del Agua, Herramienta Indispensable para la Gestion Adecuada de los Recursos Hidráulicos*, Colegio de Postgraduados, Montecillo.
Palacios, E. (1997), *Benefits and Second Generation Problems. The Case of Mexico*, Paper presented at the International Workshop on Participatory Irrigation Management, Cali, Columbia, February 1997.
Palacios, E. (1998), *Problemas de Segunda Generación que tienen las Asociaciones de Usuarios del Agua*, Colegio de Postgraduados, Montecillo.
Poder Ejecutivo Federal (1989), *Plan Nacional de Desarrollo. 1989-1994*, Poder Ejecutivo Federal, Mexico City.
PRI (1988), *Los Retos de la Modernización. Agua y Desarrollo*, PRI, Mexico City.
Repetto, R. (1986), *Skimming the Water: Rent-seeking and the Performance of Public Irrigation Systems*, World Resources Institute Research Report no. 4, World Resources Institute, Washington, D.C.
Sánchez, Martin (1998), 'La Primera Transferencia. Gestión y Administración Federal del Agua en México', in P. Avila-García (ed), *Agua, Medio Ambiente y Desarrollo en México*, pp. 139-146, Colegio de Michoacán, Zamora.
SAyF (1929), *Ley de Aguas de Propiedad Nacional*, SAyF, Mexico City.
SARH (1981), *Plan Nacional Hidráulico*, SARH, Mexico City.
SARH (1991), *Programa Nacional de Aprovechamiento del Agua. 1991-1994*, SARH, Mexico City.
Smith, B. (1985), *Decentralisation: The Territorial Dimension of the State*, George Allen & Unwin, London.
SRH (1975), *Plan Nacional Hidráulico*, SRH, Mexico City.
SRH (1976), *Politica Hidráulica en México. Pasado, Presente y Futuro*, SRH, Mexico City.
Stanford, L. (1993), 'The "Organization" of Mexican Agriculture: Conflicts and Compromises', *Latin American Research Review*, 28(1): 188-201.
Trava, J.L. (1994), 'Transfer of Management of Irrigation Districts to WUAs in Mexico', in *Indicative Action Plan and Proceedings of the National Seminar on Farmers Participation in Irrigation Management*, Water and Land Management Institute, Aurangabad, Maharashtra.
Vargas, S. (1996), 'Las Grandes Tendencias Históricas de la Agricultura de Riego', in R. Melville and F. Peña (eds), *Apropiacion y Usos del Agua. Nuevas Lineas de Investigación*, pp. 31-50, Universidad de Chapingo, Chapingo.
Wade, R. (1978), 'Water Supply as an Instrument of Agricultural Policy: A Case Study', *Economic and Political Weekly*, 13(12): A9-13.
Wade, R. (1982), 'The System of Administrative and Political Corruption: Canal Irrigation in South India', *Journal of Development Studies*, 18(3): 287-328.
Wade, R. and D. Seckler (1990), 'Priority Issues in the Management of Irrigation Systems', in R.K. Sampath and R.A. Young (eds), *Social, Economic and Institutional Issues in Third World Irrigation Management*, pp. 13-29. Westview Press, Boulder, Colorado.
Wionczek, M.S. (1982), 'La Aportación de la Política Hidráulica entre 1925 y 1970 a la Actual Crisis Agrícola Mexicana', *Comercio Exterior*, 32(4): 394-409.

World Bank (1983a), *Mexico Irrigation Subsector Survey – First Stage. Improvement of Operating Efficiencies in Existing Irrigation Systems. Volume I – Main Findings*, Report no. 4516-ME, World Bank, Washington D.C.

World Bank (1983b), *Mexico Irrigation Subsector Survey – First Stage. Improvement of Operating Efficiencies in Existing Irrigation Systems. Volume II – Annexes*, Report no. 4516-ME. World Bank, Washington D.C.

Yanow, D. (1988), 'Tackling the Implementation Problem: Epistemological Issues in Implementation Research', in D.J. Palumbo and D.J. Calista (eds). *Implementation and the Policy Process: Opening Up the Black Box*, pp. 213-227, Greenwood Press, New York and London.

Zaag, P. van der (1992), *Chicanery at the Canal: Changing Practices in Irrigation Management in Western Mexico*, CEDLA Latin America Studies no. 65, CEDLA, Amsterdam.

Chapter 4

Irrigation Development and Management Reform in the Philippines: Stakeholder Interests and Implementation

Thomas Panella

Introduction

Irrigation reform in the Philippines has been a political challenge due to tension among the interests of irrigation stakeholders, the groups who influence or are affected by irrigation policy. Politics in this chapter means not only the will and action of the state or other political units, but includes the various types of relationships among members of society at all levels that involve authority or power; from farmers and irrigation agency field staff to the offices of state leaders or international lenders. To pursue or protect their interests, stakeholders have caused technically sound or well-intentioned reforms to founder or have transformed their implementation to suit their own purposes. Stakeholder interests may vary within a single organization and stakeholder coalitions are dynamic, adapting to changing situations and interests. The Philippine case also demonstrates that policy changes adopted to meet stakeholder interests at one given moment may have long-term consequences that affect future reform efforts, or today's problems may be yesterday's solutions.

The interaction of interests among the National Irrigation Administration (NIA), international lenders and foreign aid agencies, and domestic politicians within different social, economic, and political contexts have shaped irrigation policy and its reform in the Philippines. The National Irrigation Administration (NIA) has been the vehicle for modern irrigation development and provides the focal point for this analysis. Since its inception, NIA has sought to expand or maintain resource allocation to support its existing functions as well as to increase the purview of its activities. The strategies NIA has used to achieve its objectives have varied according to its operating environment.

International lending and aid agencies have strongly influenced Philippine irrigation and NIA's development, and a strong symbiotic relationship between NIA and lenders has always existed. Although lenders' irrigation policy objectives have changed over the years, NIA has remained the primary channel through which lenders' irrigation money has flowed. This has helped to maintain NIA's power.

Domestic politicians have helped shape irrigation policy. President Ferdinand Marcos played a large role in NIA's growth, while President Joseph Estrada has had a dramatic impact on NIA's revenue through pursuing policies that undermine cost recovery. Farmers have had little independent influence over irrigation development. The small voice they are gaining is being supplied through an organization created and supported by NIA.

Modern irrigation in the Philippines is representative of the expansion of irrigation in Asia over the last generation. When NIA was formed in 1964, there were 79 national irrigation systems (NIS), 771 communal (CIS), and 2,540 private or pump irrigation system for a total of 217,500 ha, 393,000 ha, and 51,500 ha respectively. In 1999, there are 190 National Irrigation Systems, 6,692 Communal Irrigation Systems and an estimated 4,001 private/pump systems for a total of 678,549 ha, 486,066 ha, and 174,200 ha respectively. The Communal Irrigation Systems are generally smaller than National Irrigation Systems (100 to 200 ha) and farmers are supposed to manage them. National systems are larger and managed by NIA. Private systems are generally shallow tubewell and very small (NIA, 1999). This chapter focuses on the Communal Irrigation Systems and National Irrigation Systems, which have been developed by NIA and have dominated irrigation in the Philippines. These systems are mainly gravity-fed, run-of-the-river or reservoir types, yet some pump systems exist.

Three main sections in roughly chronological order comprise this chapter. The following section, Section 2, examines NIA's formation and growth into a powerful bureaucracy during the 1970s. During this period NIA exploited substantial foreign assistance and strong domestic support from President Marcos to develop into a powerful irrigation construction agency. The period was characterized by massive capital expansion and concentration of power within NIA.

Section 3 discusses policy changes during the mid-1970s and early 1980s. Due to its strong position, NIA was able to initiate significant institutional changes to enhance its financial viability, to expand its budget, staff, and to increase its bureaucratic responsibilities. During this time, NIA initiated its first participatory irrigation management (PIM) programmes to address financial and operation and maintenance (O&M) concerns in the National Irrigation Systems and Communal Irrigation Systems. The section also examines the development of NIA's early Participatory Irrigation Management efforts and how the interplay of NIA and World Bank interests affected Participatory Irrigation Management programme institutionalization in the National Irrigation Systems and Communal Irrigation Systems.

Section 4 addresses current irrigation reform issues after the Philippine's transition to democracy in 1986. Decentralization, populist politics, and the erosion of NIA's power characterize this period. This section explores the vicious cycle of irrigation management problems that are now pervasive in the Philippines National Irrigation Systems. Reform has been difficult because different aspects of the cycle have different stakeholders whose interests counter any comprehensive solutions. Previous policy changes initiated by NIA contribute to the current problems and may act as obstacles to more permanent reforms. This section focuses on the problems of the incomplete devolution of the communal irrigation systems, the trade-off between financial viability and Operation & Maintenance, problems of Participatory Irrigation Management and current irrigation management transfer (IMT) efforts, and populist politics and the irrigation service fee (ISF).

Section 5 summarizes the discussion and assembles some lessons from the past thirty years of irrigation development in the Philippines.

NIA's Growth and Evolution

Laying Foundations: Capital Expansion, Construction, and Centralization

Understanding NIA's evolution is necessary to appreciate current political impediments to irrigation reform in the Philippines. Two primary influences shaped NIA's early development: foreign assistance and President Ferdinand Marcos. The United States Bureau of Reclamation (USBR) directed NIA's technical orientation and international lending facilitated rapid agency expansion. Marcos' strong support provided agency resources and insulation from domestic political concerns that afforded NIA substantial latitude for policymaking and operations. NIA parlayed these circumstances into a period of tremendous growth to become a large and powerful irrigation construction bureaucracy. Most NIA employees remember this period as the 'golden age of NIA'.

Formation of NIA

Even prior to NIA's formal creation, irrigation development in the Philippines has always been affected by foreign development agencies and lenders. The United States has played an especially large role due to its colonial occupation of the Philippines from 1898 to 1946.[1] The Philippines' 1961 five-year plan included verbatim recommendations from a World Bank assessment. The World Bank required a survey of Philippine water resources before any lending for water resources development would be considered. The United States Agency for International Development (USAID) provided a short-term consultant to conduct water resources surveys, and the final report in 1961 included preliminary studies of seven river basins. The report proposed feasibility studies for multipurpose water resources development projects and recommended the Philippines 'should have the benefit of outside guidance in technical aspects of over-all planning for development of a water resources program' (NIA, 1977). In October and November of 1962, the Philippines government reached agreement with USAID and United States Bureau of Reclamation to 'undertake surveys, investigation, and studies and to take steps as may be necessary to formulate comprehensive programmes for multipurpose river basin developments' (NIA, 1977).

During this time, a congressman from a region where the USBR studies recommended project development introduced legislation to create a national irrigation agency.[2] The bill passed in May 1963, and in 1964 Republican Act 3601 created NIA as a corporate body and abolished and transferred the Bureau of

Public Works Irrigation Department's (BPWID) resources to NIA. The original NIA charter contained the following provisions.

- To investigate, study, improve, construct, and administer all national irrigation systems
- To investigate all available and possible water resources in the country for the purpose of utilizing the same for irrigation, and to plan, design, and construct the necessary projects to make the ten to twenty-year period following approval of this Act as the Irrigation Age of the Philippines
- To collect fees from users to cover Operation & Maintenance and to recover capital cost over 25 years (no interest charges are mentioned)
- NIA was given a 300 million pesos capitalization to be paid at 30 million pesos a year for ten years.

Ascension to Power

NIA had a faltering start, yet its fortune changed markedly with the election of Ferdinand Marcos. NIA struggled with low ISF collections during its first two years, and it received only five of the 60 million pesos intended for its capitalization. In 1965, Marcos was elected under his well-known campaign slogan, 'Irrigation is the crying need of the hour!', and NIA's rapid expansion started. In 1966, USAID and the World Bank indicated that NIA needed to reorganize, improve its capacity, and have a continuity of leadership before they would provide assistance or lend (NIA, 1990). Marcos responded by appointing Alfredo L. Juinio, Engineering Chair from the University of the Philippines, as the NIA administrator. Juinio was reluctant to assume a political position, but Marcos replied, 'You do your job, I'll take care of the politics. Don't worry about the money, but make every centavo count' (NIA, 1990). Juinio became Administrator in December 1966 and stayed until 1980. The period marked unparalleled growth.

Concomitant with the solidification of NIA's domestic support, foreign technical assistance was influencing the agency's future. After NIA's inception in 1963, the USBR provided the two advisors, one for operations and one for construction. 'All the thinking, the methods of NIA especially those pertaining to the technical side, were all drawn from USBR – designs, operating manuals, training methods and so on, all USBR' (Bagadion, 1999). To date, NIA's technical manuals are based on USBR designs. The original USAID/USBR studies commissioned in 1962 finished in September 1966. They provided the impetus to construct the Upper Pampanga River Project (UPRP) in the Central Luzon Basin and the Magat River Multipurpose Project (MRMP) (USBR, 1966b; NIA, 1986). Today, these two projects serve approximately 160,000 ha, over 25% of NIA's current National Irrigation Systems service area and dwarf the other 188 National Irrigation Systems projects (NIA, 1999). These projects developed NIA's construction capacity and established its power base as a large bureaucracy.

Massive inflows of domestic and foreign capital to NIA accompanied the two projects' development. In 1969, a NIA project received the first Asian

Development Bank (ADB) loan ever made, which was also NIA's first foreign loan. NIA's first World Bank loan funded the UPRP in 1970. The MRMP was funded with close to $200 million of World Bank and Asian Development Bank loans. Between 1970 and 1980, 48% of all World Bank Philippine irrigation funding went to these two projects, and irrigation was 20% of the World Bank's overall Philippine lending portfolio (Svendsen et al., 1990; USBR, 1966a; NIA, 1977; NIA, 1990).

After successful completion of UPRP in 1974, Philippine irrigation lending increased dramatically and peaked in 1978 but has declined ever since (Table 4.1).

Table 4.1 NIA foreign loans and capital expenditures from 1969-1986

Year	Loan	Project	Loan Amount (million US$)	Capital Releases (Million Pesos)	% Forex	% Local
1969	ADB	Cotobato Irrigation	2.50	188.6	n.a.	n.a.
1970	WB	Upper Pampanga River Project	34.0	168.0	n.a.	n.a.
1971				356.7		
1972				568.4		
1973	ADB	Angat-Magat Integrated Irrigation Development	9.6	901.3	n.a.	n.a.
	ADB	Davao del Norte Irrigation	4.2		n.a.	n.a.
1974	WB	Aurora Penaranda Irrigation Project	68.0	901.3	n.a.	n.a.
	ADB	Agusan del Sur Irrigation	5.8	901.3	n.a.	n.a.
1975	WB	Tarlac Irrigation system Improvement Project	17.0	1,889.8	n.a.	n.a.
	WB	Rural Infrastructure Project	25.0			
	ADB	Pulangui River Irrigation	13.5			
	ADB	Laguna de Bay Development	27.5			
1976	WB	Magat River Multipurpose Project	42.0	1,795.1	17.3	82.7
	WB	Chico River Irrigation Project	50.0			
	ADB	Second Davao del Norte Irrigation	15.0			
1977	WB	Jalaur River Irrigation Project	15.0	1,801.6	23	77
	WB	National Irrigation System Improvement Project I	50.0			

Year	Source	Project	Amount	Total	%	%
	WB	Mindoro River Development Project	15.0			
	ADB	Tago River Irrigation	22.0			
1978	WB	National Irrigation System Improvement Project	65.0	1,889.8	23.5	76.5
	WB	Samar River Development Project	28.0			
	WB	Magat River Development Project II	150.0			
	ADB	Allah River Irrigation	23.5			
	ADB	Second Agusan Irrigation	14.0			
	OECF	CIADP	49.3			
1979	WB	Magat River Development Project III	21.0	2,937.2	35.6	64.4
	ADB	Bukidon Irrigation	15.0			
	ADB	Third Mindanao Irrigation Study	1.70			
	ADB	Bicol River Basin Irrigation Development	41.0			
	OPEC	Bukidon Irrigation	3.5			
	IFAD	Magat River Development Project III	10.0			
1980	WB	Philippine Medium Scale Irrigation project	71.0	2,937.2	35.5	64.4
	ADB	Second Laguna de Bay Irrigation	20.0			
1981	WB	Watershed Management and Erosion	38.0	2,307.5	44	56
	ADB	Palawan Integrated Area Development	47.0			
	OPEC	Laguan de Bay Project	7.5			
1982	WB	Communal Irrigation Development Project I	71.1	2,046.7	38	62
	ADB	Third Davao del Norte Irrigation	45.3			
	IFAD	Communal Irrigation Development Project I	12.0			
1983	ADB	Special Assistance for Selected Bank Projects	30.2	1,618.9	53	47
	ADB	Irrigation Sector	36.8			
	ADB	Fourth Mindanao Irrigation Study	1.5			
	OECF	Bohol Irrigation	40.0			
1984	ADB	Allah River Irrigation (Supplementary)	27.9	810.1	83.6	16.4
1985				876.9	63.3	36.7
1986	ADB	Special Project Implementation Assistance	30.2	554.3	53.5	46.5

Year	Source	Project	Amount			
	ADB	Highland Agricultural Development	18.8			
1987						
1988	WB	Irrigation Operation Support Project I	45.0	n.a.	n.a.	n.a.
	ADB	Sorsogon Integrated Area Development	3.7	n.a.		
1989	OECF	Irrigation System Improvement Project I	17.6		n.a.	n.a.
1990	WB	Communal Irrigation Development Project II	46.2	n.a.		
	ADB	Palawan Integrated Area Development	13.9	n.a.		
	ADB	Irrigation Systems Improvement Project I	29.0			
	OECF	Malitubog – Maridagao	38.9			
1991	OECF	Pampanga Delta	68.3	n.a.	n.a.	n.a.
1992	ADB	Kabulnan Irrigation and Development Project	48.0		n.a.	n.a.
	IFAD	Visayas Communal Irrigation Program	15.1			
1993	WB	Irrigation Operation Support Project II	51.3	n.a.	n.a.	n.a.
1994				n.a.	n.a.	n.a.
1995	WB	Water Resource Development Project	54.7	n.a.	n.a.	n.a.
	ADB	Irrigation Systems Improvement Project II	30.0			
	OECF	LADP-IC	40.3	n.a.		
1996				n.a.	n.a.	n.a.
1997				n.a.	n.a.	n.a.
1998	OECF	Central Luzon Improvement Project	103.9		n.a.	n.a.
1999	ADB	Southern Philippines Irrigation Project	60.0	n.a.	n.a.	n.a.
		Total Lending for 1969 - 1999	2,001.3			

Source: Svendsen et al. (1990), World Bank (1992) and NIA (1999). This does not include 56 million dollars in grants from agencies such as USAID and JICA. Loan amounts reflect the portion of loan used for the irrigation component of multipurpose loans. Loan dollars are not time adjusted. Capital expenditure amounts are denominated in constant 1982 dollars for the data available. Note the following acronyms Asian Development Bank (ADB), World Bank (WB), International Fund for Agricultural Development (IFAD), Overseas Economic Cooperation Fund (OECF), Oil Producing Exporting Countries (OPEC).

This was consistent with overall World Bank irrigation lending, which climaxed in 1978 at almost two billion dollars and dropped precipitously to half that level by 1988 (Jones, 1995). As Table 4.2 indicates, the World Bank (47.8%) and the Asian Development Bank (31.9%) have made almost 90% of all NIA loans. Table 4.1 indicates that NIA's overall capital expenditures peaked in the late 1970s and early

1980s and have not reached similar levels since. The relative share of domestic funding for capital expenditures has declined as loans increased in the mid-1970s. Since the mid 1980s, foreign funding has been the majority of NIA's capital budget and has averaged about 75% all capital funding over the last decade (NIA, 1999).

Alignment of Stakeholder Interests

In the ten years after Marcos took office with Juinio as Administrator, NIA had become an effective irrigation construction agency, and irrigated area had doubled in the National Irrigation Systems by 1976 (see Table 4.3). At the inauguration of Pantabangan Dam in 1974, Amnon Golan of the World Bank declared that 'NIA is the finest irrigation agency in the whole of Asia and in any developing country in the World' (NIA, 1990).

Table 4.2 Foreign lending to NIA by funding agency

Lender	Percent of Total	Total Amount (million $)	Number of Loans
ADB	31.9%	637.6	27
IFAD	1.9%	37.1	3
OECF	17.9%	358.3	7
OPEC	0.5%	11.0	2
World Bank	47.8%	957.3	20

Source: World Bank (1992) and NIA (1999). Dollars are not time adjusted, which tends to understate World Bank funding compared to later OECF and ADB funding.

This high growth period reflected an alignment of stakeholder interests that shaped NIA into a powerful construction bureaucracy. USBR involvement strengthened NIA's technical capacity while at the same time influenced project selection of UPRP and MRMP. These very large and complex multipurpose schemes required NIA to develop significant construction capacity and a large centralized bureaucracy for project management. The projects had their genesis during the peak of the USBR dam building in the US when multipurpose project construction and river basin development were being aggressively pursued. Through extending foreign assistance to the Philippines, the USBR was also maximizing its bureaucratic activities and influencing development of water resources worldwide (White, 1969). USBR and USAID involvement with NIA also played a role in securing World Bank and Asian Development Bank confidence to start lending, which accelerated irrigation development and construction.

International lenders, especially the World Bank, have substantially influenced irrigation policy. The Philippines and NIA even responded to World Bank desires to strengthen domestic capacity and funding for irrigation even before any lending had commenced. The domestic irrigation agenda has been set in many cases by the concerns and ideas of the World Bank, and the selection of irrigation activities that lenders haven chosen to fund affects the overall national irrigation development. In this early period, a focus was on construction of large multipurpose gravity systems, although this is partly influenced by the USBR. The amount of money available from donors at a given time also influences the total level of investment in the irrigation sector, and the late seventies coincided with the peak of World Bank irrigation funding (1978) as well as the peak of investment in the Philippines (Svendsen et al., 1990; Jones, 1995). Finally, lending influence may be strengthened through loan conditions.

Table 4.3 Growth in national and communal irrigation systems (1000 ha)

Year	NIS	Growth	CIS	Growth	Year	NIS	Growth	CIS	Growth
1964	217.5		n.a.	n.a.	1982	549.3	6.8%	634.8	5.4%
1965	217.5	0.0%	n.a.	n.a.	1983	554.7	1.0%	648.8	2.2%
1966	217.5	0.0%	n.a.	n.a.	1984	567.2	2.3%	658.8	1.5%
1967	288.5	32.6%	n.a.	n.a.	1985	595.9	5.1%	665.1	1.0%
1968	305.7	6.0%	n.a.	n.a.	1986	600.5	0.8%	668.8	0.6%
1969	336.8	10.2%	n.a.	n.a.	1987	618.0	2.9%	673.1	0.6%
1970	349.0	3.6%	n.a.	n.a.	1988	617.0	-0.2%	684.6	1.7%
1971	349.1	0.0%	n.a.	n.a.	1989	621.0	0.7%	695.9	1.7%
1972	349.1	0.0%	n.a.	n.a.	1990	637.3	2.6%	714.8	2.7%
1973	350.2	0.3%	429.0	n.a.	1991	645.8	1.3%	724.5	1.4%
1974	355.0	1.4%	449.0	4.6%	1992	646.5	0.1%	734.1	1.3%
1975	396.3	11.6%	470.0	4.7%	1993	646.5	0.0%	741.4	1.0%
1976	435.9	10.0%	493.0	4.9%	1994	651.8	0.8%	442.0	-40.4%
1977	455.9	4.6%	519.0	5.3%	1995	651.8	0.0%	474.3	7.3%
1978	463.7	1.7%	538.0	3.7%	1996	651.8	0.0%	488.5	3.0%
1979	472.1	1.8%	552.0	2.6%	1997	662.7	1.7%	491.4	0.6%
1980	491.7	4.2%	580.0	5.1%	1998	678.5	2.4%	486.0	-1.1%
1981	514.3	4.6%	602.1	3.8%	1999	678.5	0.0%	486.0	0.0%

Source: Svendsen (1990) and NIA (1999). Note that there was a major reassessment of functioning system in 1994 that reduced the number of communals by 40%.

The World Bank loans have had greater sector influence on irrigation policy in the Philippines than other lenders. The World Bank has made fewer but larger loans than the Asian Development Bank. World Bank loans have historically integrated institutional components in addition to infrastructure. The World Bank has also been more willing to become involved with internal policy dialogue than the Asian Development Bank. Furthermore, the World Bank loans have usually focused on irrigation and not been bundled with other rural development components in smaller more complex projects. World Bank loans and have affected greater geographical area than Asian Development Bank loans, which are more targeted and may incorporate many components other than irrigation. Finally, the World Bank has also lent far more money than any other institution to the Philippines (Svendsen et al., 1990; Husain, 1999; Rosario, 1999).

Marcos' authoritarian regime provided strong irrigation support and facilitated NIA's rise in power. Heads of state have often supported large-scale water resources development projects to implement national transformational strategies or for national security in terms of agricultural self-sufficiency or as monuments to their administrations (Scudder, 1994). It is difficult to claim that Marcos' intentions were primarily to create monumental projects, however, since in addition to UPRP and MRMP, he supported substantial investment in far less monumental communal irrigation systems during the 1970s (Rosario, 1999). For Marcos, irrigation was one part of a two-pronged strategy to achieve a national goal of rice self-sufficiency in the 1970s. The second prong was a production support programme (Masagana 99) with extension, credit, input, and post-harvest services that were integrated with the irrigation projects (Ng and Lethem, 1983). Large projects and intensive irrigation development of certain 'rice bowls' areas of the Philippines was the strategy selected to most quickly achieve rice self-sufficiency (Svendsen et al., 1990; Bagadion, 1999).

Marcos was also an astute politician and in addition to keeping the rural sector pacified, irrigation and food self-sufficiency helped keep the stomachs of Filipinos full and him in power. Marcos declaration of martial law on September 21, 1972 facilitated project development with regard to right of way and domestic funding as well as insulated NIA from domestic politics (NIA, 1977, 1990). Development of numerous irrigation systems, both Communal Irrigation Systems and National Irrigation Systems also extended the presence of the government throughout the provinces. NIA flourished under Marcos' highly centralized and authoritarian regime, and this government model may have contributed to NIA's development as a highly centralized, autonomous bureaucracy.

The combination of the foreign and domestic stakeholders interests provided NIA's tremendous opportunities for growth. NIA used these opportunities to develop its own organizational culture as a large irrigation construction bureaucracy and to strengthen its own abilities to achieve its organizational objectives. For NIA, being proclaimed the finest irrigation agency in the world meant being the best hydraulic irrigation construction agency in the world, which reflected the view of irrigation development at that time (for some, this is still the prevailing view).

The historic construction role in the 1970s still seems to be at the core of NIA's organizational culture today. NIA's capital budget has always been much larger than the Operation and Maintenance (O&M), budget. Most NIA staff, including system management, are civil engineers which contributes the to the agency's construction orientation. As engineers, construction is far more interesting and high profile work than ensuring canals are desilted and turnout structures properly maintained. Most of the current staff remains from NIA's 'golden' construction period.

Construction may also involve both a legal and illegal rent-seeking element. NIA is a monopoly service provider, so costs may be padded to benefit the agency (salaries/benefits are still well above market rates). In the Philippines, most government awarded contracts involve a tacit agreement for a 10% kickback from the contractor to government project manager (SOP, 'Standard Operating Procedure', is the colloquialism for the practice). Payments may also be made to agency staff that inspect or sign-off on a final project.[3] Construction may also satisfy contractor and agency relationships since some ex-NIA employees are now involved in contracting services. Finally, as will be discussed, construction or the capital budget is critical to supporting current Operation & Maintenance activities.

The strong alignment of interests that developed in the early growth period involved foreign concerns, domestic politicians and the agency itself. The alignment of interests was similar to the classic iron triangle that leads to the rise of many public bureaucracies in democratic countries, such as the USBR in the US. Yet in this case, electoral politics and constituent pandering were not a significant factor due to marital law, although maintenance of Marcos' power provided a strong surrogate interest. Foreign lending interests, who were also satisfied through the alignment, also provide a twist on the domestic model. Yet, the opportunity of the bureaucratic NIA to parlay these interests to its own benefit is the same in either.

The First Period of Reforms: Mid-1970s to Early-1980s

Financial Reforms or Bureaucratic Enhancement?

Financial pressures within NIA lead to policy changes in the 1970s and early 1980s. From its inception until 1975, NIA received direct appropriations from the central treasury for administration and Operation & Maintenance. Throughout this time, however, NIA had continuing problems with insufficient Irrigation Service Fees (ISF) collections to cover Operation & Maintenance costs for the National Irrigation Systems. NIA's was unable to fulfil its charter as a government corporation that required viability for Operation & Maintenance (Bagadion, 1999). To address the situation, NIA pursued institutional changes for increased financial viability and for Participatory Irrigation Management (PIM). The policy changes were also consistent with NIA's objectives of maximizing agency resources and expanding operations.

Institutional Changes

NIA was instrumental in affecting five institutional changes from 1974 to 1980 to support its operating interests: Presidential Decree (PD) 552, an increase in the ISF, changes to the Philippine Water Code, the National Economic and Development Authority's Resolution 20, and PD 1702. PD 552, enacted 1974, is the most important statutory provision affecting NIA to date, and it increased NIA's financial viability and greatly expanded NIA's purview of activities. According to the former assistant administrator, PD 552 did not take much more than a month to draft because NIA had a good idea of what it wanted (Bagadion,

1999). PD 552 reflects the power and latitude of NIA during its rise in power. Box 4.1 highlights the important powers granted to NIA and provisions of Presidential Decree 552.

Box 4.1 Powers and provisions for NIA under Presidential Decree 552

Powers Granted to NIA by PD 552
1. to 'undertake concomitant projects such as flood control, drainage, land reclamation, hydraulic power development, domestic water supply, roads or highway construction, reforestation and projects to maintain ecological balance in coordination with the agencies concerned.'
2. the 'authority to supervise the operation, maintenance and repair, or otherwise administer temporarily all communal and pump irrigation systems constructed…'
3. 'to delegate the partial or full management of national irrigation systems to duly-organized cooperatives or associations…'
4. to charge and collect fees from all irrigation systems to cover the costs of operations, maintenance, and construction costs 'within a reasonable period of time.'
5. 'to recover funds or portions thereof expended for the construction and/or rehabilitation of communal irrigation systems which funds shall accrue to a special fund for irrigation development…'
6. to construct multipurpose water resources projects.
7. to plan, design, and construct drainage facilities and to recover costs for construction and operation of such facilities and the power of eminent domain.

Provisions of PD 552
1. It increased NIA's capitalization from 300 million to two billion pesos paid at 200 million per year for ten years with no obligation to pay back the capitalization.
2. It provided general appropriations for 'general administration, current operating expenses, and operation and maintenance and administration of irrigation systems'.
3. It provided NIA an additional six million pesos annually for 'feasibility studies, investigations, surveys, and plans for preparation of projects'.
4. It allowed NIA to borrow up to US$500 million with government guarantees for irrigation projects.
5. It relieved NIA from having to pay any 'direct and indirect taxes, fees, imposts, or the charges and restrictions, including import'.
6. It stated, 'All amounts collected by NIA as irrigation fees, administration charges, drainage fees, equipment rentals, proceeds from the sale of unserviceable equipment, and materials, sale of all reparation goods allocated to the defunct Irrigation Service Unit and the National Irrigation Administration, and all other incomes shall be added to its operating capital.

Source: Presidential Decree 552, 1974 from NIA (1990). Also note that the provisions of PD 552 were also in direct support of NIA's construction of multi-purpose dam projects.

On October 21, 1974, Marcos approved NIA's request to increase the ISF (see Table 4.4). New fees approximately trebled the old 1968 rate and were denominated in rice. ISF could be paid in kind or in cash equivalent based on the National Food Authority (NFA) support price.

NIA brokered changes to the Water Code of the Philippines in 1976. The revisions strengthened elements of appropriative doctrine in the Code to improve water security for irrigator associations in communal system. The code changes also made it possible for Irrigators Associations to have water rights, so they could control allocation among members (Korten and Siy, 1988).[4] This supported NIA's planned strategy for irrigator participation for the communal irrigation systems.

The National Economic and Development Authority's (NEDA) Resolution No. 20, Series of 1978 allowed NIA to impose charges to cover Operation & Maintenance costs and to recover capital costs over 50 years. This resolution, however, was conditioned on the beneficiaries' ability to pay and provided that the government would pay all interest on irrigation development (Small et al., 1989).

Table 4.4 Irrigation rate change 1968 to 1974[a]

Irrigation Type	Wet Season	Dry Season	Third Season
Prior to 1968	12 Pesos	12 Pesos	12 Pesos
1968 All Systems	25 pesos	35 pesos	30 pesos
1974 Gravity	2 cavans[b]	3 cavans	3 cavans
1974 Pump	3 cavans	5 cavans	5 cavans
1974 Reservoir	2.5 cavans	3.5 cavans	3.5 cavans

a) All rates refer to paddy.
b) One cavan = 50 kilos
Source: Raby (1997) and Svendsen (1991). The fee structure was changed 1974 to denomination in paddy to facilitate payment of the fee increase and as a hedge against inflation. The fee for Gravity Systems at the time was equal to about 250 Pesos annually after the change, so this reflected over a 300% increase or greater for all National Irrigation Systems.

Finally, Marcos enacted PD 1702 in 1980, which boosted NIA capitalization to 10 billion pesos. Importantly, PD 1702 also allowed NIA to collect a 5% management fee from all its capital projects to be used to support agency Operation & Maintenance expenditures.

The five institutional changes, especially PD 552, not only provided the possibility for NIA to shore up its Operation & Maintenance budget and support Participatory Irrigation Management, but the changes significantly expanded the range of activities that NIA could pursue.

Viability or Subsidy?

NIA operates with two budgets, a capital budget and an Operation & Maintenance budget. NIA uses the capital budget to subsidize its Operation & Maintenance activities. Operation & Maintenance revenues include ISF, equipment rental, the

5% management fee, CIS and pump system amortization, interest (when NIA had corporate savings), and the sale of assets. ISF collections have only once been more than 50% of total agency revenues and have never come close to covering Operation & Maintenance expenses (see Table 4.6 below). Both the pump and CIS amortization used for agency Operation & Maintenance are actually equity payments that should be used to repay the capital costs for these systems. The management fee on both foreign and domestic capital projects has become a vital source of operating revenue, yet is an implicit tax on new system development and system rehabilitation. Equipment rental comes primarily from NIA renting equipment to itself to undertake capital projects (Galvez, 1999). These latter two revenue sources come out of NIA's capital budget that eventually gets paid from general revenues. In this respect, they are no different than a subsidy to NIA and farmers. The management fee and equipment rental funding sources are viable only so long as the capital budgets remain healthy, and also serve as an additional incentive to pursue construction activities.

NIA's Early Participatory Irrigation Management Experience

Besides the aforementioned policy changes, NIA pursued Participatory Irrigation Management (PIM) as means to improve its financial viability and support increased agency activities. PD 552 gave NIA responsibility for the numerous, fragmented Communal Irrigation Systems and doubled the size of its irrigation responsibility. Participatory Irrigation Management represented a possible low-cost strategy to manage the Communal Irrigation Systems, and PD 552 codified and strengthened the rights of the irrigator associations for management. NIA also saw Participatory Irrigation Management as a way to reduce agency costs by divesting the smaller National Irrigation Systems (approximately 1,000 to 2,000 ha) that were not viable even at 100% ISF collection efficiency to irrigators associations.

Communal Irrigation Systems Provide a Testing Ground for Participatory Irrigation Management

While Communal Irrigation Systems responsibility significantly expanded NIA's scope of activities, irrigated area, and budget, they posed a large management challenge. NIA needed to organize and strengthen the irrigator associations (IAs), so farmers would be able to manage the systems keeping NIA's Operation & Maintenance costs down (Praline, B.P., and Mark W., Lusk, 1991). The Communal Irrigation Systems farmers were not intended to pay ISF since the Irrigator Association members were supposed to be responsible for system Operation & Maintenance after construction. Prior to NIA, Communal Irrigation Systems construction had been funded through grants from congressmen's discretionary funds, so farmer involvement was also seen as way to encourage a sense of ownership and capital repayment for the systems. Farmers were expected to contribute labor during construction and pay off the capital costs. Farmer involvement was also needed to improve Communal Irrigation Systems development since many of the systems were poorly designed and had shoddy construction that caused beneficiary rejection (Wijayaratna and Vermillion, 1994; Illo, 1999).

In 1976, NIA started two pilot projects in the Communal Irrigation Systems that used community organizers (COs – later called Institutional Development Officers, IDOs) to work with Communal Irrigation Systems Irrigator Association members during system construction. The COs helped develop technical and institutional skills for eventual system management. The pilot projects attracted significant academic attention and led to the formation of the Communal Irrigation Committee (CIC) to assist NIA develop the Participatory Irrigation Management programme. The CIC was composed of the Ford Foundation, the Institute of Philippines Culture from Ateneo de Manila University, the Asian Institute of Management, the International Rice Research Institute, UP Los Banos Economic Development Foundation, and Central Luzon State University. The CIC was able to draw on engineering, agriculture, sociology, economics, anthropology, institutional management, and training expertise to facilitate Communal Irrigation Systems development.

The CIC representatives became important stakeholders in the Participatory Irrigation Management programme through the investment of their own research and organizing efforts. The strategy of thorough documentation and wide dissemination of the experience through the academic and development community added credibility to the programme both within in NIA and to outside agencies. It is important to note, however, that the CIC members and NIA staff involved with the early Participatory Irrigation Management efforts were separate stakeholders and in no way reflective of NIA as a whole.

For some NIA staff and members of the CIC, Participatory Irrigation Management was an intrinsic goal rather than simply a means to financial viability or improved irrigation service was their motivation. With regard to the work of the Community Organizers at the time, 'it was as close to activism that one could get without going underground' since this work was taking place under martial law (Illo, 1999). There were ties between some of the Community Organizers and the New Peoples Army (NPA – a communist resistance organization). The NPA was known to have intervened occasionally with right of way disputes and other community issues during the development of the Communal Irrigation Systems (Illo, 1999). While the Participatory Irrigation Management programme may not have led to direct subversion, the programme trained farmers to confront authority in the form of NIA engineers and to plan and work together under an extremely repressive regime that regulated public meetings. It is ironic, however, that the Participatory Irrigation Management programme would not have come about if Marcos had not provided NIA resources and autonomy to pursue the programme.

The widely documented fledging Participatory Irrigation Management experience in the Philippines satisfied a plurality of financial, academic, and political interests as well as improved delivery of irrigation service.[5] Various stakeholders were able to satisfy an interest with the programme: improved Operation & Maintenance and financial viability, the purported organizational transformation of NIA and the triumph of community participation, and effective activism under an authoritarian regime. Importantly, the early Communal Irrigation

Systems programme did not threaten NIA's interests, supported NIA expansion, and complemented NIA's construction activities; thus the programme advanced.

Participatory Irrigation Management in the National Irrigation Systems

The National Irrigation Systems posed the biggest financial problems for NIA since ISF collection was only covering about one third of total Operation & Maintenance costs. The goal was to use Irrigator Associations for O&M to reduce NIA staff costs. It was assumed that Operation & Maintenance would improve with Participatory Irrigation Management since farmers have greater incentive to perform these functions, and the improved service delivery would lead to increased ISF collections (Bagadion, 1999). Based on the experience of the Communal Irrigation Systems, the CIC developed a pilot Participatory Irrigation Management programme for the National Irrigation Systems.

Prior to development of CIC's National Irrigation Systems programme, however, far more significant National Irrigation Systems Participatory Irrigation Management activities had been taking place in the Magat River Integrated Irrigation System (MRIIS, formerly MRMP). The MRIIS programme's impetus was also to reduce Operation & Maintenance costs, improve revenues, and improve Operation & Maintenance. In 1976, MRIIS field staff had started organizing farmers irrigator groups (FIGs) at each turnout. Supporting MRIIS staff was an Agricultural Development Coordinating Council (ADCC) made up of representatives from various government agricultural support agencies. Although irrigation management was an important reason for the farmers to organize, Irrigator Association members were also very interested in possible group marketing, credit, and input purchasing benefits from the collective action. Starting in 1981, 22 Irrigator Associations had signed contracts with MRIIS staff for canal clearing (Bautista, 1986, 1999).

Although initiated independently, the two Participatory Irrigation Management programmes' development and success shared common elements. Firstly, the programmes had strong champions, Ben Bagadion for the CIC and Honorio Bautista in MRIIS. Secondly, the programmes had support of top management. Juinio, the Administrator, told Bagadion, the programme director, see what approach worked best and to 'use any and all resources of NIA' (Bagadion, 1999). Thirdly, NIA provided substantial time, resources, and autonomy to develop and refine the programme. In addition to NIA's resources, the Ford Foundation donated more than one million dollars over ten years to support the CIC effort. Fourthly, with both the CIC and the ADCC, there was a collective pool of multidisciplinary resources and capacity to assist programme development existed. Most importantly, neither of these programmes initially jeopardized field staff or agency resources (Bautista, 1986, 1999).

Institutionalization of Participatory Irrigation Management

The institutionalization of Participatory Irrigation Management in the communal and national irrigation systems elucidates two important points. First, when stakeholder interests are threatened by reforms, stakeholders will transform reform implementation to support their own objectives. The greater the threat, the less likely the original intent of the reform will be maintained. Second, both NIA and the World Bank have been willing to accommodate one another to keep each other's interests moving forward.

Communal Irrigation Systems Although reluctant and uncertain at first, the World Bank's support of the Communal Irrigation Systems Participatory Irrigation Management programme was important for its institutionalization throughout NIA as well as for acceptance of Participatory Irrigation Management world-wide. A large part of World Bank acceptance for Participatory Irrigation Management was from the influential and well-disseminated work of the CIC. The World Bank's first Communal Irrigation Systems loan in 1982 did not provide funding for the institutional development officers (IDOs – formerly Community Organizers) who organized and developed the Irrigator Associations, however, the World Bank incorporated the Participatory Irrigation Management approach in the loan's programme design. Funding for the IDOs was later secured from USAID. If the World Bank had not supported the Participatory Irrigation Management approach, the concurrent use of two different programme approaches within NIA would have caused problems with communal implementation, and avoiding this situation was critical to programme's success. As a source close to the programme expressed it:

> The Bank's Seal of Approval of the participatory approach on communals also added to the credibility of the approach with some outside institutions – particularly the Asian Development Bank, which later backed the participatory approach in its irrigation loans (Korten, 1993).

The World Bank's support for the participation with the communals also helped create support for beneficiary participation within the World Bank itself (Korten, 1993; NIACONSULT, 1993b). Under the Communal Irrigation Systems, no serious interests were threatened, and the World Bank accommodated NIA's desire for the Participatory Irrigation Management approach.

National Irrigation Systems The institutionalization of Participatory Irrigation Management within the National Irrigation Systems, however, demonstrates how reform policies can be transformed as they threaten stakeholder interests. Based on the early Participatory Irrigation Management experiences, NIA Memorandum Circular 34 of 1985 established three possible contracts for National Irrigation Systems Irrigator Associations to sign, Stage I, Stage II, and Stage III. As the name implies, the contracts had progressive levels of responsibility. Stage I simply paid Irrigator Associations to do regular canal maintenance and gave them a 2% share of the ISF collections. A Stage II contract gave Irrigator Associations essentially full responsibility for Operation & Maintenance at the lateral level and a large ISF share for their own collections (50% default). Under Stage III, Irrigator Associations did not pay ISF, but managed the system and paid amortization for the construction and rehabilitation of the system. Stage III contracts were intended for the smaller non-viable National Irrigation Systems (World Bank, 1992; NIACONSULT, 1993c).

As the National Irrigation Systems Participatory Irrigation Management programme was becoming institutionalized, however, interests contrary to the programme caused it to be fundamentally altered. NIA changed the Stage I, Stage II, and Stage III contracts to be changed to Type I, Type II, and Type III contracts. The official reason given by NIA for the change in Memorandum Circular 14 of 1989 was '... in response to the commitment of the NIA to the World Bank under Irrigation Operations Support Project (IOSP I), Loan No. 2984 PH dated June 13, 1988' (NIA, 1989). IOSP I was the first large irrigation rehabilitation loan by the World Bank. Under the new contracts, Type I and Stage I were still for general canal clearing and not very different. Stage III contracts and Type III were also not markedly different, but neither was actively promoted. However, the Type II contract significantly reduced the authority and latitude given to the Irrigator Associations with regard to system management and Operation & Maintenance compared to the Stage II contract. The Type II contract focused activities on ISF collections with a considerably reduced ISF share. Various explanations exist for the changes (Galvez, 1998, 1999; Husain, 1999).

One explanation is that the IOSP I contained loan conditions for targeted increases for both annual Operation & Maintenance expenditures and ISF collections. Under the existing arrangements, NIA could not meet the targets with the Stage II contracts, which gave too large an ISF share to the Irrigator Associations. The maximum Irrigator Association ISF share under the new Type II contract was 15% for a 90 to 100% collection efficiency by the Irrigator Association and much less for lower collection efficiencies. This was far lower than the 50% default share under the previous Stage II contract (Korten, 1993).

Another explanation is that the World Bank and many in NIA management had concerns that maintenance activities were not being properly addressed under the Stage II contracts. Some Irrigator Associations under Stage II contracts focused their efforts primarily on ISF collections while maintenance was seriously neglected. These findings were part of the World Bank appraisal mission for IOSP I. Since the World Bank was funding a maintenance and rehabilitation loan, it wanted greater assurances that adequate system maintenance would be carried out. This lead to the modulated Operation & Maintenance responsibilities and latitude under the Type II contract.

Importantly, NIA's support for Participatory Irrigation Management was never unified throughout the agency and existing support winnowed. In 1985, Ben Bagadion, the early champion, left NIA, and Juinio, the supportive administrator, had left in 1980. Top management at the time of IOSP I, and the Irrigator Association contract changes did not have the same support for Participatory Irrigation Management (Bagadion, 1999). A considerable amount of resistance and opposition had always existed within NIA management with regard to irrigator participation. Apprehension was also great among NIA field staff that Participatory Irrigation Management would lead to job losses, especially under the Stage II and III contracts, which lead to staff resistance for Participatory Irrigation Management (Galvez, 1999; Rosario, 1999).

NIA also had financial concerns about extending the Stage contracts as a system-wide policy. The Stage II and especially Stage III contracts were developed to improve NIA's financial position with regard to small nonviable systems. When the National Irrigation Systems Participatory Irrigation Management programme was being developed, the pilot programme included 26 mainly smaller, less viable systems. After the programme was developed, however, NIA realized that by offering the contracts to Irrigator Associations system-wide, a significant potential for lost revenue existed from the larger viable systems. This realization may have helped NIA to welcome the change mandated by the World Bank for IOSP I (interviews NIA staff, 1999). The reduced responsibilities of the Irrigator Associations under the type contracts meant that NIA field staff no longer had to worry about losing their positions.

All of the aforementioned reasons may have lead to the change to the 'Type' system of contracts with limited shared management focused on canal clearing and ISF collections. In one opinion, 'Thus, the Bank financed IOSP I with the Irrigator Association development component has weakened NIA's earlier participatory programs and retrogressed farmer's participation from broader managerial experience on system's operation and maintenance into narrowly defined tasks' (NIA, 1993).

Stakeholder interests had aligned to change the original contracts and programme objectives. NIA accommodated the World Bank and was able to receive its IOSP I loan money. Even if some NIA management wanted to maintain the existing Participatory Irrigation Management approach, the desire was outweighed by the opportunity the loan provided to enhance the capital budget and keep construction activities moving forward, especially since capital funding was very tight at the time. NIA was still able to achieve some cost savings though farmer canal clearing and improved ISF collections with a greater share for NIA. The World Bank was satisfied with the assurances for adequate maintenance to make the loan, and NIA field staff could keep their positions, so they no longer opposed the programme.

The difference between the institutionalization of the Communal Irrigation Systems and National Irrigation Systems Participatory Irrigation Management programmes simply reflect that stakeholder interests were not threatened with the Communal Irrigation Systems programme, so the programme moved forward. In both these cases, NIA and the World Bank were willing to accommodate one another's interests to keep the financial relationship moving forward.

The Period after 1986: Irrigation Reform and the Vicious Cycle

The period from the mid-1980s to the present has witnessed many changes in NIA's operating environment and the manifestation of numerous challenges to irrigation reform in the Philippines. In 1986, Marcos's authoritarian regime was replaced with a democratic government, and NIA's power and control has eroded. The power change was accompanied by a period of decentralization, yet attempted devolution of irrigation responsibility has proven difficult. The cycle of poor irrigation management – inadequate maintenance, insufficient user participation and low ISF collection, and financial non-viability – have become endemic to the National Irrigation Systems systems. Reform has been difficult because different aspects of the cycle of poor management have stakeholders whose interests counter any comprehensive solutions. NIA's previous policy changes for financial viability and the early Participatory Irrigation Management initiatives have contributed to current problems and serve as an obstacle to more fundamental reforms. This section looks at problems of incomplete devolution of the Communal Irrigation Systems, the trade-off between financial viability and Operation & Maintenance, problems of current Participatory Irrigation Management efforts, and populist politics and the ISF.

Historical Patterns are Difficult to Overcome

NIA's capital growth started to decline in 1983. In that year, a divergence developed between the World Bank's and NIA's interests with regard irrigation development. The World Bank recommended NIA scale back its ambitious development plans and give greater emphasis to rehabilitation and smaller scale projects. The World Bank's position has been reflected in subsequent loans. Philippine plans were very different from World Bank recommendations, yet reduced funding severely curtailed NIA's ambitious expansion plans. The Philippine Government's justifications for continued irrigation expansion were the following: 1) periodic rice shortages and subsequent hoarding created by the frequent calamities (typhoons mainly), 2) negative economic multiplier effects from rice shortages, and 3) rice imports subsidized foreign farmers and drained precious foreign reserves (NIA, 1990). Rice self-sufficiency continued to be a paramount national objective. While the World Bank's recommendations may have been based on economic pragmatism, rice and irrigation in the Philippines are a political as well as an economic commodity and service.

By the early 1980s, NIA had grown into a powerful construction bureaucracy that was in a large part created by substantial foreign backing just a few years earlier. Although the World Bank desired reform, redirecting loan money is a much simpler task than scaling down an entrenched agency that is providing a very politically popular service. The situation to date demonstrates how previous lending patterns have long lasting effects that may hamper future reforms.

Change in Government

On February 25, 1986 Corazon Aquino became the seventh president of the Philippines; Marcos had fallen from power, and NIA's protection from domestic politics was gone. Federico N. Alday, a lawyer, now headed NIA. The two or three years following the revolution were very 'messy' and it was very demoralizing for NIA staff to be headed by a lawyer not an engineer (Illo, 1999). Past and present NIA staff agrees that the frequent changes of administrators after the change in government and their outsider, non-engineering backgrounds have had a negative impact on NIA.[6] Table 4.5 shows the change in administrators in NIA's history.

Table 4.5 NIA administrators

Tenure	Administrator	Education	Background
March 1964 – Dec. 1966	Tomas de Guzman	BA Civil Engineering	Chief BPWID
Dec. 1964 – March 1980	Alfredo L. Juinio	BA Civil Engineering ME Civil Engineering	Chair of College of Engineering – UP
March 1980 – Feb. 1983	Fiorello R. Estuar	BA Civil Engineering PhD Civil Engineering	Private Sector
Feb. 1983 – May 1986	Cesar Tech	BA Civil Engineering	From within NIA
May 1986 – July 1989	Federico N. Alday Jr.	BA Law	Private Sector
July 1989 – Oct. 1992	Jose B. del Rosario	BA Engineering	From within NIA
Oct. 1992 – Dec. 1995	Apalonio V. Bautista	BA Journalism	Dept. of Agriculture
Jan. 1996 – Jan 1997	Rodolfo C. Undan	PhD Agricultural Engineering	State University Official/Academic
Feb. 1997 – July 1998	Orlando V. Soriano	BS Engineering MBA Business	Military
July 1998 – Present	Manuel Arevalo	BS Military Academy MA Management	Military

Source: NIA (1990) and NIA (1999).

NIA now had to operate in a very politicized climate, and the institutional security of the last 20 years was precarious. As a former administrator stated, 'the interest of our people giving their service was diluted with establishing political connections', and 'if you try to establish political connections, how can you accomplish your priorities' (Rosario, 1999). The change of approach and diminished support of the World Bank, the instability of internal leadership and direction, and loss of political protection under Marcos greatly impacted NIA's power and resources. The 'golden age of NIA' as the finest irrigation construction bureaucracy in the world had passed.

The Local Government Code and Incomplete Devolution of the Communal Irrigation Systems

Passage of Republican Act 7160, The Local Government Code (LGU Code) of 1991 devolved substantial authority away from the central government to the local government units (LGUs, i.e. provincial, municipal, and city governments). The

LGU Code devolved *de jure* Communal Irrigation Systems responsibility from NIA to the Local Government Units. Implementation of the Communal Irrigation Systems devolution has been problematic for the following reasons: the Local Government Units have inadequate resources/capacity to develop Communal Irrigation Systems, the Local Government Units are highly politicized, foreign lending has helped maintain the status quo, and bureaucratic resilience of NIA makes it difficult to establish new modes of Communal Irrigation Systems development.

The LGU Code provided for the transfer of the personnel, assets, liabilities, records, as well as the 1992 appropriations of national government agencies to Local Government Units. NIA lost official responsibility of 734,104 ha of Communal Irrigation Systems and 517.9 million pesos from the 1992 budget allotment. Loss of the 1992 appropriations disrupted construction or rehabilitation of 259 Communal Irrigation Systems projects (18,010 ha) with 19,760 Irrigator Association members (NIA, 1993). NIA, who fifteen years earlier was able to dictate institutional changes to take control of the Communal Irrigation Systems, was caught unaware that it would lose the Communal Irrigation Systems, which was indicative of NIA's diminished stature. Part of NIA's ignorance to the coming LGU Code may have been that its attention was on lingering internal political and financial difficulties (Bagadion, 1999). The motivation for passage of the LGU Code was a reaction to the Marcos dictatorship and a desired tangible embodiment of local control and autonomy through decentralization. Current and ex-NIA staff believes the Communal Irrigation Systems were devolved because the drafters of the bill were unaware of the capacity and resources required to develop a Communal Irrigation System. An additional explanation offered is that when drafting the LGU Code, the authors were trying to maximize the monetary transfer from the central government to the Local Government Units. Communal Irrigation Systems funding represented substantial funds, and Communal Irrigation Systems devolution seemed within the spirit of the new code (Rosario, 1999 and Bagadion, 1999).

Responsibility for the Communal Irrigation Systems is still not firmly established. To settle funding and responsibility for Communal Irrigation Systems development, an agreement between the Department of Agriculture (NIA's parent agency), the Department of Management and Budget (DBM), and the Department of the Interior and Local Governments (DILG) was forged in 1992. This agreement called for 920 million pesos to go to NIA to complete the 1992 and 1993 Communal Irrigation Systems programme of work. Ultimately, the DBM rejected this because the 1992 funds had already been given to the Local Government Units (NIA, 1993). Three subsequent attempts in 1992, 1993, and 1995 also failed to finalize an agreement, and as of June 2000, there has been no final resolution. NIA is essentially still carrying out all Communal Irrigation Systems development. Although weakened, NIA's behaviour with regard to the LGU Code reflects a current agency strategy of bureaucratic inertia and selective non-response that it has used to resist intended reforms. Whether the strategy is intentional or not is not clear, yet NIA's entrenchment has made it difficult to alter the status quo.

Policy Design Problems of Devolved Communal Irrigation Systems

Devolution of resource management responsibility, especially in the irrigation sector, is often treated as a beneficial reform. This is the premise behind Participatory Irrigation Management.[7] In this case, however, responsibility was devolved to the Local Government Units and not to the Irrigator Associations themselves, and the current Communal Irrigation Systems devolution policy faces financial, capacity, and political hurdles.

Financial Problems The LGU Code's intent was that Local Government Units would fund Communal Irrigation Systems development and Operation & Maintenance out of their internal revenue allotments[8] (IRA), which are limited to 20% for general infrastructure. Local Government Unit funds increase with the population and size of the Local Government Unit, yet it is the smaller and lesser populated cities and municipalities where Communal Irrigation Systems are usually developed. A small municipality may only receive 10 million pesos per year and if only 20% of this can be used on infrastructure, the entire infrastructure budget would be devoted to one communal irrigation project at current development costs.[9] Previous Communal Irrigation Systems funding to NIA is now diluted among all Local Government Units (even non-rural areas) and not targeted for Communal Irrigation Systems development. Attempts to require portions of the IRA be used for Communal Irrigation Systems development have not materialized (NIA (1993a); De Guzman, 1999).

Capacity Problems Most Local Government Units and their staff lack capacity to implement Communal Irrigation Systems. Unlike the Department of Agriculture (DA) and many other national bureaucracies, NIA staff did not devolve to the Local Government Units. It is not entirely clear why this was the case. The inability of NIA, the DBM, and the DILG to reach agreement regarding agency resources may have hindered devolution of staff due to the uncertainty (Meiji, 1999). Because of their corporate status, NIA salaries and benefits are quite expensive compared with other government employees, and Local Government Units may have been reticent to accept NIA staff (Husain, 1999).

Political Problems NIA used technical, economic, and social criteria to select Communal Irrigation Systems project sites, while under the LGU Code, community project selection is based largely on the discretion of local officials and thus more politicized (De Guzman, 1999). Communal Irrigation Systems must compete with roads, markets, and other general infrastructure project that have wider political constituent usage. Municipal leaders are subject to re-election funding pressures, which may come from well-capitalized contractors who expect a *quid pro quo* for construction projects that are usually not small irrigation systems. Most municipal governments only have three-year terms while most Communal Irrigation Systems take five years to develop. New administrations may have a different agenda and drop projects even once it is started (De Guzman,

1999). All these political factors have limited Communal Irrigation Systems development by the Local Government Units.

The politics and intents of the LGU Code were much larger than irrigation reform. The desired devolution required comprehensive legislation affecting all sectors and government services. Due to its broad scope, the LGU Code is a blunt instrument, yet it would have been very difficult to evaluate and devise comprehensive transfer guidelines for each responsibility devolved. Thus the Code's transitioning and implementation provisions are weak, which has lead to problems with devolution of Communal Irrigation Systems responsibilities to the Local Government Units.

Bureaucratic Resilience

NIA is still responsible for most Communal Irrigation Systems development in the Philippines. In 1995, NIA received 329 million Pesos to complete the projects that were disrupted with the loss of the 1992 Communal Irrigation Systems budget allotment. In 1996, the Philippine government allocated 800 million pesos to NIA for Communal Irrigation Systems development under a poverty alleviation programme. The original intent was for the money to go directly to the Local Government Units, but the DBM thought disbursement and management would be too difficult, so the money went to NIA to manage the programme. There were two methods of programme implementation: 1) the Local Government Unit requested NIA to implement the project (similar to the old model), or 2) NIA prepared the technical designs, but the Local Government Unit implements the project if they have the ability to administer and/or complete the construction work. According to NIA, all NIA projects were completed on time, yet many of the Local Government Unit projects linger unfinished.

NIA still receives many direct requests to implement Communal Irrigation Systems projects. In some cases, NIA has worked in a quasi-contractor capacity with Local Government Units who have secured funding from other sources for Communal Irrigation Systems. NIA even lobbies directly with local politicians to help secure their discretionary funding to develop systems (De Guzman, 1999).

The fact that NIA is still implementing the Communal Irrigation Systems underscores the bureaucratic resiliency of NIA as the Philippines only irrigation alternative. Other agencies who fund or need provision of irrigation service rely on NIA since relationships and established patterns of bureaucratic interaction have been ingrained, and since the transaction costs or risks of doing anything different are too high. If reforms are to take place, credible alternative institutional arrangements have to be created for service delivery. These alternative arrangements have not materialized, so Communal Irrigation Systems devolution to the Local Government Units has not taken place.

Foreign Assistance Supports Status Quo

Foreign funded Communal Irrigation Systems development projects still pass through NIA. This includes World Bank and International Fund for Agricultural

Development (IFAD) Communal Irrigation Systems loan projects that were developed prior to the LGU Code. NIA implements the projects using the original Communal Irrigation Systems programme protocol with Institutional Development Officers (IDOs) to develop the irrigator associations. Foreign funded Communal Irrigation Systems projects now have specific budget allocations to support the IDO's work, yet elimination of NIA's domestic Communal Irrigation Systems funding in 1992 dramatically reduced the number of IDO positions. Currently, it is common practice for a domestically funded Communal Irrigation Systems to 'borrow' IDO staff from a neighbouring foreign funded project, and foreign projects fund over 70% of the IDO requirements for all Communal Irrigation Systems development (De Guzman, 1999).

The Overseas Economic Cooperation Fund (OECF) and the World Bank both support projects that are part of the Philippine Comprehensive Agrarian Reform Program (CARP). In addition to irrigation, CARP projects provide inputs, post-harvest facilities, and other infrastructure under a more comprehensive rural development approach. The IFAD Communal Irrigation Systems project uses a similar integrated approach. NIA is the implementing agency for these projects' irrigation component. Therefore, one outcome from the Communal Irrigation Systems devolution is that a more comprehensive approach to Communal Irrigation Systems development is being employed than simply providing irrigation, yet less total irrigated area is being developed in the Philippines. These projects may provide greater benefits and sustainability in the long-run than simple irrigation projects. On the other hand, comprehensive programmes develop far less irrigation area than straightforward irrigation development loans/programmes, and they are far more difficult to coordinate successfully (Husain, 1999). The trade-offs and costs and benefits of these two approaches warrant further study for Communal Irrigation Systems reform.

The Communal Irrigation Systems case demonstrates how aligning lender and borrower interests, policies, and a programme design is critical to move reforms forward. As long as foreign funding supports most irrigation development and this money is channelled through NIA, Local Government Units it will find it difficult to develop capacity or take control of the Communal Irrigation Systems. A serious effort has not yet been made, however, to develop capacity on behalf of the Local Government Units or to transfer the resources from NIA. This is partly due to NIA's inertia and/or resistance. Since foreign funds help keep NIA in the position of developing Communal Irrigation Systems, it helps to reinforce the bureaucratic resiliency of NIA with domestic agencies who need irrigation services. Like domestic agencies, however, foreign lenders need to have assurances of credible institutional arrangements and capacity to manage responsibilities before a loan is made. Since foreign loans are almost always made with national governments, the loans always support centralization to some degree since they must pass through a national conduit agency. This arrangement presents an inherent contradiction for any loan programme meant to support reform efforts based on devolved authority and management. To establish reform policies, lenders and borrowers need comprehensive joint planning to develop reliable alternative institutions for service delivery with a time horizon that extends throughout the loan period and beyond.

The Politics of Communal Irrigation Systems Rehabilitation

A bigger problem than new system development, however, is Communal Irrigation Systems rehabilitation and Operation & Maintenance. Much of the current Communal Irrigation Systems money has been targeted at new system development, since NIA involvement for Operation & Maintenance is supposed to end after construction. The lack of resources directed at Communal Irrigation Systems Operation & Maintenance and the inability of Irrigator Associations to maintain the systems adequately has led to significant deterioration with many systems no longer functional. System rehabilitation after natural calamities (primarily typhoons) is a also problem for the Communal Irrigation Systems. While National Irrigation Systems get immediate attention from NIA, the Communal Irrigation Systems need to apply for assistance since Local Government Units may not have adequate funding or capacity for repair. Irrigator Associations often do not collect enough funds for Operation & Maintenance, and some Communal Irrigation Systems Irrigator Associations may use a strategy of deferred maintenance to let NIA rehabilitate systems. Only 20% of the Communal Irrigation Systems pay their required amortization to NIA (NIA, 1997; De Guzman, 1999).

In response to a letter from the National Confederation of Irrigator Associations (NCIA – discussed later), NIA conducted a 'Study to Find Ways and Means of Reducing the Burden of Farmer-Beneficiaries in Communal Irrigation Systems' (NIA, 1997). NIA recommended that it immediately receive increased funding to augment staff and operations to address the needs of the Communal Irrigation Systems. In the medium-term, the study recommended that the LGU Code be amended so the Communal Irrigation Systems come back under NIA's purview and budget and be allowed to convert to National Irrigation Systems systems, which still receive NIA Operation & Maintenance support. In the long-term, the study recommended eliminating the distinction between National Irrigation Systems and Communal Irrigation Systems and the Type I (canal clearing) and Type II (collections) contracts be used for both systems.

The Department of Agriculture (DA) regards the Communal Irrigation Systems reverting to NIA as a step backwards, yet recognizes that the Communal Irrigation Systems have national importance that must be addressed (Serrano, 1999). The DA has just funded a new rehabilitation programme with a negotiated cost share between the Local Government Unit, the DA through NIA, the Communal Irrigation Systems Irrigator Association, and local congressman if he/she is interested. The Local Government Unit would pay 20 to 50% of the cost, the Irrigator Association would pay 20 to 30% and NIA would pay the rest out of a 100 million-peso fund. The Irrigator Association must present a viable management plan with adequate collections to cover Operation & Maintenance costs before rehabilitation funds are granted (DA, 1999). The World Bank is considering a similar programme.

One conclusion of the current Communal Irrigation Systems situation is that the original NIA Participatory Irrigation Management intention for farmers to fully operate, maintain, and manage the systems after construction has failed if NIA's departure could cause such a rapid deterioration of facilities. NIA's recommended solution for the Communal Irrigation Systems to revert to NIA's control is attractive to the agency since it would be able to increase its budget and staffing. This solution may also appeal to the Local Government Units who likely do not want to expend resources for Communal Irrigation Systems activities. For farmers, the recommendations increase an Operation & Maintenance subsidy and provide additional support. If the Communal Irrigation Systems revert back to NIA and are converted to a shared management as under the National Irrigation Systems, this is a set back for greater cost-recovery and would expand the co-dependant situation between farmers and NIA that currently exists in the National Irrigation Systems. An Operation & Maintenance crisis with regard the Communal Irrigation Systems exists, however, and the expediency of simply having the Communal Irrigation Systems revert to NIA as a quick solution may have myopic political appeal to those under pressure to do something. If new institutions are to be forged to address the Communal Irrigation Systems situation, it will take stronger leadership, a larger commitment of resources, and more creativity than has been demonstrated thus far to overcome NIA's bureaucratic resiliency for Communal Irrigation Systems responsibility.

The Vicious Cycle - Financial Viability and Operation & Maintenance

This is the first of three sub-sections that looks at the current interrelated problems of irrigation management in the National Irrigation Systems: Financial Viability and Operation & Maintenance, Irrigator Participation, and the Irrigation Service Fee (ISF). Addressing these problems must be done comprehensively, yet each of the different problems has stakeholders whose interests impede reforms. NIA's policy changes of the 1970s and 1980s are also compromising current attempts to pursue more permanent structural reforms.

Viability Incentives and Short-lived Solvency NIA implemented several internal policies to improve financial performance during the early 1980s. On the revenue side, these measures included: incentives to ISF and amortization collectors, increasing the NIA staff eligible to collect fees, reallocation of Operation & Maintenance staff and resources to support ISF collection, paying expenses of NIA staff for ISF collection, and improved incentives to encourage payment of back accounts. The Viability Grant Incentive Program was instituted in 1984 and that programme provided direct incentives to personnel of viable systems, regions, and provincial offices. On the cost side, NIA embarked on cost cutting through: expanding the responsibilities of existing field staff, incentives to encourage early retirement of personnel, leaving unfilled position vacant, and attrition. NIA also restricted expenditures for facility upgrades and Operation & Maintenance (Svendsen et al, 1990; NIA, 1990).

These measures along with the creation of the additional revenue sources (management fee, etc.) allowed NIA to operate without a subsidy for the first time in 1982 (Svendsen et al., 1990; NIA 1990). Few other irrigation agencies have been able to achieve this. As discussed, however, NIA's Operation & Maintenance viability was partly based on revenue generated from the capital budget. In spite of these efforts, NIA's Operation & Maintenance viability was short lived. NIA has been running deficits for much of the 1990s. In June 1994, President Ramos approved an Operation & Maintenance subsidy that equalled 50% of the difference between estimated Operation & Maintenance expenditures and projected ISF collection at a 60% collection efficiency. NIA has been able to have the subsidy extended for 5 years (Ramos, 1994 and NIA, 1999). From 1995 through 1999 NIA's operating losses ranged from 20 to 73%.

Table 4.6 Collection efficiencies and ISF as percentage of O&M expenses

Year	Current Account (CA) (Millions of Pesos)	Back Account (BA) (Millions of Pesos)	Collection Efficiency BA-clct./BA	Collection Efficiency CA-clct./CA	Collection Efficiency BA+CA/CA	ISF as % of O&M Expense
1980	100.58	395.45	5.5%	52.6%	74.3%	53.8%
1981	132.76	318.54	4.2%	34.8%	44.9%	21.5%
1982	134.73	389.42	4.0%	45.0%	56.5%	27.5%
1983	136.77	432.92	3.7%	41.5%	53.0%	39.7%
1984	210.78	496.77	4.7%	36.8%	47.8%	37.9%
1985	306.84	534.23	5.4%	39.9%	49.2%	47.1%
1986	341.31	807.95	4.4%	40.3%	50.8%	50.7%
1987	337.93	1,092.35	3.3%	40.5%	51.3%	41.8%
1988	343.14	1,316.42	2.6%	39.1%	49.2%	39.4%
1989	379.34	1,537.62	2.9%	41.4%	53.3%	45.9%
1990	483.49	1,769.44	3.0%	43.3%	54.4%	42.0%
1991	568.40	2,118.49	2.6%	51.0%	60.6%	50.3%
1992	552.41	2,394.33	2.4%	47.1%	57.3%	42.9%
1993	609.40	2,664.53	2.3%	44.5%	54.4%	48.2%
1994	652.60	3,191.13	2.1%	43.6%	53.8%	47.3%
1995	604.79	3,254.15	2.5%	44.0%	57.5%	39.6%
1996	720.22	3,563.20	3.2%	44.9%	60.6%	41.1%
1997	813.06	4,095.25	3.1%	47.3%	62.7%	41.8%
1998	677.32	4,426.29	2.0%	36.0%	49.0%	26.8%

Source: Raby (1997); Galvez (1998) and Tables 4a-d of this document; NIA (1998).

Viability Leads to Revenue Enhancement at Expense of Maintenance NIA's financial viability policies have focused NIA staff efforts on ISF collections and have led to and under-investment in maintenance, particularly Operation & Maintenance equipment (Galvez, 1999; Panella, 1999). Even though many Irrigator Associations have been engaged to perform ISF collections under Type or Stage II contracts, they often do not fulfill their contracts. NIA field staff, whose salaries are dependent on ISF, have strong incentives to pick up the slack. ISF collections, which occur twice a year (wet and dry season), consume significant staff time and resources. In some systems prior to the collection period, management substantially redeploys field staff and other agency resources for collections. In some National Irrigation Systems, NIA field staff may spend up to 40% of their annual work on collection activities. In one district in MRIIS, collection expenses are a line item and have been 12% of non-salary expenses over the last few years. The policy of ISF payment-in-kind significantly increases the required resources for collection, and equipment such as dump trucks are displaced from maintenance functions to collect paddy. NIA loses about 10-15% of its gross receipts in collecting, storing, and selling paddy (World Bank, 1996; Raby, 1997; Panella, 1999).

In spite of the efforts, collection efficiency has remained low. As Table 4.6 indicates, ISF collections have never come close to covering Operation & Maintenance expenditures. 1980 was the collection efficiency peak and the only time ISF exceed 50% of expenditures. Table 4.6 also shows that there is a significant amount of money owed to NIA from delinquent accounts due to the low collection rate.

Table 4.7 NIS irrigation performance 1993 to 1998

	Service Area (ha)	Wet Season	Dry Season	Total	Cropping Intensity	Benefited Area	Per cent Benefited
1993	661,716	460,129	404,118	864,247	130.6%	785,580	118.7%
1994	651,812	460,386	401,991	862,377	132.3%	797,824	122.4%
1995	651,812	457,440	399,955	857,395	131.5%	735,901	112.9%
1996	651,812	474,476	407,832	882,308	135.4%	785,020	120.4%
1997	651,812	465,201	406,455	871,656	133.7%	811,408	124.5%
1998	669,767	458,964	371,427	830,391	124.0%	665,918	99.4%

Source: Galvez (1998).

The emphasis on financial viability has led to an under-investment in maintenance. As a general policy, new equipment, especially heavy equipment, is not purchased for system Operation & Maintenance. All backhoes, graders, dump trucks, and other equipment needed for Operation & Maintenance come from the project's original construction or other capital projects in the region where the budget provided for new equipment. A severe shortage of maintenance equipment plagues most systems. Extant Operation & Maintenance equipment is old, requires significant maintenance, has substantial down time while being repaired and waiting for parts (perhaps 50%), and is improperly sized or designed for Operation & Maintenance work, generally too large, since it was purchased for construction. When equipment repairs are made, they are usually the cheapest, most remedial measures so the equipment breaks down frequently. The oversized equipment also increases fuel and maintenance costs.

As mentioned, NIA rents the limited Operation & Maintenance equipment for contracting to generate revenue; mainly renting to itself for its own projects. Equipment rental has become the second largest source of revenue for the agency averaging between 15 and 20% over the last 5 years. The rental limits the time the equipment is available for Operation & Maintenance, which provides no revenue. In addition to agency revenue generation, anecdotal evidence suggests that equipment is used and rented for NIA staffs' personal gain, again reducing its availability for Operation & Maintenance (Panella, 1999).

As Table 4.7 indicates, the National Irrigation Systems have a very low cropping intensity and low benefited area (benefited areas produce more than 2000 kg per ha). Although it may not be the only cause, inadequate maintenance results in some areas of the systems receiving inadequate water supply, which contributes to the low cropping intensity. Many farmers, especially at the tail end, do not pay their ISF because of low production from a lack of water and dissatisfaction with the service.

Operation & Maintenance Viability at Any Cost? Attempted financial viability has come as a trade-off with focusing field staff attention on collections away from Operation & Maintenance and farmer development and with cutting Operation & Maintenance expenses. In spite of these actions, NIA has still required Operation & Maintenance subsidies. Although contrary to most reform agendas, perhaps increasing the subsidy for Operation & Maintenance is viable strategy if it results in better maintenance, higher ISF collections, and improved productivity. The question then becomes how much of an additional subsidy is needed, how it should be targeted, the expected returns, and criteria for its funding source.

Even if the Operation & Maintenance subsidy increases, it is not clear as to how significantly Operation & Maintenance would improve. Operation & Maintenance does not appear to be a priority for the agency, and it may take more than financial resources to reorient the NIA's priorities to focus on Operation & Maintenance. Construction is still perceived as NIA's core activity for reasons already mentioned. The current financial shortfalls, however, do not help the situation and provide a legitimate excuse to eschew maintenance activities further.

Rehabilitation and Deferred Maintenance The under-investment in Operation & Maintenance activities may be part of a strategy of deferred maintenance with rehabilitation. Deferred maintenance causes premature deterioration of the system, but NIA can use the capital budget expenditures for rehabilitation projects. A NIA staff person in one system estimated that half of the rehabilitation costs under the ongoing World Bank rehabilitation programme was due to deferred maintenance. Capital expenditures for rehabilitation contracts also provide the same rent seeking incentives as new construction, while Operation & Maintenance provides no rent seeking opportunities. Long-term system rehabilitation, however, is necessary in the Philippines, which has heavy rains and numerous natural calamities. This makes it more difficult to discern what is deferred maintenance and what are required rehabilitation expenditures (Rosario, 1999; Galvez, 1999; NIA Staff, 1999).

Rehabilitation funding is consistent with current lending patterns, yet rehabilitation raises difficult issues. World Bank and Asian Development Bank lending has shifted to rehabilitation in the National Irrigation Systems since the cost is perhaps one third to one half of constructing new systems and yields a better return on investment (David, 1996 and Husain, 1999). However, if much of the rehabilitation is for deferred maintenance and poor management, then rehabilitation loans reward and encourage this practice. Rehabilitation lending represents a clear moral hazard problem, yet no clear answer exists as to the best policy to remedy the situation. Additionally, from an equity perspective of rain-fed farmers, rehabilitation is spent on the same farmers who have already received irrigation once, but may have provided inadequate Operation & Maintenance assistance. Finally, rehabilitation creates no new irrigation area to increase agency revenue generation.

Vicious Cycle – Irrigator Participation

Deficiencies of the Current National Irrigation Systems Participatory Irrigation Management Approach NIA's current National Irrigation Systems Participatory Irrigation Management programme is implemented by the Institutional Development Department (IDD) and Participatory Irrigation Management has become institutionalized as a part of NIA's National Irrigation Systems management strategy. A main function of IDD has been to organize and sign contracts (Type I or II) with Irrigator Associations. Although independent research has substantiated better ISF collections and Operation & Maintenance in some systems with Irrigator Association contracts, many problems still afflict Irrigator Association performance. Along with shirking ISF collections, many Irrigator Associations fail to perform their canal clearing obligations, or may have weak management practices (World Bank, 1996; Panella, 1999).

The Irrigator Association shortcomings led to IDD's development of the Irrigator Association functionality survey in 1994 to assess Irrigator Association performance, yet functionality is difficult to measure. The survey grades Irrigator Associations based on operation and maintenance, organization, financial performance, and organizational discipline (meetings etc.), and Table 4.8 indicates, Irrigator Association performance has been low. This has led IDD to focus on 'Irrigator Association strengthening', or institutional development for Irrigator Association officers that is accomplished through training on basic leadership development, financial management and cost reconciliation, and systems management (NIA, 1998). Training, especially for Irrigator Association officers, has become a primary IDD activity along with organizing Irrigator Associations and signing contracts. These activities are easy to monitor and the statistics demonstrate IDD activity in the NIA annual reports. However, the fundamental problem is that the number of Irrigator Associations registered, number of contracts signed, and the number of Irrigator Association trainings and their attendance have become proxies for irrigator/farmer capacity development and actual farmer participation.

The Irrigator Association officers and boards of directors (BODs) may not be well connected to farmers in their area, and the average farmer may have limited incentives to participate.[10] This results in any irrigation participation being limited to the Irrigator Association officers and boards of directors (BODs). Irrigator Association officers may change yearly, and Irrigator Associations themselves may undergo frequent reorganization increasing the difficulty of developing a cohesive and active Irrigator Association membership. The more basic unit for irrigator organization (both within NIA's programme and in general) is at the turnout level where water flows from the lateral canal to the farmers' fields. Each turnout service area (TSA – an Irrigator Association may be compose of between 5 and 10 TSAs) is supposed to have a chairman who is member of the Irrigator Association BOD. Under current IDD activities, programmes are not targeted at the TSA or farmer level and there are few assurances that NIA staff or Irrigator Associations BODs actually affect this level or individual farmers. Many TSAs are termed 'inactive' and may have no direct contact with the Irrigator Association leadership. Irrigator Association BODs/officers or TSA chairmen do not always communicate information from meetings or NIA trainings back to the membership. This may be due to a lack of time or resources (snacks for meetings) or intentional for control purposes, yet little interactions may take place with the farmers. Because of these factors, farmers may feel little involvement with or responsibility for the system Operation & Maintenance or the Irrigator Association (Bagadion, 1999; Galvez, 1999; Bautista, 1999; Hassall and Associates, 1998).

Table 4.8 NIA functionality rating of irrigator associations 1994 to 1998

IA Rating	1994		1995		1996		1997		1998	
	CIS	NIS	CIS	NIS	CIS	NIS	CIS	NIS	CIS	NIS
Very Functional (80 to 100%)	0%	0%	8%	23%	7%	11%	7%	2%	18%	10%
Moderately Functional (60-79%)	0%	26%	69%	55%	63%	51%	60%	62%	64%	63%
Not Functional (<60%)	0%	75%	23%	22%	30%	38%	33%	36%	18%	27%
Total Number of IAs	2,512	1,753	2,605	1,802	2,643	1,830	2,784	1,911	2,979	2,012
Per cent Surveyed	4%	6%	22%	22%	47%	46%	44%	71%	41%	73%

Source: NIA (1999).

Since Irrigator Associations BODs may have little support from their farmer members, they become reliant on NIA to help them with contract functions. This has led to dependency between NIA staff and the Irrigator Association leadership. Under the current programme, however, little accountability exists among NIA, Irrigator Associations and farmers, so Irrigator Associations and farmers may shirk contract obligations under shared management, and NIA has a strong incentive to assist with collections and even other duties.[11] Greater farmer involvement, thus Irrigator Association strength, may be achieved if development activities are targeted at the turnout and/or farmer level rather than focusing on the Irrigator Associations organization level. To do this, however, is time and resource intensive since it requires much more groundwork, meetings, and other organizing activities by agency staff.

The weakness of the current Participatory Irrigation Management programme is compounded by the fact that the Irrigator Associations and farmers have limited system responsibility and authority beyond collections and canal clearing. The current programme that was started in 1980s has stagnated. Without a progression to greater responsibilities and exercise of real authority, farmers may not perceive adequate benefits to warrant their participation. Little opportunity exists for Irrigator Associations and farmers to maintain and advance the social capital that might be developed through more substantial responsibilities, authority, and perceived benefits. This limits the Irrigator Associations viability and the effectiveness of the Irrigator Association development programme (Bagadion, 1999; Galvez, 1999; Bautista, 1999; Hassall and Associates, 1998).

The problems of the current Participatory Irrigation Management programme led consultants for the World Bank's Irrigation Operations Support Project II (IOSP II – the second major National Irrigation Systems rehabilitation loan) Program to make the following recommendation in 1997, 'Rather than continue the ongoing modification and refinement of the present organizational approach, the Consultants recommend complete reformulation in the context of a clearly articulated and vigorously implemented programme to transfer the management of national irrigation systems to Irrigator Associations' (Hassall and Associates, 1998). Current Asian Development Bank and World Bank rehabilitation loan programmes contain institutional components that are based on much greater farmer participation through joint system management (JSM) or irrigation management transfer (IMT). The Asian Development Bank programme uses the term Joint System Management and the World Bank uses Irrigation Management Transfer. Although the terms differ, there is little difference in the memorandum of agreements (MOA – contracts) that define the roles and responsibilities of NIA and the Irrigator Associations (or CIAs) under each programme.[12]

Agricultural and Fisheries Modernization Act – Farmer Management Mandate
Domestic recognition of need to change and improve irrigation management also exists. The Philippine government has mandated greater farmer participation in the operation of National Irrigation Systems through The Agricultural and Fisheries Modernization Act. The Agricultural and Fisheries Modernization Act (AFMA), passed in 1997, requires Irrigation Management Transfer in the National Irrigation

Systems. The current DA Administration and Secretary supported the legislation, and it exhaustively addresses most activities under the DA's purview. Implementation of AFMA, however, has been very difficult.

Chapter 4 (Irrigation) of AFMA directly addresses NIA and Irrigator Association involvement in system operation. Rule 30.2 states that, 'The NIA, in consultation with the Department (of Agriculture), IAs and other relevant entities shall accelerate and complete the turnover of the Operation and Maintenance (O&M) of secondary canals and on-farm structures of the National Irrigation Systems to the management and O&M by IAs.'

Complete turnover to Irrigator Associations at the lateral level was supposed to have happened by June 30, 1999, but several provisions can justify a delay. Irrigator Associations must be deemed ready to manage the system and transition measures should be provided for any personnel who may be adversely affected. The Rule requires NIA to create an Irrigator Association capability-building programme (including technical assistance, logistical support, and training) for approval by the DA on or before October 30, 1998. As of June 2000, the plan had not yet been drafted.

Rule 31.1 states that, 'The Department shall, within five (5) years from the affectivity of this Act devolve the planning, design, and management of Communal Irrigation Systems, including the transfer of NIA's assets and resources in the relation to the Communal Irrigation Systems to the LGUs...turnover shall include responsibilities related to financing, planning, management, design, operation, and maintenance, relevant assets and resources, and transition measures for affected personnel.' NIA is required to develop a turnover proposal as well develop a capability building programme for the Local Government Units by December 30, 1998. As of June 2000, this had not yet been implemented, which is indicative of NIA's ability to resist change. AFMA simply reiterates what should have happened under the LGU Code six years earlier.

Rule 35 states that, 'Upon affectivity of this Act, the NIA shall immediately review the ISF rates and recommend to the Department reasonable rates within six (6) months from the affectivity of this Act.' The ISF review was to be submitted to the NIA Board and DA by 10 August 1998. The review was submitted 4 August 1998 by NIA and recommended an annual rate increase of 60% for diversion systems and 41% for reservoir systems (NIA, 1998). Compliance with this rule, which serves to enhance NIA resources, and NIA's non-compliance for the other rules, which threaten agency resources and power, simply reflects that NIA does not respond to initiatives that run contrary to agency interests.

Rule 111.5 specifies that for the first year of implementation of AFMA, 30% of the recommended 20 billion-peso budget should go to towards irrigation. Additionally, 5% of the appropriation was recommended 'for capability-building' of farmers, fisherfolk and Local Government Units. Instead of receiving 20 billion pesos for its first year of implementation, however, AFMA received no funding. AFMA provides a legislative mandate for Irrigation Management Transfer, yet NIA's resistance to change and adequate funding are challenges that AFMA has yet to overcome.

Current Irrigation Management Transfer Status within the National Irrigation Systems As mentioned, World Bank and Asian Development Bank Irrigation Management Transfer initiatives are taking place within the Philippines. In addition to these new programmes, some National Irrigation Systems exist that have adopted 'full turnover' under Stage III or Type III contract under NIA's existing Participatory Irrigation Management programme. In 1997, out of the 1,769 Irrigator Associations with contracts, 43 were Type or Stage III. However, only five National Irrigation Systems (with 14 total Irrigator Associations) existed where all of the Irrigator Associations in the irrigation system were Stage or Type III. The other Irrigator Associations are mixed within systems where other Irrigator Associations have Stage/Type I and II contracts, so that NIA operates the system as it would operate any other system. Even in the system with Type/Stage III contracts, some NIA staff is still present for operations (Galvez, 1999; Mejia, 1999). Some of the systems that have completely Type/Stage III Irrigator Association contracts are also covered under foreign rehabilitation loans so NIA channeled assistance is still prevalent. Importantly, the Type/Stage III contracts have never included elements that are currently considered necessary to a successful Irrigation Management Transfer programme such as water rights, crop selection freedom, scheduling control, or infrastructure rights (Vermillion, 1997). Because of these factors, the term 'full turnover', as used in the Philippines case, has always been inaccurate.

The Asian Development Bank's Second Irrigation Systems Improvement Project (ISIP II, 1995 – $30 million) involves system rehabilitation along with management turnover for nine systems on the island of Leyte. The systems range from about 650 to 1,900 ha and together cover about 12,500 ha. ISIP II has a large institutional component to develop the Irrigator Associations and the programme includes post-harvest facilities, a pilot credit scheme, agricultural extension, and a watershed management component. As of June 2000, the pilot areas are just being completed and little data is available concerning programme outcomes.

The World Bank's Improved Operation Support Project II loan (IOSP II, 1993 – $51.3 million) and Water Resource Development Project (WRDP, 1995 – $54.7 million) share the same programme approach. The loans couple management turnover with system rehabilitation and modification through proportional weirs or reduced pipe sizes for more equal water distribution. As previously noted, the programme consultants recommended dropping the ongoing Participatory Irrigation Management approach with a new Irrigation Management Transfer programme, and the conditions of the loan were changed to reflect this. IOSP II and WRDP cover 18 National Irrigation Systems systems for a total of 96,000 ha, about 1/5 of the National Irrigation Systems operational service area (World Bank, 1993).

The World Bank's IOSP II and WRDP loans will likely influence the form of any eventual national Philippine Irrigation Management Transfer programme. The programmes' Irrigation Management Transfer provisions already cover large portions of the National Irrigation Systems. The programmes do not contain agro-extension and post harvest facilities, so they are less expensive to implement and more simple to coordinate than the Asian Development Bank programme. The

World Bank's Irrigation Management Transfer programme mirrors the management requirements of AFMA and provides an easier model to replicate nationwide.

Change of Roles and NIA-World Bank Dependency There is some irony in the fact that the contract developed for the current World Bank mandated Irrigation Management Transfer programme under ISOP II (and Asian Development Bank programme) is similar to the old Stage II contract that was changed to the Type II contract for the initial IOSP I loan. In the first instance, it was a NIA developed programme/contract that the World Bank wanted changed as a loan condition. NIA accommodated the World Bank through a change from the 'Stage' to the 'Type' Irrigator Association contracts, although other internal factors contributed to NIA's policy change. Under the current IOSP II loan, Irrigation Management Transfer is a World Bank developed programme that is departure from the NIA's ongoing programme, which is based on the World Bank's previous desired position with regard to Irrigator Association participation in the National Irrigation Systems. Again, NIA has accommodated and is implementing the Irrigation Management Transfer programme per the loan requirements.

The World Bank's current change in position and support of greater participation in the National Irrigation Systems is due to some case specific and more general factors. Specifically, the World Bank's WRDP loan appraisal mission in 1995 recognized the problems of limited responsibility and programme stagnation under the NIA Participatory Irrigation Management approach. The World Bank's consultant's findings under ISOP II substantiated this and recommended greater farmer participation and a change of approach with the NIA programme. The current Asian Development Bank and World Bank Irrigation Management Transfer initiatives in the Philippines, however, also reflect a broader change in lenders attitudes to include greater emphasis on participation by beneficiaries. Although perhaps not articulated in precise operational directives, a participatory component is integrated into most current World Bank and Asian Development Bank loan programmes.

Various general factors have stimulated greater lender support for participation. Policies at the World Bank and Asian Development Bank are partly shaped in response to the constituent countries that influence their funding. These policies reflect the values of the lenders' funding countries, and end up as contingencies (loan conditions) for continued financial support to lenders.[13] Various constituencies within the lending institutions have also pushed for changes in approach, and lenders have come to realize that incorporating client participation can yield more effective and sustainable projects. Ironically, part of the World Bank acceptance for participation may have been generated by the early Philippine Participatory Irrigation Management experience in the 1980s. Current World Bank support for greater integration of participation into current loans has now come back to the Philippines to help push forward the initial, partial Participatory Irrigation Management efforts that have stagnated. Some of the stagnation, however, may in part be due to the Irrigator Association contract changes required by the World Bank in the 1980s when it was less embracing of participation.

Irrigation Management Transfer Concerns – Old and New In spite of the new acceptance of greater participation, however, the current World Bank Irrigation Management Transfer programme raises several concerns, some that are not new.

Inadequate Maintenance:
As before with the Stage II contracts, adequate maintenance has been a significant World Bank concern with the new Irrigation Management Transfer programme. After one year under the new pilot Irrigation Management Transfer programme, a significant maintenance improvement in the Irrigation Management Transfer programme areas had not been witnessed, and Council of Irrigator Associations (CIA) resources devoted to maintenance are not substantial (Husain, 1999; Panella, 1999).[14] During the early stages of the programme, the farmers have concentrated on collections since there is a 50% ISF share between NIA and the CIA, and adequate revenues are necessary to the organizations' viability. The focus on collections at the expense of Operation & Maintenance is what helped cause the change of original Stage to Type contracts. Part of the problem with the CIA focus on ISF, however, may be due to the fact that NIA has portrayed improved ISF collection and financial viability as the primary impetus of the programme. NIA may be reluctant to dwell on the fact that current agency maintenance efforts have been inadequate. Most of the CIAs that have signed Irrigation Management Transfer contracts already had Type I contracts for canal clearing and minor maintenance, and CIAs perceive little need to increase maintenance activities. Without a change in farmer attitude or mechanisms to ensure greater maintenance attention, the status quo will hold (Panella, 1999).

Long Term Rehabilitation and Major Repairs:
Responsibility for long-term system rehabilitation is not clearly defined in the Irrigation Management Transfer agreement, and rehabilitation does not show up in NIA Operation & Maintenance costs. Since much of the rehabilitation budget may be for deferred maintenance, current Operation & Maintenance expenses provide no real measure of system Operation & Maintenance costs. Actual ISF may need to be higher as indicated in the NIA report for AFMA. Current Irrigation Management Transfer contracts are also written so that the farmers understand that NIA will pay for 'major repairs,' which are defined as those over 100,000 pesos or that require heavy equipments (there is a similar provision in the Asian Development Bank's programme's contract). If the CIAs believe that NIA or the government will support rehabilitation and major repairs, then they have an incentive to defer maintenance. An incentive exists for both NIA (additional construction work and ability to keep control of heavy equipment) and farmers (free maintenance) to support this contract provision.

Capacity versus Dependency Development:
The current World Bank Irrigation Management Transfer programme in MRIIS grants a one to two year 'transition period', which is problematic. During the transition period, NIA field staff are still working even while duplicating CIA responsibility, and NIA provides temporary CIA managers who are NIA

employees. The situation is symptomatic of the 'vicious cycle of institutional development', which has plagued NIA's previous Participatory Irrigation Management efforts. In order to prepare farmers to manage the system, a need exists to increase agency staff for adequate farmer training, yet the goal is to create farmer self-sufficiency so that staff can be pulled out. Once staff is hired to with farmers, they would also like to keep their jobs. At the same time farmers grow dependent on the agency staff not only in terms of technical assistance, but often in terms of power relations. Without intervention, staff keep their jobs and farmers are never forced to create the skills or social capital necessary to manage the system. This is expressed in staff statements such as, 'the farmers are not ready yet' or 'the farmers need more training.' In order for the farmers to develop the capacity necessary to manage the system, increased staff for capacity development is necessary, yet development staff must be withdrawn at some point. If the CIAs do not have the capacity to manage the system when staff are withdrawn, however, the programme and irrigation system management will collapse. This has been one of the most difficult problems to resolve with regard to farmer management turnover.

Staff Displacement:
As before, NIA field staff has significant concern with displacement, and employees would like an early retirement package to ease displacement. Philippine Civil Service Code protects workers and generally provides one-month severance pay for every year of service under retirement. The NIA Employee Association is bargaining for two and one half months severance for each year of service. They believe this would be adequate to cause sufficient exodus to facilitate Irrigation Management Transfer implementation, however, less than this amount may be sufficient (interviews NIA Staff, 1999; Rosario, 1999). The ability to reach a severance agreement will help dissipate lingering dissention and non-cooperation with programme implementation, yet funding any agreement that causes early retirement is difficult. The Operation & Maintenance budget is struggling and already almost 80% is for personal expenses. Over the last several years, NIA has also often been in arrears with payment of retirement benefits, which is an added disincentive for employees to retire since an employment cheque is more certain than retirement.

Although the AFMA and the World Bank require NIA to formulate a policy to protect the involved staff, displacement is still unresolved as of June 2000. It is very difficult to expect full staff cooperation to implement an effective Irrigation Management Transfer training programme if they know it will lead to their displacement. Displacement from Irrigation Management Transfer is also perceived to have a perverse impact affecting the more competent and dedicated employees first. In the case of MRIIS, the programme is selecting the best Irrigator Associations for the initial Irrigation Management Transfer areas. Irrigator Association ability to implement Irrigation Management Transfer is partly related to the quality of the agency field staff in the area. As one NIA staff member commented, because they did their job well, they will be the first to be displaced.

Weak Assurances and Limited Authority:
Some of the aforementioned concerns could be addressed through changes in current programme implementation, yet more fundamental problems exist. Like the Type III contracts, the current Irrigation Management Transfer contracts do not provide water rights, crop selection freedom, scheduling control, or infrastructure rights. This may make it difficult for farmers to fully accept and carry out the responsibility for system operation. Importantly, CIAs cannot set their own irrigation service fee to cover cost of service. If the current rates, which are set by the government, are not able to cover the cost of providing adequate service and the CIAs cannot easily avail of government subsidies like NIA, then it will be impossible for them to execute Operation & Maintenance. The current Irrigation Management Transfer contract is an expanded service contract where a large motivation to participate is the 50% share of ISF that goes to the CIA while much of the authority for system operation still rests with the NIA. The current Irrigation Management Transfer contracts contain few assurances for both the CIAs and NIA to carry out their obligations under the agreement. If the programme fails, and farmers eventually reject the Irrigation Management Transfer responsibilities, there will be significant problems especially if NIA field staff has already been removed.

Slow Irrigation Management Transfer Adoption The Asian Development Bank and the World Bank projects are the only two substantial programmes where Irrigation Management Transfer is moving forward in the Philippines although it is required by AFMA nationwide. As of June 2000, less than one half of the Magat system and one smaller system had signed the Irrigation Management Transfer contract under the World Bank programme, although some preparations had taken place in the other systems covered by the loans. There is little indication that NIA is developing a comprehensive system-wide plan to pursue Irrigation Management Transfer. One reason for the delay given by NIA is the limited funding provided by AFMA to support the transition, yet even minimal compliance with AFMA to produce a transfer plan has not occurred. Without external pressure, Irrigation Management Transfer will not move forward on the agency's own accord. Inadequate funding is a limitation, and an effective Irrigation Management Transfer capacity development programme requires time and a large commitment of resources. If additional resources are committed and staff hired to develop farmers for Irrigation Management Transfer, however, a plan and timetable to withdraw those resources is needed, or the agency may simply absorb the resources and co-opt the programme into its ongoing operations without an real management turnover.

The current Asian Development Bank and World Bank Irrigation Management Transfer programmes are tied to over $150 million in rehabilitation funds, which helps explain any irrigation agency change thus far. A large risk exists, however, from implementing a token effort. If Irrigation Management Transfer is poorly executed and system management degrades, farmers will become resistant to Irrigation Management Transfer programmes and the agency has a convenient excuse to delay system-wide implementation.

134 The Politics of Irrigation Reform

Vicious Cycle – Irrigation Service Fee (ISF) and Populist Politics

Proposed Elimination of the ISF The change to democratic government in the 1980s brought not only decentralizing tendencies, but also a rise in populist politics. The current President's, Joseph Estrada's, position on ISF has significantly complicated the National Irrigation Systems reform process since he promised free irrigation water during his campaign. In his first State of the Nation Address on July 27, 1998 the President said, 'We are condoning irrigation fees. *Ibibgay natin nang libre ang patubig sa magasasa*ka' (Philippine Daily Inquirer, 1998). The Presidential staff maintains that the president was speaking only of condoning the back accounts, and the English is sufficiently ambiguous to imply this. The Tagalog is quite straightforward, however, stating free water will be given to farmers (Ignacio, 1999).

Upon review of the policy implications for free irrigation water, Administrative Order Number 17 (AO 17), a compromise, was issued as a NIA resolution in September 1998, AO 17 written in response to the President's desires. The Order reduced ISF for the majority of farmers by about one half and differentiated ISF rates based on farm holdings (see Table 4.9). AO 17 also reduced Communal Irrigation Systems equity payments.

Table 4.9 Socialized ISF from AO 17

Rates in Cavans (1 Cavan = 50 Kilos)	Diversion		Reservoir	
	Wet	Dry	Wet	Dry
Old Rate	2	3	2.5	3.5
0 to 2 ha	1	1.5	1.5	2
2 to 5 ha	2	3	2.5	3.5
5 ha and above	3	4.5	4	5

Source: NIA (1998). The rates for pump systems remained the same and other crops are not included in the table.

Abolition of the ISF has been a pervasive political gambit throughout the 1990s, and as early as 1989, congressmen from rice producing regions had proposed legislation to eliminate ISF and/or condone the back accounts owed by farmers. NIA has issued several position papers addressing the negative impacts on NIA operations from eliminating ISF and has argued that removing ISF and increasing the irrigation subsidy takes money that could be used for more needed poverty alleviation since farmers with access to irrigation are not the poorest farmers. However, NIA has often requested additional subsidy, so its motives for supporting ISF are likely more related to financial viability than welfare concerns, yet the welfare arguments are still valid (NIA, 1997).

Impact of AO 17 Although AO 17 reduced the ISF by half for most farmers, the immediate impact of the President's actions was that collection efficiency dropped by almost a quarter, from 47.3% in 1997 to 36.0% in 1998 (see Table 4.6 above). The decreased ISF collections and lower rate reduced NIA revenue from 511 million pesos in 1997 to 360 in 1998. ISF collections went from 38.5% of NIA's total Operation & Maintenance revenues in 1997 to 27.0% in 1998, so NIA struggled even more with its Operation & Maintenance budget. The infusion of a 298 million-peso subsidy made operations viable in 1998 (See Annex 1). The NIA Employees Association has petitioned the President for increased subsidies to cover ISF reductions since staff salaries have been delayed. AO 17 has further increased the amount of time NIA staff spends on collections further limiting time for system Operation & Maintenance.

The President's announcement for free irrigation and the resulting lower ISF collections from AO 17 impacted the World Bank's Irrigation Management Transfer programme. Many CIA Board of Directors and farmers like the new rate, yet realize that the CIA and farmers share has been cut in half. This caused some CIA reluctance to sign the Irrigation Management Transfer contracts. Under the Asian Development Bank Irrigation Management Transfer programme, however, some Irrigator Associations decided to maintain the previous (pre-AO 17) ISF rates. The additional money from charging old rates will be to support Operation & Maintenance and other agricultural activities of the Irrigator Association. The programme has met with limited success, however.

The fact that ISF rates decreased, yet collection efficiency still declined due to the president's announcement demonstrates that willingness and ability to pay ISF are very different issues that vary with farmers perceptions and expectations. Most farmers were aware of the President's statements and expected free irrigation, although many farmers agree the old rate was not burdensome. AO 17 has lead to many farms being subdivided on paper since the new rate uses a graduate tariff based on size of holdings. Some large land-owners simply refuse to pay the new rate, and NIA has little recourse but to accept the old rate (Galvez 1998; Panella, 1999).

Larger ISF Concerns The government, not NIA, sets ISF rates, which is a problem.[15] Even though ISF was originally based on cost of service provision and indexed through denomination in rice, real per unit Operation & Maintenance costs have increased and the price of rice has not kept pace with inflation (Svendsen et al., 1990; World Bank, 1992). The ISF rate had not changed from 1974 to 1998 and raising ISF to sufficient cost-recovery levels is more difficult given the recent ISF decrease and presidential announcement. NIA is stuck with a corporate charter to recover Operation & Maintenance costs yet cannot price its own services. This situation is responsible for NIA seeking alternate revenue sources in the 1980s as well as for the recent required subsidies. The situation has also led to proposed legislation to raise NIA's management fee on capital projects from 5% to 10%, which would cover the ISF revenue loss from AO 17 (Galvez, 1998). Understandably, NIA supports this measure since the capital budget looks robust for the next few years, and this revenue is much easier to collect and more assured than ISF.

The recent ISF developments have set back irrigation management reform; Irrigation Management Transfer becomes far more difficult since ISF funds may not be sufficient to cover CIA Operation & Maintenance costs, and NIA's Operation & Maintenance budget suffers. Neither of these bodes well for improved service delivery. Not surprisingly, the ISF situation is also troubling to international lenders. After the President's announcement, the Asian Development Bank put all agricultural loans on hold. It took the DA several months to repair relations and unfreeze the loan money. In late 1999, the Asian Development Bank awarded a technical assistance to examine the ISF issue (ADB, 1999).

The National Confederation of Irrigator Associations and Farmer Representation
Farmers have had little role in shaping irrigation policy development in the Philippines. Most farmers are poor and politically weak to demand better service or influence the policy process. Especially under Marcos, public influence was squelched, yet during this time irrigation was a national priority backed with resources. Irrigation development, however, has also been heavily subsidized so farmers with irrigated lands may be complacent since they are far better off than most rain fed farmers even if service delivery is poor (many farmers also do not pay their irrigation service fees). Asymmetric power relations between NIA and farmers may also inhibit farmer demands for better service. To date, there have been few if any demand driven irrigation policies. Effective stakeholders representing farmers' interests to ensure irrigation policies and farmers' needs are aligned have not been present, which helps explain why many irrigation policies have been shaped and co-opted by NIA, lenders, and other political interests.

The National Confederation of Irrigator Associations (NCIA) recently formed to represent farmers. Each Administrative Region in the Philippines (12 total) has a Federation of Irrigator Associations that provides representation to the NCIA. The first national NCIA congress (meeting) was held in June 1997, and in February 1998, permanent BODs and officers were installed. According to their by-laws, the mission of the NCIA is the following: to represent farmers in all government agencies and policy making bodies; to represent farmers at international symposia and public hearing; to channel technical assistance to Irrigator Associations; to hold regular meetings of the regional presidents; and promote collaboration among Irrigator Associations, NIA, other government agencies, and NGOs. NCIA officers are already sitting on advisory boards for agricultural programmes. NIA supported and facilitated the NCIA's development and is paying almost all costs for BOD and officer trainings, the national congress, and quarterly BOD meetings (Dulig, 1999; NIA, 1999).

Politics of the NCIA and NIA Currently, the NCIA's and NIA's interests are aligned. The NCIA advocates policies that appeal to farmer interests such as continued subsidies to NIA to ensure operations, especially in light of the lower ISF rates. The NCIA has supported unification of the Communal Irrigation Systems and National Irrigation Systems Operation & Maintenance policies, so that the Communal Irrigation Systems would receive the same subsidized Operation & Maintenance support under the existing Type I and II contracts. The

NCIA is supportive of Irrigation Management Transfer, yet sees no reason to displace NIA field staff so long as the service area can keep increasing, and continued expansion of irrigated area is a shared interest of NIA and the NCIA. The current NCIA President is the past General Manager for MRIIS and for NIA Region II (interviews NIA Staff, 1999).

The relationship between the NCIA and NIA advances both of their interests. Through its stated role in representing farmers, the NCIA has legitimacy to intervene in policy and fills a current void for farmer stakeholders. NIA support has propelled the NCIA into a position of greater power in various policy-making arenas. The NCIA is viewed as an independent advocate for policies that maintain NIA's budget and staff and subsidized service to farmers. Prominent elected politicians, including the vice-president, have spoken at NCIA congresses. The NCIA plans eventually to run candidates as sector representatives for the House of Representative under the Philippines' proportional representative system. Since the NCIA has approximately 750,000 members through all of the individual Irrigator Associations, a realistic possibility exists that candidates can be elected (NCIA, 1999). So long as interests remain aligned, this will be productive partnership for the NCIA to gain power, farmers to maintain or increase subsidy, and for NIA to maintain agency resources. If NCIA power grows and interests diverge, however, then NIA may find itself in a difficult position with a formidable foe.

The NCIA may be able to play a constructive coordinating role to facilitate Irrigation Management Transfer, yet the organization is still not able to function independently and its motives and the policy positions of the leadership are not clear or cohesive. In addition to assisting with any transfer of management, the NCIA and its leadership may also be well positioned to receive a share of the wealth and power that is transferred away from NIA towards farmers under a national Irrigation Management Transfer programme. At least 50% of the ISF collection and an eventual higher share will be transferred to the Irrigator Associations if national Irrigation Management Transfer moves forward. As has been the case with other irrigation reform policies in the Philippines, interested parties have creative abilities to transform the policy process toward their own ends. Some of the NCIA officers have large agro-business and NIA contracting concerns, so their interests in the policy agenda are likely different than small tenant farmers in irrigation systems. It is too early to tell what role the NCIA will play, what policy positions they will adopt, or whether the will represent the average farmer, yet it is gaining political influence (Dulig, 1999; NIA, 1999).

Summary and Conclusions

Progression of Irrigation Policy

Irrigation policy, reform oriented or not, has moved forward in the Philippines when stakeholder interests are aligned and has been thwarted or transformed when they are not. The alignment of foreign technical assistance, international lending, the domestic political agenda, and NIA resource-maximizing interests led to the centralized, construction oriented entrenchment of NIA during the 1970s and early 1980s. NIA's capital build up increased financial management pressures that led to the policy changes of the late 1970s and early 1980s. The management fee and equipment rental, which have supported Operation & Maintenance viability, threatened no interests and have survived to this day. The early National Irrigation Systems Participatory Irrigation Management initiatives which sought system turnover, however, did not have full support of NIA top management, threatened NIA field staff and agency revenue, and were contrary to World Bank interests at that time. Accordingly, the Participatory Irrigation Management contracts were transformed into shared management to protect these interests.[16]

Irrigation management reforms in the Philippines need to address the legacy of the gravity systems that have fallen victim to the vicious cycle of irrigation management. The policy changes from the 1970s and 1980s have complicated the situation. Although ISF is not sufficient to cover Operation & Maintenance, NIA's creation of other revenue sources lessens the severity of the financial situation and masques a direct subsidy. This has helped to prevent a crisis situation that may trigger more substantial changes. NIA's emphasis on viability has led to extreme cost cutting, a severe under-investment in Operation & Maintenance, staff orientation towards collection at the expense of Operation & Maintenance work, and reliance on deferred maintenance. Deferred maintenance, however, not only satisfies a financial objective, but rehabilitation increases the capital budget and generates construction, which is more in line NIA's interest. NIA's previous Participatory Irrigation Management activities have fostered dependency between Irrigator Associations and NIA that may be more of an obstacle to implementing Irrigation Management Transfer than if there had been no previous Participatory Irrigation Management activities.

Many of the institutional factors are in place to address the inter-related irrigation management problems. There is 1) a history of irrigator associations with legal standing; 2) a statutory basis and a history of paying an irrigation service fee; 3) a mandate to cover cost of service and even capital costs; 4) current legislation that mandates Irrigation Management Transfer in the National Irrigation Systems. Different aspects of the problem, however, have different political constituencies who are impeding reform efforts. Irrigation Management Transfer provides a possible solution to address Operation & Maintenance and viability issues, yet threatens NIA field staff, resources, and status. A successful Irrigation Management Transfer programme requires sufficient revenue generation to support the cost of service provision by the Irrigator Associations. However, Irrigator Associations do not have the ability to set ISF rates and populist politics have reduced the ISF rate to unsustainable levels. All of these conflicting interests make reforming irrigation very difficult, and lasting reforms will require significant changes across numerous parameters due to the interconnectedness of the problems.

NIA's build-up into a powerful centralized bureaucracy during its early period and continued support from foreign lenders still allows it to thwart decentralizing reform activities. NIA's selective non-responsiveness to reform mandates, resilient patterns of bureaucratic interaction to provide irrigation service, the government's inability to create viable new institutional arrangements, and political expediency have helped maintain the status quo rather than make difficult long-term changes.

International Lenders and Reform

The politics of irrigation reform are often the economics of irrigation reform. Development banks and other foreign lenders make loans and governments borrow. From the creation of NIA to the present, foreign lenders have played a large role in shaping irrigation policy through the types of loans, loan conditions, and lending totals. Aligning lender and borrower interests, policies, and programme designs is critical to move reform efforts forward. Developing irrigation management reforms takes comprehensive planning, time for change, and commitment from both lenders and borrowers.

Foreign lenders can facilitate the reform process. They can provide expertise to help host government create reforms strategies. Lenders provide longevity of tenure to help keep reform processes on track, which is useful in situations of desultory government policy making. The loan process provides an ongoing financial incentive that transcends domestic political office. Finally, foreign lenders can also play useful political scapegoats for difficult domestic reforms, such as staff displacement.

The World Bank and Asian Development Bank are the only stakeholders that have had any success in pushing irrigation reforms forward. Without greater domestic support from an agency outside of NIA that has leverage to influence NIA or more forceful use of the tools at the lenders' disposal, it is uncertain that reforms such as ISF reform or Irrigation Management Transfer will continue to more forward. Loan conditions are a commonly used strategy to influence policy, yet there are limits to how much foreign lenders can influence domestic policy if strong political opposition to reforms exists. Lenders are less subject to political constraints and have greater flexibility to change policy direction with sector priorities. Frequent changes in lending policy or a lack of resources, incentives, and long-term commitment to support policy changes, however, will not yield desired reforms.

Conversely, agencies such as NIA have complied with loan conditions, yet in NIA's case, compliance may have more to do with resource maximization and construction desires, than support for reforms. Compliance may be more cosmetic than structural. Borrowing countries also have lending options, so as World Bank loan priorities have moved away from new construction towards irrigation rehabilitation, NIA has also switched lenders to suit its purposes. The Overseas Economic Cooperation Fund lending (now Japanese Bank of International Cooperation), which imposes less conditions and is more in line with NIA's interests, has grown significantly in the 1990s and has supported NIA's most recent National Irrigation Systems expansions (See Table 4.1).

While international lenders may favor reform, a troublesome contradiction exists with lenders initiating reforms that attempted to facilitate decentralization or attenuation of national agencies' power or resources. Large international loans have tended to keep national bureaucracies centralized and strong since funds are channelled through central agencies. In NIA's case, foreign funding is the majority of the capital budget, which presents obstacles to devolve management responsibility whether it be to farmers or to Local Government Units. Even if power is devolved, reliance on the central bureaucracies for programme funding such as long-term rehabilitation may compromise reform efforts.

Final Lesson

The experience in the Philippines demonstrates that to reform the vicious cycle of irrigation management or change any other irrigation policy, stakeholder interests must be addressed. If stakeholders' interests are not represented, such as farmers, they will not influence policy development, which will likely hinder final policy adoption. Prior to an investment in devising technical solutions, it is necessary to ensure that political and economic interests are aligned to support reform policies and stakeholders are represented. One conclusion of this is that there needs to be an open, inclusive process that allows irrigation interests to be debated and accommodated for policy development prior to policy implementation. The problem still remains, however, that implementation can be transformed to support more narrow stakeholder interests. Assurances, feedback mechanisms, and incentives need to be in place to address this concern. Yet experience in the Philippines has shown this is very difficult since stakeholders will not support or implement policies that run contrary to their interests. This behaviour is not necessarily conscious, insidious strategizing to thwart initiatives, but rather a natural response on behalf to stakeholder whose interests are threatened.

While this chapter describes many problems with irrigation reform, it is admittedly rather weak on solutions to overcome the political obstacles to moving reforms forward. These obstacles do not have simple solutions, yet their explicit recognition is necessary so that the process of developing institutional arrangements and policies for improved irrigation reform can address them.

Notes

1. Passage of Act No. 1854 in 1908 established the BPWID in the Philippines to construct and manage national irrigation systems. The BPWID was based on the USBR that had been created six years earlier, and it hired a USBR employee to help structure the new agency (Svendsen et al., 1990 and NIA, 1990).
2. It is unclear whether this legislation was introduced specifically to facilitate the development of the USBR projects or from a general perceived need to increase irrigation development.
3. These comments do not imply that all government employees engage in these practices, yet discussions with agency staff, contractors, and local businessmen indicate that the practices are more common than not.
4. In the Philippines, only communal systems' IAs are given water rights. In the NIS, NIA retains the water rights.
5. Several studies (Korten, 1988; Svendsen, 1991) provide empirical evidence regarding improved performance of the CIS that integrated PIM in their development versus

Irrigation Development and Management Reform in the Philippines 141

those that did not have participation. As will be seen in the next chapter, however, the sustainability of the CIS based on a purely PIM strategy has failed.

6   Even if the leadership skills of the new NIA administrators were excellent, this general negative perception would make effective leadership difficult.
7   More precise terms for transferring power horizontally and vertically exist, i.e. devolution v. decentralization v. deconcentration etc. Devolution is used generically here.
8   LGU government funding comes through dispersing of national taxes. The national government remits funds to the LGUs through the internal revenues allotment (IRA).
9   NIA staff knows of only two CISs funded from the IRA, and these had matching funds from other sources.
10 IAs are typically about 150 to 300 ha, and the average farm size in the NIS is 1.35 ha.
11 There is an anecdote of a Water Resources Field Technician (WRFT) who paid workers out of his own pocket to clear canals even though an IA was supposed to perform this work. The clearing was prior to system inspection and the WRFT would suffer a bad evaluation if the canals were not cleared. IA contract failure for maintenance, however, can partly be explained by the fact that recently NIA is often in arrears with IA payments.
12 This paper uses the term IMT since this is the more widely used in the literature although JSM is more reflective of what actually takes place.
13 Gender issues and environmental concerns are two other areas where changes in lender policies may be partly in responses to concerns of the affluent nations who influence lender funding. This does not insinuate that these are spurious concerns, yet without the awareness and pressure to incorporate these concerns (NGOs play a large role here), it is uncertain that the borrowing countries and the lenders themselves would have generated these policy changes as quickly or comprehensively.
14 Under the current World Bank program IAs are organized into a Council of Irrigator Associations (CIA) that are made up of approximately five to seven IAs around a lateral canal. The CIA is the organization that signs the IMT memorandum of agreement (MOA) with NIA.
15 The last two ISF rate changes were Presidential orders.
16 No clear indication exists, however, that these early transfer efforts would have been successful since the program still lacked many features such water and infrastructure rights.

References

Bagadion, Benjamin (1991), 'Farmer Participation in Irrigation Management: The Philippines Experience', in Bradley W. Parlin and Mark W. Lusk (eds) *Farmer Participation and Irrigation Organization*, Westview Press, Boulder, CO.

Bagadion, Benjamin (1999), Personal Interview in North Fairview, Metro Manila, September 24, 1999. Former Assistant Administrator for NIA, until 1985.

Bautista, Honorio (1987), *Experiences with Organizing Irrigators Associations: A Case Study from the Magat River Irrigation Project in the Philippines*, IIMI Case Study no. 1, International Irrigation Management Institute, Sri Lanka.

Cruz, Conception J., Luzviminda B. Cornista and Diogenes C. Dayan (1987), *Legal and Institutional Issues of Irrigation Water Rights in the Philippines*, University of the Philippines Los Banos, Agrarian Reform Institute, Los Banos, Philippines.

David, Wilfredo P. (1996), *Acceleration Transformation to Irrigation Agriculture: Policy*

Issues and Recommendation, University of the Philippines Los Banos, College of Engineering and Agro-Industrial Technology, Los Banos, Philippines.

David, Wilfredo P. (1999), *Summary of the Finding of the 1997-98 Inventory of Shallow Tubewells and Low-lift Pumps in Region I*, Department of Agriculture Memorandum, 25 August 1999, Quezon City, Philippines.

De Guzman, Rodrigo (1999), Personal Interview NIA, Quezon City, 23 September 1999. Project Director in charge of foreign funded communal projects.

Del Rosario, Jose B. (1999), Personal Interview at Hydroterre Inc., Quezon City, 27 September 1999. NIA employee from 1964 to 1986. He became NIA Administrator from 1989 to 1992.

Easter, William K. (1993), 'Economic Failure Plagues Developing Countries Public Irrigation: An Assurance Problem.' *Water Resources Research*, 29 (7): 1913-22.

Galvez, Jose (1999), Personal Interview in Munoz, Nueva Ecija, 28 September 1999. NIA employee from 1971 through 1994, Assistant Administrator from 1989 through 1994.

Hassall and Associates (1998), *Completion Report – Second Irrigation Support Project – WB Loan No. 3607 – PH*, In association with Primex, Manila.

Husain, Syed (1999), Personal Interview at the Makati City Shangri La Hotel, 25 September 1999. Senior Economist with the World Bank and is responsible for the Improved Systems Operation Project II and Water Resources Development Project Loans to NIA.

Illo, Jean Francis I. (1999), Personal Interview at her residence in Quezon City on September 24, 1999. Former member of Communal Irrigation Committee and contributing author to *Transforming a Bureaucracy*.

Jones, William O. (1995), *The World Bank and Irrigation*, The World Bank, Washington D.C.

Korten, Frances F. (1993), Personal Correspondence sent to Dr. Bhuvan Bhatnagar of the World Bank, March 17, 1993.

Korten, Frances F. and Robert Y. Siy. (1988) *Transforming a Bureaucracy – The Experience of the Philippines National Irrigation Administration*. Kumarian Press, West Hartford, Connecticut.

Ng, Ronald and Francis Lethem (1983), *Monitoring Systems and Irrigation Management: An Experiences from the Philippines*, Monitoring and Evaluation Case Studies Series, The World Bank, Washington DC.

NIA (1977), *Completion Report – Upper Pampanga River Project, Loan No. 637 PH*, NIA, Quezon City, June 1977.

NIA (1989), *Memorandum Circular #14 of 1989 – Revised Criteria in the Shared-Management (turn-over) of National Systems or Part Thereof to Irrigators' Associations Under three (3) Types*. NIA, Quezon City.

NIA (1990), *A Comprehensive History of Irrigation in the Philippines*, NIA, Quezon City.

NIA (1992), *Devolution of Certain Irrigation Activities to Local Government Units – Internal Position Paper prepared by Corporate Planning*, Quezon City, July.

NIA (1993a), *Locally-Funded Communal Irrigation Projects – What Happened After Its Devolution to LGUs – Internal Position Paper prepared by Corporate Planning*, Quezon City, May.

NIA (1993b), *Corporate Plan: 1993 - 2002 – prepared by Corporate Planning*. Quezon City, June.

NIA (1995), *Devolution of Locally-Funded Irrigation Projects to Local Government Units: the Later Developments – Internal Position Paper prepared by Corporate Planning*, Quezon City, August.

NIA (1997), *Study to Find Ways and Means of Reducing the Burden of Farmer-Beneficiaries in Communal Irrigation Systems*, National Irrigation Administration,

Quezon City, December.

NIA (1998a), *Position Paper on the Proposal to Abolish Irrigation Service Fees*, National Irrigation Administration, Corplan, Quezon City.

NIA (1998b), *Year-End Report to the President: 1998*, National Irrigation Administration, Quezon City.

NIA (1999), Training materials for the IMT transfer. Presented for D2B series IAs of the Magat River integrated Irrigation System, Echague, Philippines, June 2, 1999.

NIACONSULT (1993a), *An Evaluation of the Impact of Farmers' Participation on the National Irrigation Systems' (NIS) Performance*, NIACONSULT, Quezon City.

NIACONSULT (1993b), *Farmers' Participation in the National Irrigation Systems in the Philippines: Lessons Learned*, NIACONSULT, Quezon City.

NIACONSULT (1993c), *The Role of the World Bank in NIA's Participatory Programs*, NIACONSULT, Quezon City.

Panella, Thomas (1999), *Irrigation Management Transfer in the Philippines*, Ongoing research for Ph.D. dissertation, University of California, Berkeley.

Pascua, Domindor (1999), Personal Interview 22 September 1999, NIA, Quezon City. Manager of the Corporate Planning Staff for NIA.

Philippine Daily Inquirer (1998), Transcript of President's State of the Nation Address. Philippine Daily Inquirer, Manila, July 27, 1998.

'Presidential Decree 552' (1990), in NIA, *A Comprehensive History of Irrigation in the Philippines*, NIA, Quezon City.

Raby, Namika (1997), *Participatory Irrigation Management in the Philippines: National Irrigation Systems*, Economic Development Institute of the World Bank, Washington D.C.

Ramos, Ferdinand (1994), Memo From the Office of the President for The Provision of Financial Assistance for the operation and Maintenance of National Irrigation Systems, Malacanang, Manila, June 22 1994.

Scudder, Thayer (1994), 'Recent Experiences with River Basin Development in the Tropics and Subtropics.' *Natural Resources Forum* 18(2): 101-113.

Small, Leslie, Marietta S. Adriano, Edward D. Martin, Ramesh Bhatia, Young Kun Shim, and Prachanda Pradhan (1989), *Financing Irrigation Services: A Literature Review and Selected Case Studies from Asia*, International Irrigation Management Institute, Colombo, Sri Lanka.

Svendsen, Mark (1992), *Assessing Effects of Policy Change on Philippine Irrigation Performance*, Working Papers on Irrigation Performance 2, International Food Policy Research Institute, Washington D.C.

Svendsen, Mark, Marietta Adriano, and Edward Martin (1990), *Financing Irrigation Services A Philippine Case Study of Policy and Response*. International Food Policy Research Institute, Washington D.C.

United States Bureau of Reclamation (USBR) (1966a), *A Report on Upper Pampanga River Project*, USBR, Washington D.C.

United States Bureau of Reclamation (USBR) (1966b), *A Report on the Cagayan River Basin Luzon Island Philippines*, USBR, Washington D.C.

Vermillion, Douglas (1997), *Management Devolution and the Sustainability of Irrigation: Results of comprehensive versus Partial Strategies*, Draft paper presented at the International Workshop on Participatory Irrigation Management (PIM): Benefits and Second Generation Problems, Cali, Colombia, 9-15 February 1997.

White, Gilbert (1969), *Strategies of American Water Management*, the University of Michigan Press, Ann Arbor.

Wijayaratna, C.M. and Douglas L. Vermillion (1994), *Irrigation Management Transfer in the Philippines: Strategy of the National Irrigation Administration*, Short Report Series on Locally Managed Irrigation, Report no. 4, International Irrigation Management Institute, Sri Lanka.

World Bank (1993), *Improved System Operation Project II - Staff Appraisal Report*, The World Bank, Washington D.C.

World Bank (1993a), *Water Resources Management*, The World Bank, Washington D.C.

World Bank (1993b), *Philippines Irrigated Agriculture Sector Review – Volume I*, The World Bank, Washington D.C.

World Bank (1993c), *Philippines Irrigated Agriculture Sector Review – Volume II*, The World Bank, Washington D.C.

World Bank (1995), *Water Resources Development Project - Staff Appraisal Report*, The World Bank, Washington D.C.

Chapter 5

From Voice to Empowerment: Rerouting Irrigation Reform in Indonesia

Bryan Bruns

Robert Repetto's 1986 paper, 'Skimming the Water: Rent-Seeking and the Performance of Public Irrigation Systems' analyzed how rent-seeking interests have dominated irrigation development in countries around the world. Irrigation agencies, political leaders, farmers, and construction interests benefit from and seek to expand subsidized construction. Farmers receive valuable water while paying little or nothing for it. Economic rents, the difference between the cost of water and its marginal value to farmers, become capitalized into land values, with farmers seeking to keep and expand the benefits of subsidized irrigation services. Within irrigation agencies, power and money are concentrated in construction, offering the best opportunities for career advancement, and personal enrichment (Chambers, 1988). As long as centralized technical agencies control infrastructure operation and maintenance, they have little incentive to carry out maintenance (Ostrom, Schroeder and Wynne, 1993). Under-budgeting of maintenance perpetuates a cycle of premature degradation and subsequent rehabilitation, reinforcing construction interests. The pragmatic response for officials in such agencies is to accept the pattern of deferred maintenance and rehabilitation projects (Levine, 1986). Proposals to solve apparent economic inefficiencies through cost recovery, volumetric water wholesaling and irrigation fees paid to government frequently fail to recognize the dynamics of the institutional interests in irrigation (Moore, 1989; Small and Carruthers, 1991).

In Indonesia as in many countries, international donors funded centralized technocratic development of irrigation, in contrast to the local control of irrigation that prevails in much of the U.S., Europe and Japan. Centralized approaches were convenient for donors and national government agencies, and fit easily into technocratic, state-led development. Under the 'New Order' regime, authoritarian rule and political demobilization meant that the politics of irrigation were primarily the internal politics of the government bureaucracy, and particularly the Ministry of Public Works, which controlled irrigation investment. The alignment of institutional structures through which investments were made promoted and reinforced rent-seeking, not merely in the sense in which rent-seeking is sometimes misunderstood or disingenuously used as a euphemism for corruption, but more importantly in the sense of who would have power over and profit from 'rents' embodied in agency budgets, bureaucratic authority, political power and patronage in the provision of subsidized services. Availability of international finance enabled agencies to subsidize projects and avoid the accountability that might have come with greater dependence on local resources. This process often overwhelmed and displaced local efforts to invest in improving irrigation. The participatory reforms initiated in 1987 offered Indonesian farmers greater 'voice' in irrigation management (Paul, 1992; Hirschman, 1970), but as this paper shows, little choice to exit from the dominant patterns of agency-controlled development in irrigation.

The first section of this paper outlines Indonesia's 1987 irrigation reforms through which the government, with support from international donors, sought to increase farmer participation in irrigation management. The discussion focuses on the case of irrigation management transfer under World Bank-funded projects. The second section analyses how the set of interests that supported and benefited from irrigation construction reshaped reform efforts, and the persistent dynamics of a construction-driven approach to irrigation. Sustainability of reforms was endangered by the one-sided emphasis on reducing old government roles, and inadequate attention to developing the dynamics necessary to sustain new institutions. Beginning in the late 1990s, a second round of efforts to reform irrigation and water resources management was introduced in the context of ambitious efforts to restructure Indonesia's governance institutions. The third section describes the key principles of the second phase of reforms and how they sought to escape problems that derailed earlier reforms.

Context: The 1987 Irrigation Operation and Maintenance Policy Reforms

Since the beginning of Indonesia's New Order government in the mid-1960s, national irrigation policy focused first on rehabilitation, and then on expansion and construction of new schemes (Varley, 1989). Oil revenues and international loans financed government intervention to increase agricultural production. Projects controlled by the central government were the major vehicle for government intervention in irrigation. From 1968-1993 about $10 billion was invested in irrigation, about 70% of which was financed by external loans, building and improving public schemes irrigating about 5 million hectares (Varley, 1999). Donors made money available for construction and rehabilitation projects executed by Directorate General of Water Resources Development (DGWRD) in the Ministry of Public Works (MPW).

Irrigation development was a core part of efforts to achieve national self-sufficiency in rice production. Even in relatively well-watered western areas, such as Sumatra and west Java, irrigation helped expand second and third season crops, as well as reducing losses due to dry spells in the wet season. On parts of Sumatra, Kalimantan and Sulawesi, as well as in drier areas of eastern Indonesia, construction of irrigation schemes was also part of transmigration policies intended to move people from the islands of Java and Bali to other parts of the country.

The Ministry of Public Works and provincial irrigation agencies managed newly-built or improved schemes, including older locally-managed schemes incorporated into new schemes and rehabilitation works. Legally, government responsibility in government-built irrigation schemes extended all the way down to tertiary outlets and fifty meters beyond. Government's role in irrigation management expanded steadily even though by international standards most Indonesian irrigation schemes are relatively small, mostly less than 5,000 hectares in size, as shown in Table 5.1. Even these categories are somewhat misleading since designated 'irrigation areas' are often composed of multiple adjoining sub-units with separate headworks, often linked by complex networks of supplementary canals. In addition to about 4.8 million hectares of government irrigation schemes, 'village schemes', managed by local irrigation institutions and village authorities, cover an additional one to 1.5 million hectares of small systems.

Table 5.1 Sizes of public irrigation areas

Size	Systems		Area	
	(number)		(million ha.)	
Small <1,000 ha	5,783	86%	1.65	34%
Medium 1,000-5,000 ha	814	12%	1.57	33%
Large >5,000 ha	133	2%	1.57	33%
Total	6,730		4.78	

Source: CID (1998), based on 1996 DGWRD Inventory.

The fiscal crisis caused by the collapse of oil prices in the mid-1980s highlighted problems with construction-oriented policies. That strengthened the voice of those in government concerned with the need for increased attention to operation and maintenance (O&M). Those involved included irrigation officials directly responsible for Operation & Maintenance as well as the Ministry of Home Affairs and provincial governments, concerned with broader objectives in public administration, and the National Planning Board (BAPPENAS) and the Ministry of Finance, concerned with efficiency of public expenditure. The crisis occurred in a context where the cost of constructing new schemes was rising, budget allocations for operation and maintenance were low, and the performance of schemes in terms of area actually irrigated and benefits to farmers was often well below expectations. Degradation and need for rehabilitation much sooner than the expected lifetime of the infrastructure signalled the inadequacy of operation and maintenance.

International donors had become increasingly concerned about the lack of adequate operation and maintenance. The government's fiscal crisis increased the leverage of the World Bank and Asian Development Bank in policy dialogue, which resulted in the government's 1987 Irrigation Operation and Maintenance Policy (IOMP), issued as a precondition for the first Irrigation Subsector Project (ISSP-I) funded by the World Bank. As was usual at the time, preparation of projects and of the IOMP, was conducted as a secret (confidential) process between government and donors, without public consultation and with little publicity. The IOMP was designed as an integrated package of reforms to ensure adequate funding for operation and maintenance and improved irrigation management. The government committed to increase budget allocations for Operation & Maintenance, turn over small schemes to water user associations, establish irrigation fees, reform property tax administration, and mobilize more resources from beneficiaries. Implementation of the new policies was supported by the World Bank's Irrigation Subsector Project (ISSP-I) from 1998 to 1991, IISP II from 1991 to 1995 and then the Java Irrigation and Water Resources Management Project (JIWMP) from 1995 to 2001, as well as projects funded by the Asian Development Bank and other donors. Table 5.2 gives an overview of the principal programmes carried out to implement the IOMP: turnover, irrigation service fees and efficient operation and maintenance.

Turnover The IOMP envisioned that over a period of about fifteen years all schemes smaller than 500 hectares would be turned over to water user associations (WUA). Initial emphasis was on schemes smaller than 150 hectares. Water Users Associations were to be formally established and registered with district government. A modest level of funding, initially about US $100 per hectare, was provided for repairs and improvements, with designs to be prepared in consultation with farmers (Bruns and Atmanto, 1995; Vermillion et al., 2000).

Irrigation Service Fees Irrigation service fees were introduced through a participatory approach, involving water user associations in identifying and prioritizing Operation & Maintenance needs and in collecting fees. In the approach developed at pilot sites, farmers had a voice in determining the Operation & Maintenance needs through joint walkthroughs, and through district-level consultative bodies composed of selected heads of Water Users Associations federations (Gerards, 1995). Based on the scheme Operation & Maintenance budget, individual Irrigation Service Fees charges were to be calculated by a formula based on the level of service received. Fees were to be paid into a bank account held by the district government, and used to pay for Operation & Maintenance done by the irrigation agency. Whereas before this budget was fully under the irrigation agency and its projects, the new arrangement created a potential 'check and balance' intended to promote greater accountability (Paul, 1992).

Efficient Operation & Maintenance The Efficient Operation & Maintenance programme (EOM) was supposed to ensure effective water delivery and adequate preventive maintenance, breaking the cycle of neglect and premature rehabilitation. Maintenance budgets for government-managed schemes were to be based on field analysis of needs. Donor-funded projects financed 'special maintenance' to improve scheme infrastructure and facilities to the level needed for EOM. Loan funds subsidized Operation & Maintenance budgets in EOM schemes for a five-year transitional period, while loan conditions required maintaining adequate funding levels after loan subsidies ended. Improved Operation & Maintenance procedures were developed and initially applied in schemes selected as 'Advanced Operation Units' (AOU). A large Water Users Training Project (WUTP) disseminated information about Water Users Associations and Operation & Maintenance to WUA leaders and government officials at all levels.

Table 5.2 Project activities to implement the 1987 Irrigation Operation & Maintenance Policies

	Institutions	Finance	Construction
Large schemes >500 ha	WUA formation and development
Joint walk-throughs to assess O&M needs
District consultative bodies approve use of fees for O&M
Strengthening of agency O&M procedures (starting in 'Advanced Operations Units')
Water User Training Programme for officials and farmer leaders	Central government subsidizes 'special maintenance' (at a rate of about $200/ha)
Salaries of permanent staff in provincial budget
Central government subsidies to provincial O&M budgets (~$20/ha)
Irrigation service fee (~$20/ha expected)
WUA still responsible for O&M of tertiairies
Strengthening of land and property tax for local government revenue	Central government projects construct 'special maintenance' works to improve canals and headworks, and O&M facilities (transport, communications, and buildings)
Small schemes <500 ha	WUA formation and development
Consultation about design and construction of 'special maintenance' works
Training on O&M (in the field, and sometimes formal)
Transfer of management	Central government subsidizes 'special maintenance' (~$100/ha)
WUA contributions to construction (mainly earthworks)
WUA responsible for subsequent O&M.
Possible government aid for later repairs if 'beyond farmers' capacity',
Strengthening of land and property tax for local government revenue	Central government projects construct 'special maintenance' works to prepare schemes for turnover

Misadventures on the Road from Policy to Practice

Attempts to institute participatory reforms challenged the network of interests among individuals, groups and organizations that benefited from construction-oriented approaches to irrigation development. This section analyses several examples of how reforms were redirected during the process of translating new policies into practice.

Detours: Construction before Turnover

The Disappearance of Category A The turnover programme was formulated to prepare for management transfer by carrying out design and construction of improvements in parallel with development of Water Users Associations. For the working group of agency and donor officials that designed the programme, minor construction works offered a way to interest farmers and give them meaningful participation in decisions. In the original formulation of the turnover programme, schemes that had been included in the register of government irrigation schemes, were supposed to be gradually inventoried and classified into three categories.

- Schemes that had no government-built infrastructure. These could be 'turned over' by just revising administrative records, or, at most, organising a Water Users Association to receive formal responsibility for the scheme.
- Schemes with government-built infrastructure that had recently been improved, and so would need Water Users Association organization, but no physical improvements, before turnover.
- Schemes that had government-built infrastructure, and which needed physical improvements as well as Water Users Association organization, before turnover.

As the turnover procedures were developed in more detail in 1988 and 1989, DGWRD officials argued that all schemes should receive consideration as to whether they needed physical improvements before turnover. It was asserted that schemes which had never received assistance were often in worse condition than those were that already had government-built infrastructure. Assessments of scheme 'condition' emphasized physical hardware such as need for permanent dams and lined canals. Aiding schemes that already had more elaborate infrastructure, while not helping those with fewer permanent structures of stone and concrete, was portrayed as unfair.

Many schemes appeared to fall into category A as initially defined. Often farmers were unaware that the government considered them to be 'government schemes'. According to law, if there had been any government investment, then legally this made a scheme a 'government' scheme. Since Operation & Maintenance budgets were based on the area of 'government schemes', irrigation agencies had a strong incentive to record as much area as possible as government schemes. Many areas were included which had never received government investment. In other cases, although government had funded construction works at some time in the past, the schemes were otherwise fully farmer-managed. All these schemes could have been rapidly 'taken off the books', formally recognizing what was already true in practice, that they were de facto managed by farmers.

However, turnover would potentially reduce the flow of Operation & Maintenance subsidies, which in practice were mainly used for other schemes, or for occasional construction aid, not routine Operation & Maintenance (Murray-Rust and Vermillion, 1989). The National Planning Board (BAPPENAS) had blocked DGWRD from involvement in project assistance to village irrigation schemes. Transferring category A schemes therefore also might have deprived DGWRD of the opportunity to be involved in future project assistance to these schemes.

In the end, procedures were revised so that schemes were first assessed according to their need for improvement. In all cases of schemes without government-built infrastructure (the original category A), the conclusion was that improvements were needed. Even recently improved schemes (the original category B) were almost always found to be inadequate or incomplete, so that construction was needed. The outcome was that all schemes were judged to require construction. World Bank officials supervising the project acquiesced in this, seeming to feel they lacked the ability to influence this redefinition. The policy that turnover without construction was still possible in theory made it harder to argue against the change than if there had been an explicit decision that all schemes would have construction. Acceptance of the change may also have been due to the changes in the World Bank personnel involved with the project, lesser leverage in an ongoing project and the fact that this change would not threaten the disbursement of construction funds or achievement of formal project targets. The original design of the turnover process had involved a range of agencies, and donor organizations. However once implementation began, most decision-making was concentrated within DGWRD as the executing agency, and decisions came to reflect a narrower set of priorities. In hindsight this redefinition to make all schemes eligible for construction was a fateful first step in transforming a programme ostensibly directed at strengthening Operation & Maintenance and local participation into a construction programme.

Diversions There was substantial contention about the appropriate level of improvement to be carried out before turnover. This was a major point of discussion among central and provincial officials implementing projects, and of discussion with World Bank project supervision missions. In many cases farmers were accustomed to building and repairing temporary headworks on their own to divert water into canals. However, many engineers in DGWRD and provincial irrigation agencies felt that permanent headworks were needed before turnover. They felt that it would be excessively burdensome to turn over schemes that required frequent reconstruction of temporary weirs. Counter arguments offered by some of the central officials involved in the programme, consultants (including the author) and Bank missions, stressed that these were functioning schemes, and that usually farmers had been mobilizing to build and rebuild temporary weirs since the schemes were first built, in some cases generations ago. Arguments for building permanent weirs were framed in technical terms, usually with little attempt to compare costs and benefits.

Farmers wanted permanent weirs, particularly if the alternative they faced was between the possibility of getting a government-subsidized permanent structure for free, or continuing to mobilize their own resources for frequent repair of temporary weirs. Farmers and agency staff had little problem agreeing on the desirability of using project loan funds, which neither had to repay, to build permanent weirs. After substantial debate, the conclusion was that the turnover programme would generally not fund new permanent headworks, only repairs to existing structures or construction of semi-permanent (gabion) structures. In part this was driven by budgetary constraints, since the project lacked enough budget to fund many new headworks. Setting limits on budget items and declining to agree to changes was one of the few points of control available to the donor agency and central officials. Funding was still allocated according to a per hectare average, roughly $100 per hectare. Sites requiring more than this were either excluded, or required additional justification through an economic analysis.

Cost sharing Farmer cost-sharing was another point of debate during the World Bank's supervision missions for ISSP-I. Inclusion of a scheme in the programme was basically a unilateral decision by government, not a choice by farmers or something negotiated based on their readiness to invest in improving the scheme. The issue of cost sharing was framed in terms of farmer 'contributions', not as a matter of joint investment by farmers and government (Murray-Rust and Vermillion, 1989). A commonly expressed view on the part of government and donor officials was that construction was an incentive to persuade farmers to accept turnover. Many agency officials felt requirements for local contributions would be difficult to implement and impose an unfair obligation on farmers, who were already being 'burdened' with future responsibility for Operation & Maintenance. The conclusion was that farmers were expected to do earthworks. These local contributions were monitored, though attention to even this level of local 'contribution' dwindled over time.

Construction Before Turnover During implementation, farmers did gain a voice in offering their suggestions about priorities and design of improvements (Bruns and Atmanto, 1995; Mott-MacDonald, 1993). Opportunities were increased for farmers to sell their labor and local materials during construction. Water Users Associations were set up that the government recognized as speaking for farmers. However the process was largely controlled by the agency, with little empowerment of farmers. The decision to carry out physical improvements before transfer constrained the pace and scope of turnover to the rate at which construction could be accomplished. Control of funds, and management of the design and construction process was still in the hands of the agency, not farmers. Much of the time of senior government officials and World Bank staff supervising the project was focused on debating the need for permanent headworks, per hectare ceilings of the amount of construction to be done, and farmer contributions to construction, which reduced the time available to attend to institutional issues.

Shortcuts to Nowhere: Targeting WUA Establishment

The turnover pilot programme worked out procedures for organising water user associations in pilot areas. These were supposed to start from the level of quarternary blocks, groups of 10-30 farmers. This approach was intended to build on local social relationships and be closely linked to farmers' involvement in prioritizing ways for schemes to be improved in preparation for turnover. Progress in Water Users Association development was measured by the steps of formulating the WUA constitution and by-laws, choosing a slate of officers, and formal recognition of the organization by village, sub-district, and district level authorities. Much of this process appeared to work successfully during the pilots and initial implementation during ISSP I. However, organization of formal Water Users Associations was easily disconnected from substantive development of management capacity for irrigation operation and maintenance. Water Users Association development often skipped the bottom-up organising process, with Water Users Association leaders instead appointed because of their relationship to village authorities. Water Users Associations' constitutions and by-laws followed pre-printed examples, filling in a few blanks in these forms, rather than being based on thorough local discussion and genuine consensus. Project targets were met, in terms of organizations formed on paper. However, such paper organizations quickly became inactive. Commonly the essential tasks of pre-season maintenance and rotational water distribution during periods of shortage were still carried out by persons authorized by village officials, or informal mutual aid among farmers, rather than by the formal Water Users Association.

Experience in the Irrigation Service Fees programme was similar. Initially the consultants guiding the pilot projects recruited community organizers who worked in the field to facilitate a bottom-up process of forming Water Users Associations. There was a stress on identifying local management problems and developing solutions, which often could result in more equitable water distribution and better mobilization of fees. However as the project expanded it focused on targets, particularly the number of Water Users Associations formally established. Success was measured in terms of payment rates for Irrigation Service Fees, with little attention to whether there were improvements in irrigation performance.

Water Users Association federations were formed which sometimes played a role in joint walkthroughs, but mainly were just a means to channel Irrigation Service Fees to the district treasury. There was significant opposition within government to the idea of 'community organizers' and deliberate avoidance of any approach which might foster large self-governing organizations (such as that of Dutch polder management districts), or otherwise allow significant organization above the village level. This fit the New Order regime's approach of depoliticizing rural people and preventing the growth any strong organizations outside of the administrative command hierarchy, which ran all the way down to the village level. No forums were established at the scheme level for involving farmers in management. Formal arrangements for farmer voice were put into place, at least on paper. Over time even these tended to be neglected. It was argued that annual walkthroughs were unnecessary. District level consultative bodies often had little or no role in decisions. Later, interpretations of new government regulations resulted in Irrigation Service Fees funds being incorporated in the regular district budgets, and often not returned to the schemes where they were collected. Irrigation Service Fees collections declined and only weak efforts to were made implement fee collection in areas nominally included under ISF.

Who Steers? Projects and Agencies

Initially, the irrigation officials responsible for Operation & Maintenance were given responsibility for implementing the turnover programme. This combined their regular structural position as part of the provincial irrigation service with functional 'project' responsibilities as subproject managers. They had knowledge about Operation & Maintenance problems, and better incentives to seek lasting changes, since they would be around to deal with problems later on. By contrast, most 'project' staff had no responsibility after construction was complete, and usually soon moved on to other positions.

This policy changed during ISSP II, after a new Director General took charge and DGWRD was reorganized. Functional 'project' responsibilities were split from structural positions, based on the principle that one person should only hold one position. This led to a situation where most project managers came out of backgrounds in managing large construction projects. The cadre of officials who had been involved in starting the turnover programme were excluded or had much less authority over implementation. This exacerbated the tendency to focus on easily measured targets of financial and physical progress of construction works, rather than institutional development. Managers for the provincial irrigation 'projects' were appointed by DGWRD and oriented towards the national level, rather than the provincial irrigation agency. Project managers were in the driver's seat, not structural officials, and certainly not farmers.

Fuel: Money by the Hectare

Funding for special maintenance works was planned on a per hectare basis, initially averaging about US $200 per hectare for larger schemes being prepared for EOM, and $100 per hectare for turnover schemes. While some relatively simple estimate of costs was needed for project preparation, planning based on per hectare rates was often carried through to the design process. It was usually not difficult to find ways to spend all the available funds within a scheme or package of several schemes. There was a matching tendency not to prepare designs costing more than the average rate or ceiling. In theory more expensive works could be funded, so long as an economic analysis was done and submitted to the World Bank for review. However, it was much simpler to keep under the limit, even if this meant parts of schemes were left incomplete or in poor condition. Farmers and structural officials in provincial irrigation agencies had a voice in identifying needs for improvement, but decisions were in the hands of projects, which controlled design, and construction works.

The term 'special maintenance' itself had been created for ISSP I. Ostensibly there was only a need for relatively minor works to improve systems to be ready for efficient operation and maintenance. In practice the term avoided donor reluctance to fund rehabilitation, or to acknowledge that earlier construction had left many schemes unfinished and incomplete. As semantic manipulation this was quite successful. However the scope of activities was often little different from that covered by rehabilitation, just with less money. The package of IOMP reforms was also used to justify loan funding for Operation & Maintenance subsidies, albeit supposedly during a transition as ISF was phased in. However, ISF expansion was slow, and collection problematic. In practice, generous central government subsidies for Operation & Maintenance gave a contradictory signal that there was little need to be serious about collecting Irrigation Service Fees, undermining the urgency of reforms. The centre made up for declining allocations from the loan by increasing subsidies from its own resources. The central government exerted little pressure to increase local resource mobilization from Irrigation Service Fees. Financial mechanisms continued to fuel old patterns of construction driven-development, and did not create new choices for farmers.

Potholes: Maintenance Neglected

In combination, needs-based budgeting as part of EOM and the Irrigation Service Fees programme's participatory planning for Operation & Maintenance might have transformed how maintenance was planned and financed. A key attempted strategy was to build on farmers' interests in improving irrigation performance, either by transferring management to locally controlled Water Users Associations, or by using Irrigation Service Fees payments and participation in identifying and prioritizing Operation & Maintenance activities to make management of larger schemes more accountable. However the approach was still primarily one of giving greater voice for scheme level staff and farmers to express the needs they perceived. Control over expenditures was not shifted from the hands of irrigation project officials. While Water Users Association representatives sat on the consultative bodies that authorized use of Irrigation Service Fees funds for Operation & Maintenance, this was done at the district level, not for individual schemes, and did not include any authority over other parts of the Operation & Maintenance budget. While the Ministry of Finance may have been interested in reducing government expenditures on Operation & Maintenance, this was not an imperative for the other government agencies, which still benefited from controlling the flow of funds for Operation & Maintenance.

Over half the Operation & Maintenance funding was usually spent on staff, with a large contingent working on annual contracts. Of the remaining funds, most were spent on minor civil works to repair and upgrade schemes, e.g. lining canals, rather than on any systematic programme of preventive maintenance. During ISSP I, procedures were developed and implemented for needs-based budgeting, inventorying the condition of irrigation infrastructure and compiling estimates for maintenance and repair work. However, after budget requests were proposed, the centre and provinces still allocated budgets on the old per-hectare basis. This discouraged those who had put in time and effort to prepare detailed estimates. After a few years of trying needs-based budgeting with few useful results from their efforts, those responsible for preparing budget proposals usually reverted to the older, and much simpler per hectare basis. Provincial irrigation project officials retained their discretion over how to expend Operation & Maintenance budgets, including the tendering of construction contracts.

Needs-based budgeting was done separately from Irrigation Service Fees introduction, not as part of an integrated package of reforms. The definition of needs made 'needs-based budgeting' problematic. There were no clear standards for saying how much maintenance was adequate, or how to optimize maintenance expenditures in economic terms. Attention tended to focus on apparent needs for repair, and improvement, such as canal lining, rather than the more mundane tasks of routine and periodic maintenance. 'Needs-based' estimates included the backlog of neglected maintenance, not just routine annual and periodic maintenance. This often resulted in estimates far in excess of available Operation & Maintenance funds. Even though schemes were already supposed to be in suitable condition for EOM, canal networks were often incomplete, many gates were in poor condition, and schemes often lacked equipment, such as motorcycles for transport, radios for communications, and other facilities.

Operation & Maintenance budgets were not controlled at the scheme level. Neither farmers nor scheme staff had power to choose what would be done. Operation & Maintenance staff at this level could make proposals, but the usage of Operation & Maintenance funds was managed through functional 'O&M projects' set up at the provincial level. Sometimes the first time that scheme Operation & Maintenance staff knew that works had been approved was when contractors showed up to begin work. What was built often did not match priorities as perceived by staff within schemes. Often funds were concentrated into minor construction works, while more routine maintenance activities were neglected. This was both more convenient for those handling the budget and may have also allowed more opportunities for personal gain in awarding contracts.

During JIWMP, funding channels were changed so that central government subsidies for Operation & Maintenance were included in block grants provided to provinces, rather than going directly to budgets controlled by provincial irrigation projects. Most central funds intended for Operation & Maintenance were placed under a general budget item, which provinces formally had authority to allocate to whatever sector they chose. Provinces decided they had higher priorities in other sectors, so that actual Operation & Maintenance funding at the field level declined, even though on paper national budget allocations were being maintained. Maintenance was still treated as a 'project' and transformed into minor construction. More mundane needs for routine and periodic maintenance still tended to be neglected in favour of adding canal lining, new division structures and other works that could be contracted out.

Routes to Renewal

The 1987 IOMP reforms did not change the underlying dynamics of irrigation development. Government continued its role as an operator, directly implementing activities. Farmers gained some means for greater voice, but the system remained trapped in old patterns. Government sought to impose bureaucratic forms of Water Users Association development, and was largely unable to tap the energies of local initiative. Construction was highly subsidized, with funds channelled through irrigation bureaucracies. Technical, financial, legal and organizational services to aid Water Users Associations were largely absent, except for government's own target-oriented programmes directly delivering conventional, classroom-based training. Under such conditions it was hard to expect Water Users Associations to survive, let alone thrive.

During the eighties and nineties various studies, workshops and seminars provided forums for discussing problems with Water Users Association development. University academics and NGO workers helped suggest new approaches. The Ford Foundation supported many of these activities. Other donor projects also helped introduce new approaches, such as Dutch-funded efforts in the Cidurian scheme, the Madura groundwater projects supported by the United Kingdom, and the Small-Scale Irrigation Management Project in eastern Indonesia, funded by the United States Agency for International Development (USAID) and the Japanese Overseas Economic Cooperation Fund (OECF).

While substantive shortcomings in institutional reform became more apparent over time, irrigation projects continued to largely achieve formal targets as measured in terms of physical construction, financial disbursements, Water Users Association formation, preparation for scheme turnover and introduction of Irrigation Service Fees. International development agencies had strong incentives to continue to 'move money', and maintain good working relationships with the government, so their scope for pushing for major changes was limited, although their own criticisms and dissatisfaction grew over time.

In many cases, government officials supported new ideas, and in private were highly critical of their own programmes. They usually felt unable to publicly voice ideas that might upset their superiors, though sometimes outsiders could function as a channel for raising criticisms and alternatives. Over time views evolved. By the time of the 1996 national seminar on participatory irrigation management, there was substantial consensus among mid-level and many senior level officials about the shortcomings of the IOMP approaches, the need to effectively apply those principles which had been already been put into policy, and the importance of making more fundamental changes (Bruns and Helmi, 1996). At an intellectual level, many battles had been won, even though in practice implementation during this period was often retrogressing, with institutional changes neglected in favour of more construction-oriented approaches. Despite disappointing events on the ground, the conditions had been created for more drastic reforms, if senior officials were willing to consider such changes.

Changes in leadership of DGWRD and the Ministry of Public Works in 1998 opened up the possibilities for discussing institutional reforms in ways which could go beyond lip service. Advocates of top-down approaches were on the defensive, and possibilities appeared for major changes in irrigation management. Over the period of a dozen years, farmer empowerment went from something that could only be whispered about privately to something that was to be embraced as a major goal of national policy.

Reformation and Decentralization

The end of the New Order regime in mid-1998 expanded the possibilities for institutional change. Passage of laws on regional autonomy in April 1999 laid the foundation for a major shift in power and money from the central government to districts. Criticism of previous waste, corruption and abuse of power brought wide acceptance of the need for greatly increasing transparency and accountability in the implementation of government programmes. Change brought more willingness to put money and decision-making into local hands, though efforts to put this into practice also brought awareness of the risks of local corruption and abuse of power.

In 1997-1998 the Asian Development Bank had funded a technical assistance study on Options for Sustainable Irrigation Development in Indonesia (1998). This clearly signalled the concern of this major financing agency for reviewing the direction of irrigation sector activities in light of difficulties experienced in carrying out the IOMP reforms. Beginning in late 1997 (i.e., after the onset of the Asian Financial Crisis but before the change in regime) the World Bank began increased dialog about the need for major reforms in the irrigation and water resources policy and institutions, as a prerequisite for any future lending in the sector. These efforts helped to draw high-level attention from BAPPENAS, the Minister of Public Works and other senior officials. The results of the Asian monetary crisis and the new pro-reform atmosphere brought an increased sense of urgency to some of those involved.

The economic crisis provided leverage for the World Bank's offer to provide a sector adjustment loan to support reforms. Funds from the sector adjustment loans would be used for general budgetary support, but were tied to specific sector policy changes. In late 1998 and early 1999, an inter-agency working group led the formulation of ideas for reforms in irrigation and water resources policy. This drew on many ideas from earlier discussions and studies, notably a series of seminars, organized by BAPPENAS and funded by the Ford Foundation, which had included government officials, NGOs and academics. University experts helped the working group to synthesize and clarify key principles for irrigation reforms. The working group carried out public consultation meetings in Jakarta and several provinces. The outcome of this process was agreement on a major programme of reforms. In

April 1999, a Presidential Instruction and speech laid out the key principles of an irrigation reform programme.

New Directions: Reform Principles

Presidential Instruction Number 3 issued on April 26, 1999, laid out five principles for irrigation reform: redefining irrigation institutions, empowering Water Users Associations, transfer and joint management, farmer-managed fees, and irrigation sustainability. These points had already been spelled out in more detail in a longer document written in Indonesian, which was used as a basis for reform discussions including officials from various agencies, universities, NGOs and donors. Reforms were also described in the formal documents for the Water Resources Sector Adjustment Loan (WATSAL), between the Government of Indonesia and the World Bank, intended to support these policy changes, which was agreed in April 1999. The rest of this section discusses the five principles, corresponding elements of WATSAL, and some of the challenges involved in putting such reforms into practice.

Enabling Institutions The first principle of the Presidential Decree emphasized the need to rearrange the government agencies and farmer organization involved in irrigation management, so that farmers become decision-makers. Broader discussions about government reformation had emphasized the shift in government's role from provider to enabler, 'from rowing to steering'. The WATSAL Policy Matrix indicated that change would occur at all levels, with agency roles focused on water delivery and Water Users Association support services. One of the main issues in the discussion about this point concerned the need for a continuing government regulatory and advisory role in technical audit. This would address one of the shortcomings of the previous reforms, the lack of a supportive environment for Water Users Associations after turnover, including availability of specific support services and suitable regulation of water resources.

Empowering Water Users Associations The Presidential Instruction outlined principles for a paradigm shift toward Water Users Associations that would not be imposed instruments of government, but instead autonomous, self-reliant and based in the local community. The Policy Matrix emphasized autonomous governance and financial authority for management. Subsequent points highlighted the transfer of management authority and new financial arrangements. This empowerment included formal legal status, the ability to enter into contracts and open bank accounts. Water rights would enable Water Users Associations to have a clear claim to water, and a basis to negotiate over water allocation, within schemes and at the basin level.

Management Transfer and Joint Governance The proposed reforms covered all schemes, not just those under 500 hectares in size. Management transfer was supposed to be 'selective, phased and democratic'. Not all structures would necessarily be transferred, e.g. major headworks might stay under government management, secondary canal areas might be turned over before the main system, different districts might move at different speeds, and the transfer should be chosen by the community of irrigators, not imposed. In that sense, this was not a 'big bang' policy. However, areas not yet transferred would be put under joint governance, not left for 'business as usual'. Scheme management forums for joint management would enable Water Users Association representatives to make decisions about activities still implemented by agency staff. The reforms would cover all schemes, not just a subset singled out for special attention.

In the longer run, transferring governance authority over irrigation schemes would create the possibility of a choice among service providers (Vermillion, 1999). While the irrigation network itself may be a natural monopoly, the management services to deliver water and maintain facilities could be delivered by a government agency, farmer-governed organization, or a private corporation. Management transfer thus could move beyond voice to empower farmers to control the services, including gaining a choice about whether they wanted to continue to obtain the services from a government agency, hire their own manager and staff, or perhaps contract for management services from a private concessionaire.

Fees from, by, and for farmers. The core of the new approach to Irrigation Service Fees was to empower farmers to set, collect and manage fees themselves. This was intended to enhance legitimacy, make collective action more effective, and improve efficiency in the use of funds. Irrigation fees would be determined by Water Users Associations in each scheme, collected and managed by Water Users Associations. Payment would be obligatory nationwide, with enforcement supported by appropriate legislation at the national and district level. As discussed in the Letter of Sector Policy, further incentives for payment would be created by making eligibility for government aid conditional on satisfactory levels of local resource mobilization through Irrigation Service Fees. Rather than just giving farmers a voice in a programme still controlled by government, this approach would put choices, and responsibilities in the hands of farmers.

Farmers' responsibility for financing irrigation would include not just operation and maintenance, but also construction, rehabilitation and improvement. In discussions there was growing recognition that confining farmers' responsibilities to Operation & Maintenance would discourage preventive maintenance and undermine incentives for farmer investment in improving irrigation. The laws on regional autonomy passed in April 1999 envisioned radical fiscal decentralization. The challenge was then to institute new mechanisms for financing irrigation infrastructure, rather than just reproducing the same flawed, top-down, project dynamics at a lower level. An incremental approach to infrastructure improvement, shifting from infrequent lumpy rehabilitation to smaller, more frequent investments of a scale more easily managed by Water Users Associations, could be used to facilitate local control over construction (Bruns, 1998). District level irrigation improvement funds were proposed as a basis for a demand-driven approach, responding to competitive Water Users Association proposals, transparently allocating from a limited pool of funds (i.e. a hard budget constraint), and with Water Users Association empowered to control decisions in design and construction.

Irrigation sustainability. This point linked the irrigation reforms with other efforts to improve basin water resource management, improving institutions for managing water quantity and quality, including user participation in basin water management, water rights, control of water pollution from urban areas, and other changes. Rapid conversion of agricultural land to urban and residential use was reducing the area available for growing rice. There were no requirements or mechanisms for developers to compensate government for the value of irrigation infrastructure taken out of use by land-use conversion. Water was being transferred from agriculture to urban and industrial users without effective means for consultation or compensation for the affected farmers. WATSAL included commitments to establish a system of water rights covering irrigation and other agricultural water use, and improved regulation of water quality.

Experience over the twelve years between the 1987 Irrigation Operation and Maintenance Policy Statement and the 1999 Declaration on Irrigation Policy renewal illustrates some of the challenges faced by efforts to increase participation in the form of voice within agency-controlled irrigation development. New policies for Water Users Association empowerment seek to institute a paradigm with very different assumptions about the role of the state and local organizations, promoting devolution, demand-driven development and sustainability. Regime change, democratization, decentralization and development of more transparent, accountable and participatory governance institutions have opened new possibilities for reform and space for countervailing interests. New policies, laws and regulations may create an enabling environment that renders reforms possible. The combination of transferring governance authority to empower Water Users Associations and rerouting financing through decentralized irrigation improvement funds could restructure incentives and foster support for a new approach. However putting changes into practice will face many of the same challenges from the network of interests that has benefited from construction-driven irrigation development in the past.

Furthermore, devolution of governance authority and consequent empowerment of Water Users Associations would not eliminate all problems of misaligned and inadequate incentives. In their analysis of turnover programme impacts, researchers in the International Water Management Institute study noted the tendency of Water Users Associations to under-invest in maintenance after transfer (Vermillion et al., 2000). Even if water user associations and federations gain not just voice, but do actually take over more power to choose and decide, nevertheless escaping the attractions which have tended to trap irrigation in wasteful cycles of paternalistic development, deferred maintenance and inefficiently subsidized rehabilitation will not be easy.

Conclusion: The Route from Voice to Choice

Indonesia's 1987 Irrigation Operation and Maintenance Policy introduced a series of efforts to transfer management of small schemes to water user associations, and to institute irrigation service fees and improve the efficiency of irrigation operation and maintenance in larger schemes. The first Irrigation Subsector Project introduced participatory procedures for turnover of management to water user associations in small schemes. For larger schemes, the project piloted participatory collection and use of irrigation service fees, and introduced procedures for more efficient operation and maintenance. Farmers gained a voice in the design and construction of improvements made to prepare for turnover of small schemes. In larger schemes they were involved in identifying priorities for maintenance to be funded from irrigation service fees. The difficulties of expanding and institutionalizing reforms became clearer during the Second Irrigation Subsector Project. The Java Irrigation Improvement and Water Resource Management Project sought to continue the earlier reforms in irrigation operation and maintenance, amidst increasing concern about the need for more fundamental changes.

The 1987 reforms threatened the set of interests that supported and benefited from irrigation construction. These interests rerouted reform efforts, revealing the persistent dynamics of a construction-driven, rent-seeking approach to government activity in irrigation development. Donor-funded projects succumbed to a continuing bias towards construction. The decision to carry out physical improvements before turnover constrained the pace and scope of turnover to the construction programme. Institutional development tended to focus on easily measurable targets for physical works and formal registration of Water Users Association, in ways that often undermined the more fundamental objectives of reform. Project organization and agency career paths kept control in the hands of project managers oriented towards construction. Mechanistic development of water user associations produced paper organizations that quickly became inactive.

Sustainability of reforms was endangered by the one-sided emphasis on reducing old government roles, and on fixing up schemes as a prerequisite for further changes. Without rights and power, irrigators' organizations lacked incentives and capabilities to organize themselves for better irrigation management. In the absence of new arrangements for financing civil works, subsidized patterns of rent-seeking and patron-client interaction between farmers and the bureaucracy persisted. Government withdrawal, without adequate availability of technical services, and with little regulatory oversight, invited neglect and declining performance.

In 1999, amidst dissatisfaction with the results of earlier efforts, the Indonesian government proclaimed a new irrigation reform policy, initiating major changes in strategy. These efforts to reform irrigation and water resources management were formulated in the context of ambitious efforts to restructure Indonesia's governance institutions. The Indonesian government committed itself to strategic reform principles for redefining agency roles; empowering Water Users Associations; joint governance and management transfer; fees from, by and for farmers; and new approaches to financing irrigation. Reforms to empower farmers with genuine choices, not just a voice in centralized projects, sought to open new routes for effective change.

In terms of the transport metaphor used in this paper, the earlier approach to irrigation reform can be compared to the top-down construction of a centralized railroad system, monopolized by a single operator, powerful but restricted to a few destinations, dependent on detailed planning, and all too easily derailed. Reforms to empower farmers could be analogous to developing a flexible network, with many routes to reach diverse destinations, and a variety of vehicles; creating a process that could be driven by farmers, empowered with far more choices about directing their own development.

Acknowledgement and note

The paper draws on the author's experience as a consultant on various irrigation projects in Indonesia since 1988, as well as reports and publications listed in the references. Among other assignments, the author worked on the turnover programme from 1988 to 1991 as Institutional Advisor (funded by the Ford Foundation) for LP3ES' activities in the turnover programme, undertook short-term assignments during the preparation and implementation of the Java Irrigation and Water Resources Management Project, and participated in World Bank missions for JIWMP project supervision and preparation of the Water Sector Adjustment Loan, The ideas presented here benefit from discussions with many people concerned with irrigation reform in Indonesia, including farmers and government officials at national, provincial, district and village levels, and many others, including Helmi, Sigid Supadmo, Ganjar Kurnia, Sudar Dwi Atmanto, Saleh Ali, Bambang Adinugroho, Douglas Vermillion, Scott Guggenheim, Robert Varley. and Theodore Herman. Views expressed in the paper are the author's responsibility and do not represent those of any institution with which he is or has been affiliated. Comments are invited to: BryanBruns@bryanbruns.com.

A longer version of this paper was presented at the December 1999 workshop on the Politics of Irrigation Reform in Hyderabad, Andhra Pradesh, India and is available at the author's website: www.bryanbruns.com. That version includes additional description of the agencies and projects, as well as more detailed analysis of the institutional dynamics of the IOMP reforms and proposed alternatives. Appendices to that version provide the text of the 1987 Irrigation Operations and Maintenance Policy Statement, and relevant sections from the Letter of Sector Policy and Policy Matrix for the Water Resources Sector Adjustment Loan.

References

Bruns, Bryan (1998), *Incremental Rehabilitation: Restructuring Incentives for Irrigation Maintenance*, Paper presented at the Fourth International Seminar on Participatory Irrigation Management, Denpasar, Bali, Indonesia.

Bruns, Bryan and Helmi (1996), *Participatory Irrigation Management in Indonesia: Lessons from Experience and Issues for the Future*, Background Paper for the National Workshop on Participatory Irrigation Management, November 4-8, 1996, Jakarta, Indonesia. International Network for Participatory Irrigation Management, World Bank Economic Development Institute and the United Nations Food and Agriculture Organization.

Bruns, Bryan, and Sudar Dwi Atmanto (1995), 'How to Turn Over Irrigation Systems to Farmer? Questions and Decisions in Indonesia', in *Irrigation Management Transfer: Selected Papers from the International Conference on Irrigation Management Transfer*, edited by S.H. Johnson, D.L. Vermillion and J.A. Sagardoy: International Irrigation Management Institute, Wuhan, P.R. China.

Chambers, Robert (1988), *Managing Canal Irrigation: Practical Analysis from South Asia*, Oxford and IBH, New Delhi.

CID-Consortium for International Development (1998), *Options for Sustainable Irrigation Development in Indonesia*, Asian Development Bank, Jakarta.

Gerards, Jan L.M.H. (1995), 'Irrigation Service Fee in Indonesia: Towards Irrigation Co-management with Water Users' Associations Through Contributions, Voice, Accountability, Discipline and Hard Work', in *Irrigation Management Transfer: Selected Papers from the International Conference on Irrigation Management Transfer*, edited by S.H. Johnson, D.L. Vermillion and J.A. Sagardoy, International Irrigation Management Institute and the Food and Agriculture Organization of the United Nations, Rome.

Hirschman, Albert O. (1970), *Exit, Voice, and Loyalty*, Harvard University Press, Cambridge, MA.

Moore, M. (1989), 'The Fruits and Fallacies of Neoliberalism: The Case of Irrigation Policy', *World Development,* 17 (11): 1733-50.

Mott-MacDonald International Ltd. UK. (1993), *Turnover Evaluation Report. Irrigation O&M and Turnover Component, Irrigation Subsector Project II*.

Murray-Rust, Hammond, and Douglas Vermillion (1989), *Efficient Irrigation Management and System Turnover. Volumes I-III*, International Irrigation Management Institute, Colombo, Sri Lanka.

Ostrom, Elinor, Larry Schroeder, and Susan Wynne (1993), *Institutional Incentives and Sustainable Development: Infrastructure Policies in Perspective*, Westview Press, Boulder, CO.

Paul, Samuel (1992), 'Accountability in Public Services', *World Development*, 21 (7): 1047-60.

Repetto, Robert (1986), *Skimming the Water: Rent-Seeking and the Performance of Public Irrigation Systems*, World Resources Institute, Washington D.C.

Small, L.E. and I. Carruthers (1991), *Farmer-financed Irrigation: The Economics of Reform*, Cambridge University Press, Cambridge.

Varley, R.C.G. (1989), *Irrigation Issues and Policy in Indonesia, 1968-88*, Harvard Institute for International Development, Cambridge, MA.

Varley, R.C.G. (1997), 'Irrigation Issues in Indonesia - The Next 25 Years', *Visi-Irigasi Indonesia*, University of Andalas, Padang, West Sumatra, Indonesia, 13 (7): 59-86.

Vermillion, Douglas L. and Juan A. Sagardoy (1999), *Transfer of Irrigation Management Services: Guidelines*, FAO Irrigation and Drainage Paper 58, International Water Management Institute (IWMI), Deutsche Gesellschaft Fur Technische Zusammenarbeit (GTZ) and Food and Agriculture Organization (FAO) of the UN.

Vermillion, Douglas L, Madar Samad, Suprodjo Pusposutardjo, Sigit S. Arif, and Saiful Rochdyanto (2000), *An Assessment of the Small-scale Irrigation Management Turnover Programme in Indonesia*, Research Report 38, IWMI, Colombo.

Chapter 6

Irrigation Policy Discourse and Practice: Two Cases of Irrigation Management Transfer in Zimbabwe

Alex Bolding, Emmanuel Manzungu and Conrade Zawe

Introduction

Irrigation management reform invariably involves the handing over by government of authority and responsibility of management and financing of schemes to farmers. This process, known internationally as turnover, takes many forms. This paper is about turnover in Zimbabwe. The paper discusses the state of irrigation management turnover in Zimbabwe at the end of the twentieth century. It also puts forward some recommendations to improve the situation.

The paper notes that irrigation management turnover in Zimbabwe is highly variegated: it has not occurred, and does not occur, according to any recognisable framework. This *ad hoc* approach, it will be argued, is a symptom of the absence of an appropriate policy to guide the process. There has been no effort to define and conceptualise what irrigation management turnover is and what it implies. This creates gaps when it comes to implementation. For example, turnover is conceived first and foremost in terms of cost recovery. Little attention is paid to the rights of farmers under a new dispensation, while these might be considered a central element of the turnover concept. In other words, no conceptual framework has been elaborated against which the 'policy' can be evaluated. This paper attempts to provide such a conceptual framework.

In addition, the paper observes that despite the absence of a clear policy framework on management of smallholder irrigation schemes, the actors on the ground have devised their own strategies to tackle the issue of irrigation management. In some cases this has led to a *de facto* process of turnover of management responsibilities from government agencies to smallholder farmers. Whilst such events take place in the field, policy makers and top government officials are busy on their own, formulating new irrigation policy initiatives and implementing new project based approaches (turnover experiments) funded by bilateral donor agencies. The two strands of processes (the local versus the central), however, hardly link up to form a coherent irrigation policy practice.

The focus of the discussion is on the smallholder irrigation sub-sector, which is funded directly and indirectly by the government. The sub-sector entirely depends on the state for capital development, and partially or fully on the state for operation and management. The smallholder irrigation schemes are reported to face a number of problems. There are reports of low agricultural output and poor water supply, both of which have necessitated heavy government subsidies (Peacock, 1995; Manzungu and Van der Zaag, 1996). There is a growing realisation that these subsidies are not sustainable and that they need to be reduced, and in some cases eliminated altogether. This has been brought to the fore more prominently because of the Economic Structural Adjustment Programme (ESAP) initiated in 1991, which, among other things, stresses reduction in government expenditure.

Turnover of irrigation schemes to farmers is – thus – regarded as a way of doing away with subsidies, in response to adjustment policy related budget cuts in the last decade. However, it is interesting to note that the turnover debate in Zimbabwe started over half a century ago, and has been underwritten by different policy perspectives. Notwithstanding this long history, the picture regarding turnover today is that only a small percentage of the schemes (less than 10% of total irrigated area) have been turned over, despite repeated government commitment to farmer managed irrigation. The fact that farmer managed smallholder schemes have been found to perform better than government managed schemes (see Makombe et al., 1998)[1] has not helped to increase the degree and the pace of the turnover process. One of the aims of the paper is to explain the slow pace of turnover.

The plan of the paper is as follows. Section 2 presents a conceptual framework, which is used to evaluate irrigation management turnover in Zimbabwe. A distinction is made between policy-making discourse and policy practice in order to be able to analyse the irrigation turnover reality in Zimbabwe. This is followed by an overview of the turnover discourse and practice in the colonial and post-colonial era in section 3 and 4 respectively. Section 5 presents two cases of turnover: by default and by donor initiated experiment. The final section presents conclusions and issues for debate.

Conceptual Framework: Policy Discourse and Practice

First and foremost it is important to be clear on what is meant by policy, as this concept has many definitions. Anderson (1997) defines policy 'as a relatively purposive course of action followed by an actor or set of actors in dealing with a problem or matter of concern'. Anderson emphasises that the focus of policy analysis should be on *what actually is done* instead of *what is proposed or intended*. That is to say policy has more to do with action than rhetoric.

According to Howlett and Ramesh, policy is a set of interrelated decisions taken by a political actor or group concerning the selection of goals and the means of achieving them within a specified situation (Howlett and Ramesh, 1995: 5). Two critical issues emerge here. Firstly, policy is not spontaneous but is a deliberate effort taken over time. Secondly, we need to remind ourselves that we are talking about public policy, which is the preserve of the government. This means that when we have a policy position, which is linked to a (donor-funded) project, we have to be circumspect and doubt whether this can be taken as policy.

The second major task, apart from defining policy, is to examine what constitutes policy. According to Anderson (1997) a public policy:

1. must have goals;
2. should consist of patterns of action taken over time by government officials (rather than separate and discrete decisions)
3. emerges in response to claims for action or inaction on some public issue made by other actors (i.e. there are policy demands)
4. involves what actually governments do, not what they intend to do or what they say they are going to do;
5. must either be positive or negative, meaning that governments may choose to adopt a *laissez faire* attitude to an issue or be proactive and
6. is based on law and is authoritative (i.e. a public policy should have an authoritative, legally coercive quality that policies of private organisations do not have).

The evidence that we will present in sections 4 and 5 will illustrate that irrigation management turnover as it is practised in Zimbabwe does not fully meet the criteria listed above.

The third major task in elaborating a policy analysis framework is to examine the conditions that guarantee the success of a policy. Kerr (1976) advocated that there must be willingness by the relevant government organ(s) to take a particular action within certain conditions (they set for themselves). Perhaps more importantly, the relevant people must be aware of the policy (if it is not known to exist, it is not considered policy). Applied to our discussion this means that it is just not enough that governments have a policy position. Farmers and relevant government staff must be aware of that policy position. This brings us to the next point: under what circumstances might policy fail? Kerr (1976) suggests that there are three types of policy failure. Policies can fail due to:

1. *implementation* failure when the set targets cannot be met, because they are unrealistic or because of inability on the part of the implementing agency,
2. *instrumental* failure when the policy design does not fulfil or match the purpose or purposes of the policy and
3. a lack of being *normatively justifiable to the society*, meaning that while a policy might be successfully implemented and is effective as an instrument for achieving the agent's policy, it is labelled a failure when it does not meet the 'expectations of the society'.

The latter remark brings us to the issue of the politics of (public) policy. Since every society consists of various actor groups that represent different interests, one may wonder whether it is possible to come up with a policy that addresses the 'expectations of society', even if the policy is public policy assumed to address the general interest of the public at large. Long and Van der Ploeg (1989), Grindle and

Thomas (1991) and Mollinga (1998) have pointed out that policy formulation and implementation is not a matter of instrumental and implementation technicalities. These authors propose to analyse policy making and implementation as a continuous process, which takes place in different arenas, where various actors having a stake in the issues at hand apply different strategies to realise their political agenda(s). Public policy thus represents a balance of negotiated interests rather than the 'expectations of society'.

In some cases the policy process can result in policy outcomes that are cemented in law, policy statements or departmental guidelines for implementation. In other cases, like for instance irrigation management turnover in Zimbabwe, the outcomes of this process have not resulted in tangible legal statements or government directives. A policy evaluation framework that works by assessing policy outcomes against stated policy objectives does not apply in such a case. Neither does a framework that describes how the process of policy formulation and implementation *should* work, and then assesses how the policy process of formulation and implementation actually works. Since tangible policy statements and implementation guidelines on smallholder irrigation have been lacking since independence (1980), and the policy process is much more 'fuzzy' than a sequential structure of formulation and implementation, a different approach is needed. This paper investigates how the lack of a clear policy framework for smallholder irrigation turnover can be explained, given a context in which there is a lot of policy discourse on this issue, and within which numerous policy initiatives exist. Rather than formulation and implementation, the analysis of policy practices in this paper focuses on (non-)emergence and experiments.

Table 6.1 Structure of Zimbabwe's irrigation sector (1997)

Sub-sector	Area (ha)	Area (%)	# schemes	# farmers
Large scale commercial	126,000	73%	1,500	1,500
Parastatal (ARDA)	13,500	8%	20	n.a
Small-scale (out-growers)	3,600	2%	n.a.	n.a
Communal and resettlement	10,000	6%	300	20,865
Informal/micro-scale	20,000	11%	n.a	n.a
Totals	173,100	100%		

Source: ROZ (working paper 1), 1997: 5-9.

Irrigation and the Politics of Segregated Development

Because of the country's colonial legacy, access to the productive land and water resources is inequitably distributed. The balance tips in favour of about 4,500 white commercial farmers who control a good proportion (75%) of the best agricultural land ahead of 70% of the country's population that lives in the rural areas. This duality in the agrarian sector as a whole is also reflected in the structure of the irrigation sector (Table 6.1). The large scale commercial sector comprises private farms and company estates covering a large chunk (73%) of the total irrigated area, mostly employing modern irrigation technologies (sprinkler and drip) in combination with privately owned dams to produce high value crops (sugarcane, tobacco, citrus, flowers). On the other hand, smallholder irrigation schemes in communal and resettlement areas are mostly government managed, run of the river canal systems with a food crop production orientation (maize, wheat). It is estimated that smallholders use less than 10% of the total agricultural water. Less than 1 per cent of smallholder farmers have access to an irrigated plot in the schemes.

In order to understand the dual nature of the sector it is necessary to take a historical perspective of irrigation development and management.

Segregated Agrarian Development by the Settler State

When the white Rhodesian settlers in the early twentieth century failed to find the mineral wealth that had lured them into occupying the territory now known as Zimbabwe, their attention shifted to agriculture. The nascent Rhodesian settler state pursued a policy of segregated development, which legitimised state control over African development in so-called Reserves on the basis of the supposed superiority and advanced stage of development ('civilisation') of the European race (Phimister, 1988; Bolding, 2003). This led to the passing of legislation and the implementation of policies that fostered development and growth of white settler agriculture on a commercial freehold basis, whilst controlling and rationalising land and water use in African agriculture on account of its being inefficient, wasteful and environmentally destructive. The Land Apportionment Act of 1930, the linchpin of segregationist politics, set aside most of the best land for European farming, whilst designating the remainder of the land to Africans on a communal tenure basis. Its counterpart on the water front was the Water Act of 1927, establishing a water rights doctrine of prior appropriation (first come, first served) and proven beneficial use. The Act and its amended successor (1976) effectively ruled out Africans from acquiring their piece of the water cake, since they did not own title deeds and thus could not apply for water rights (except when located in a government project). In the 1930s growing concerns over the need for conservation, especially in the over-populated African Reserves, led to further legislation (i.e. Natural Resources Act 1941) which forced African land users to protect their land (contour ridges) and eradicated African irrigation practices of streambed cultivation and wetland (*dambo*) cultivation (Beinart 1984, Drinkwater 1991). The same legislation facilitated the release of government funds for the construction of water conservation works in, or nearby, European areas. Thus the white farming sector from 1936 onwards increasingly benefited from state development of dams and irrigation infrastructure, either through direct investment by government, loans at favourable interest rates or subsidies (up to 50%) for the construction of small and medium sized dams in European farming areas (Bolding et al., 1999: footnote 4, p. 248-49). Combined with the tobacco boom of the 1950s, these policies gave birth to a vibrant commercial farming sector relying on irrigated agriculture, heralded by many as a beacon of private entrepreneurship, despite the fact that most irrigation infrastructure had been financed by the state.

African Irrigation and State Control

The evolution of colonial irrigation policy for Africans is characterised by an ever-increasing degree of state control over the development and management of irrigation operations in the Reserves (Table 6.2).

Table 6.2 Evolution of smallholder irrigation policy in Zimbabwe: 1912-1980

Period	Policy objectives
1912-27	Farmer-initiated furrow irrigation with help from missionaries and settler farmers. Government watches from a distance.
1928-34	Government provides services and helps farmers develop irrigation schemes. Farmers retain control over the schemes.
1935-45	Government takes over management and development of African irrigation scheme.
1946-56	Racial segregationist land laws are reinforced. African people are moved to native reserves. New irrigation created to resettle African people.
1957-65	Government curtails development of irrigation schemes because of cost ineffectiveness.
1966-80	Government policy of separate development for Africans and Europeans revived. Irrigation schemes conceived as 'population concentration centres' around rural growth points based on irrigation. Production maximization is achieved by strict government control over farming and management operations.

Source: Rukuni and Makadho (1994: 130).

The table shows that government took over the control of schemes in the early to mid 1930s. This take-over and subsequent developments was directly related to the conservationist politics of the time. It was believed that proper development of African irrigation could only be done under technical guidance by the settler government. The conservation and water legislation of the 1930s curbed any farmer initiated irrigation furrows that had developed with the aid of benevolent missionaries and white settler farmers (Bolding et al., 1996: 194-97). Many of the government irrigation schemes in the dry and famine stricken Save Valley were constructed during the 1930s and 1940s with three main aims in mind: (1) provision of food security (famine relief); (2) maximisation and concentration of agricultural production to relieve population pressures in other areas, (3) modernization of agricultural production by introducing cash crops (Alvord, 1958; Roder, 1965). Whilst the emphasis during the early days of government controlled African irrigation was on the first objective, the emphasis during the 1950s, 1960s and 1970s shifted towards the latter two goals. African irrigation schemes provided a means to realise the policy of segregated development, creating islands of modern African production and rural industrialisation based on irrigation. Thus it was hoped future African demands for more land could be curbed and a loyal class of African yeomen producers created. To make this policy a reality and to legitimise further government expenditure on African irrigation, the government introduced strict control over the technical, managerial and socio-political aspects of irrigation, thus complicating the possible turnover of these schemes, later.

The irrigation sites were generally poorly designed, as they were based more on a political than technical rationale. The lack of technical capacity of government departments responsible for these schemes (see Table 6.3) contributed to the problematic turnover process after independence. The poor design of the majority of smallholder schemes implies that farmers are now being called to manage expensive and complex outfits that were never meant to be run by farmers.

Table 6.3 Institutions in smallholder irrigation

Period	Responsible government institution
1932-1944	Ministry of African Affairs
1945-1963	Internal Affairs African Administration
1961-1963	Department of Native Agriculture (Ministry of Agriculture)
1964-1968	Department of Conservation and Extension (Ministry of Agriculture)
1969-1978	Ministry of Internal Affairs
1979-1981	Devag (Ministry of Lands, Resettlement and Rural Development)

Source: Rukuni (1985).

Technical and economic imperatives resulted in a situation where the government exercised tight managerial control over virtually all irrigated farming operations. The state dictated the crops to be grown, when they would be grown and how they would be disposed (e.g. through commercial outlets). Irrigation was conceived as a technical operation, where centralised management synchronised the operations of hundreds of plot holders so as to ensure maximum productivity. The resultant management regime was so excessive that Reynolds (1969) observed that farmers were being treated as 'children'. This was quite a poignant point for those farmers who had constructed their own schemes, which had subsequently been appropriated by the state (see Manzungu, 1995). Hughes (1974: 213-220) likens the managerial order created by white irrigation managers on African irrigation schemes to forms of 'despotic management control' described in Karl Wittfogel's influential analysis of irrigation based societies (1957). Irrigation management was based on alienating farmers from exercising any management, since they were considered unqualified for the job.[2] Farmers had no rights, thus facing high levels of uncertainty, since they could be evicted at any time. Of particular importance in this respect are the Control of Irrigable Areas Regulations that came into force in 1970. Irrigation farmers (plot holders) were requested to sign three annually renewable permits (one to reside, one to depasture stock and one to cultivate). The Irrigation Manager who was tasked with supervision and control over all irrigation and farming operations, could withhold issuing of permits when the plot holder was found to be failing in paying water rates, timely application of seed and fertilizer, timely weeding, timely application of irrigation water, and other recommended practices issued by management. Also irrigation plot holders were not allowed to own any businesses, cultivate any dry land or engage in gainful employment. They were to be full-time irrigators. Many plot holders were evicted during the early 1970s for failing to comply with any one of the conditions.

Socio-politically the government had more subtle ways of controlling farmers. One was to ensure that farmers felt somehow that they were 'represented', thus aiding in the effectiveness of irrigation operations. It is an example of control by co-optation whereby farmers were accorded nominal management roles so that they could in turn control themselves.[3] The representation was constructed around traditional leaders.[4] Thus kraal head furrow committees, like the one in Nyanyadzi, played a purely advisory role, reporting directly to the white irrigation manager. Still, during the brief period of open African Nationalist politics (1957-1964), when African political parties were allowed to emerge and draw members from the populace, irrigation schemes became political 'hotbeds'. The relatively affluent African irrigators demanded better services and majority rule, whilst protesting against strict government regulations curbing their opportunities to accumulate wealth on the basis of agricultural production (Ranger, 1985, Weinrich, 1975). In Mutambara irrigation scheme, the very traditional leaders that were supposed to discipline their fellow irrigators, turned out to be the 'agitators' responsible for the ultimate retreat of the government and subsequent closure of the scheme (Manzungu, 1995).

The Politics of Cost Recovery

From as early as 1935 farmers needed to shoulder a proportion of the operation and maintenance costs by paying a water rate. This rate was gradually increased over the years, but the total water rates collected never paid for more than 50% of the total maintenance and running costs. A government-employed economist studied the cost effectiveness of African irrigation ventures and found that most schemes were unprofitable and heavily subsidised (Hunt, 1958). Hunt's findings led to a halt in the construction of new African irrigation schemes from 1957 to 1965. The Irrigation Policy Committee (1961) found that public resources for African development could be more profitably availed in other sectors than irrigation, whilst any new irrigation ventures should be based on repayment of operation, maintenance and capital development costs (Roder, 1965: 135-136). Most of the policy debate surrounding cost-effectiveness of African irrigation was devoted to the issue of profitable marketing of cash crops and the need to increase agricultural productivity by imposing strict government control over farming operations. The revival of segregated development policies during the Smith regime (1965-1980), led to the construction of large irrigation estates with African tenants. In these estates, presently managed by the parastatal Agricultural and Rural Development Agency (ARDA), deductions from crop proceeds were taken to recover running costs. Inspite of frequent calls by officials from the Ministry of Internal Affairs, full recovery of operation and maintenance costs from plotholders in ordinary government schemes was not considered politically feasible and therefore not imposed.

Irrigation Management Turnover Policy after Independence

One of the important features of smallholder irrigation development in Zimbabwe has been the absence of a clear policy in black and white. Many commentators have stated that policy in the smallholder irrigation sub-sector in Zimbabwe was never adequately developed (see Roder, 1965; Mupawose, 1984; Chabayanzara, 1994; Chitsiko, 1995). The director of Agritex, Makadho (1994: 20), made it more explicit when he stated that, 'irrigation policy is not in black and white: it is only understood.' What follows is an account of some attempts at smallholder irrigation policy formulation after independence. However, first of all we will provide an overview of the role of the post-colonial state in irrigation. This might help explain the lack of a smallholder irrigation management policy.

The Role of the Post-Colonial Government in Smallholder Irrigation in Zimbabwe

Rehabilitation with Donor Funds and Continuation of Colonial Agrarian Policies
With independence in 1980, the ZANU PF, 'people's government' adopted as its leading ideology, 'Scientific Socialism'. In economic terms this ideology was translated into the pragmatic 'Growth with Equity' policy (GoZ, 1981). This two-pronged policy entailed growth of the white and foreign owned capitalist sectors of the economy, whilst the equity component consisted of state administered policies aimed at alleviating the inherited Rhodesian inequalities. The latter comprised an expansion of previously denied state services in communal areas (education,

extension, marketing support), as well as a massive land resettlement programme aimed at acquiring land together with water resources from the commercial farming sector for redistribution to landless peasants and refugees. The public service was expanded and Africanised, though the latter did not imply the introduction of new attitudes and practices as many of the promoted black civil servants received their training within the Rhodesian state (Alexander, 1994: 326). The new government, by mouth of the Riddell (1981) and Chavunduka (1982) commissions, did not challenge the beliefs and practices which had informed the technical and segregated development policies during the Rhodesian era. This continuity was reflected in the reproduction of the same assumptions on the wasteful and environmentally destructive qualities of African farming practices, that were held by a now expanding bureaucracy, still largely unaccountable to representative institutions (Alexander, 1994: 323-333; see also Drinkwater, 1989, 1991). Resettlement and irrigation schemes were envisaged as self contained islands of modernisation, where settler cultivators 'were expected to sever all social and cultural ties with their past lives in order to achieve new levels of productivity under the tutelage of the state' (Alexander, 1994: 334). During the first decade of independence the United Nations High Commission of Refugees (UNHCR) and United States Agency for International Development (USAID) funded the rehabilitation of irrigation schemes destroyed during the liberation struggle, whilst little new irrigation development was taken up.

Fragmented Bureaucracy By the late 1980s, emphasis had shifted from rehabilitation to management and new irrigation development projects to fully utilise the available water resources. Also the government adopted a policy of equal access to development funds by both blacks and whites for the establishment of new irrigation projects. However the mandate for irrigation development and management had not been in the hands of a single government department. Irrigation development responsibilities were split among three departments, Derude, in the Ministry of Lands, Resettlement and Rural Development, Agritex in the Ministry of Agriculture and DWD in the Ministry of Water and Natural Resources. Derude was responsible for the management of irrigation schemes through the irrigation manager. Agritex was responsible for agricultural extension, while the DWD was responsible for water delivery to the irrigation schemes. The divided responsibility brought some problems in the development of smallholder irrigation. 'Co-ordination of the departments was poor and their co-operation inadequate. Staff members of the departments were confronted with the problem of divided loyalties. Personality clashes between management and extension staff at some schemes have not helped matters either' (Chitsiko, 1988: 70). As a result effort and debate centred more on clarifying the roles of the different organisations than on an actual policy on smallholder irrigation management. It was only in July 1987 that the irrigation component of Derude was transferred to Agritex, effectively placing both the extension and management functions to a single department (Chitsiko, 1988). However the development of the water source and the subsequent delivery of the water to the irrigation schemes remained the responsibility of the DWD. As a result the development of irrigation policy under this situation remained fragmented with the DWD and Agritex developing separate policies.

Donor Support and Invisible Farmer Rights During the period 1980 to 1997, the smallholder irrigation sub-sector enjoyed massive donor assistance. The government departments mandated with irrigation development found it difficult to deal with numerous funding institutions. The result was confusion at grassroots level where officers were not sure what to tell the farmers as the correct funding procedure and why that procedure was used and not another to establish a neighbouring irrigation scheme. The farmer's rights were therefore never clearly spelt out. Yet, the command area under smallholder irrigation steadily expanded from 4,270 ha in 1983; 4,572 ha in 1990 to 9,958 ha in 1997 (RoZ, 1997: 9). With help of the continued assurance of donor funds the irrigation engineering division within Agritex grew both in influence and number of staff employed. The adoption of the economic structural adjustment programme in 1991 led to decreased government funding for operation and maintenance activities on existing schemes. This resulted in irrigation management turnover by default in the majority of the schemes towards the end of the 1990s, when government could no longer honour its financial obligations. Meanwhile donor funds were used to experiment with new forms of smallholder irrigation development and management. However, political events in the late 1990s entailing massive land invasions by war veterans and state orchestrated oppression of opposition parties, precluded continued funding by both international and bi-lateral funding agencies. The involvement of the users in financing the running costs of smallholder irrigation schemes became the development catch phrase.

Attempts at Policy Making

After sketching the context of irrigation management turnover in post-colonial Zimbabwe in general, we will now look at attempts that have been made to elaborate a policy of irrigation management turnover.

The Derude Effort (1983) In 1981, the Department of Rural Development (DERUDE) took over the responsibility for the construction and management of smallholder irrigation schemes from the Department of Agricultural Development (DEVAG). In April 1983 DERUDE published a policy paper on small-scale irrigation schemes.[5] Coming three years after independence this was a welcome attempt to give a new direction to the smallholder irrigation sub-sector. However, the document was a mixed blessing, since it beset the smallholder irrigation sector with a host of, sometimes contradictory, objectives. These ranged from provision of food security, the generation of rural employment, the production of export crops, the provision of a relief of population pressures on the land, and the triggering of a rural industrialisation process. Thus the question of either maximising settlement and food security (by decentralising decisions on cropping strategies to as many plot holders as possible) or maximising production (by centralising management in the hands of the irrigation manager), was left open for local interpretation. The same contradiction was reflected in the management strategy that the policy proposed. The fact that the irrigation manager could withhold three annually renewable permits to plot holders on account of lacking performance, suggested a continuation of the production oriented, disciplinary colonial irrigation policy. On the other hand, the policy proposed a gradual turn-over of operation and maintenance responsibilities to farmers by increasing the farmers' financial contribution and setting up Irrigation Management Committees in schemes with the aim of achieving self-management (see Box 6.1).

Box 6.1 Aims of Irrigation Management Committees

> According to the assistant director DERUDE, later director of Agritex, the aims of the irrigation management committees were:
>
> (1) To enhance farmer participation in management and decision making at the local level;
> (2) To prepare the farmers for a complete take-over of management functions that are currently being carried out by government;
> (3) To create a responsible attitude and a sense of belonging to the scheme so that farmers could view the schemes as theirs and not simply a government project;
> (4) To introduce a self-regulatory, self-disciplining machinery at the irrigation scheme in order to enhance maintenance of discipline, cropping patterns and recommended agronomic practices (Pazvakavambwa, 1984: 423).

The document is silent on the farmer's rights. For example, ownership and user rights of irrigation infrastructure after hand over are not clarified. Also, no time schedule or criteria have been provided to indicate when and how irrigation schemes qualify for turnover. Despite the fact that the DERUDE policy has never been endorsed as official policy, it is considered the most definitive statement of smallholder irrigation policy in Zimbabwe (Meinzen-Dick, 1993: 35).

The ambiguities contained in the policy soon led to inter-departmental conflicts over its implementation. Top level DERUDE staff insisted that the IMC approach was necessitated by the lack of preparedness of plot holders to co-operate with old style Irrigation Managers, manifested by the defiance of irrigation rules and non-payment of water rates in many schemes during the first years after independence. The careful and difficult process of strengthening IMCs was meant to ultimately obviate the need for strict government supervision and continued subsidies on operation and maintenance of the schemes. Local DEVAG and Agritex staff on the ground, however, interpreted the policy as a continuation of the colonial management style, whereby IMCs could be used to enforce discipline on the plot holders and thus ensure sustained productivity of the schemes. The latter view won the day when Agritex replaced DERUDE as manager of smallholder irrigation schemes in 1987.[6]

National Farm Irrigation Fund (NFIF) Effort (1985) The aim of the NFIF was to provide smallholder farmers with access to cheap money for irrigation development. It was basically a loan facility to smallholder farmers in which as a group they could borrow money for the purchase of irrigation in-field equipment at low interest rates. The communal land tenure system precluded the use of land as collateral, necessary to finance main system irrigation equipment, which therefore remained the responsibility of the government. Still, it was hoped to reduce government spending on the development of smallholder irrigation schemes by involving the users in financing at least part of the irrigation development costs. The loan facility was also meant to ease irrigation management turnover to the farmers. However, the ownership rights of the irrigation equipment were not clear. The effort came to naught, as the smallholder farmers did not make use of the loan facility. Electoral promises by the state president and minister of Lands, Agriculture and Water Development to provide each district with a dam and smallholder irrigation system free of charge, as well as the availability of donor support to smallholder irrigation development at no charge to the ultimate users, severely undermined the policy.

The FAO Initiative (1990-1994) In 1990 FAO and GTZ (a German development agency) financed the development of an irrigation policy and strategy in Zimbabwe. The status of the resulting 1994 report is disputed. Some claim that this is the policy that is in use, while others feel it is not. At best it has been described as a semi-official document because government never endorsed it as policy. The document calls for improvement of the prevalent low levels of water use efficiency by adopting sprinkler and drip technology and handing over responsibilities for operation and maintenance to the farmers. To this end effective Water User Associations have to be established, replacing the largely ineffective Irrigation Management Committees. The policy also proposes stringent financial reforms consisting of a water pricing policy that reflects the scarcity of the commodity water and the recovery of operation and maintenance costs directly from the beneficiaries. Any subsidy of O&M costs should be justified and targeted on a case by case basis. Whilst it is proposed that development of irrigation infrastructure on state land remains government responsibility, private sector investment in irrigation is encouraged.

Decentralisation Policy (1991-1993) The goal of the policy is to promote a system of local government, based on decentralisation of authority and involving people's participation in all governance processes so as to generate sustainable development at all levels. Early attempts at setting up local governance structures had resulted in the formation of representative bodies at village (VIDCO), ward (WARDCO) and district level (District Council) in 1984. These elected bodies were meant to replace Rhodesian governance structures that had been dominated by traditional leaders.[7] At the same time central government hoped to establish control over local party committees that had managed local development in liberated areas. However, civil servants of various ministries representing different technical expertise dominated the decision-making bodies[8] of this system of local governance. Ministries regarded local authorities primarily as policy implementing, not formulating agencies. People's participation, in their view, meant assent and compliance, thus easing the implementation of centrally formulated government policies (Alexander, 1994: 329-331). The Rural District Council Amalgamation Act of 1991 (implemented in 1993) triggered a more serious attempt to reduce central government field officers involvement in the administration, operation and management of local projects including smallholder irrigation projects. Supported by heavy donor funding, Rural District Councils were expected to enhance their professional capacities and replace government experts with their own staff. However, line ministries resisted the devolution of authority, fearing the dominance of party politics over administrative and technical competence. Local bureaucrats became highly legalistic in their orientation, resulting in a lot of confusion at local level as to who was responsible for local affairs (Roe, 1995: 839). Council officials felt inexperienced in dealing with their new authority in, among other things, irrigation management, in most cases leaving day-to-day management in the hands of line ministries (see Nyanyadzi case study below).

The National Water Policy and Strategy (1995-1998) This policy aimed at redistributing access to water, privatising government departments involved in water management and involving water users in the development and management of water on a river basin scale. The water reforms that were in part triggered by this policy led to the adoption of a new Water Act (1998), and privatisation of the DWD into the Zimbabwe National Water Authority (ZINWA), a parastatal agency providing services on a cost recovery basis. Catchment councils are being established in the country's 8 major river basins, comprising all stakeholders and involving them actively in allocation of water permits and development of new (irrigation) infrastructure. The immediate effects of the 1995 policy were increases in the national blend price charged for water drawn from government dams, and the removal of electricity subsidies on pump operated smallholder irrigation schemes. This seriously affected the operation of many smallholder irrigation schemes, accelerating the *de facto* turnover of financial operation obligations from the government to the farmers (see also Nyanyadzi case study below).

The Zimbabwe Agricultural Policy Framework (1996) The Zimbabwe Agricultural Policy Framework was launched in August 1996 and embodies the current national objectives and policies of the agricultural sector as a whole. The policy provides only a general framework reflecting several ongoing policy trends without specifying implementation guidelines. It was formulated to facilitate the release of international funding for the implementation of World Bank mediated Public and Agricultural Sector Investment Plans. With regard to the smallholder irrigation sector the policy calls for increased commercialisation of irrigated production, sustained growth of irrigated area, establishment of an efficient institutional structure, and equitable and efficient use of scarce water resources through the establishment of a water pricing structure. This framework has provided a guide to most of the present reform efforts on irrigation management under implementation by international and national financing agencies. These include the FARMESA project financed by FAO and SIDA (a Swedish donor agency) seeking to develop methodologies for irrigation management transfer in government run irrigation schemes (Manzungu, 1998), and the Smallholder Irrigation Support Programme (SISP) financed by IFAD and DANIDA (a Danish donor agency) that aims to test and develop turn-over models, formulate an overall plan for irrigation management transfer, and strengthen the organisational capacity of rural district councils, irrigation service institutions and farmers by means of training. The latter programme also seeks to rehabilitate some 1,300 hectares of existing schemes before turnover, whilst developing some 700 hectares of new smallholder irrigation schemes under farmer management (RoZ, 1997).

In conclusion, we argue that at best some policy discourse has developed. This policy discourse stresses the need for irrigation management turnover for reasons of supposed better performance of irrigation schemes under farmer management, and most importantly for reasons of improved cost recovery of operation and maintenance expenditures required to keep the irrigation schemes running. The discourse is silent on the rights and responsibilities of the farmers, and has not been cemented in official law, public policy statements endorsed by cabinet, or implementation guide lines for use by the staff of the relevant departments and other actors involved in irrigation development. Rather than a diminished role of the government in irrigation management and development, the post-colonial state in Zimbabwe can be characterised by a strong continuity of colonial agrarian policies that put the onus on the state. Ample donor support has led to the expansion of government services in smallholder irrigation, whilst an expanding state bureaucracy has actively resisted attempts at decentralisation and devolution of authority to locally elected bodies. It was only due to severe budgetary pressures towards the end of the 1990s that a process of financial devolution from the government to the farmers has occurred.

Two Cases of Turnover Policy in Practice

Despite the lack of a clear, unambiguous irrigation policy on turnover and the non-existence of departmental guidelines, turn-over of O&M responsibilities has taken place in small-holder irrigation schemes during the late 1990s. This section presents two different case studies of turnover policy in practice. The Nyanyadzi case exemplifies the loss of control over irrigation operations by the post-independence state. A process of turn-over by default was triggered by (1) increasing budget deficits resulting in a *de facto* retreat of the state agencies from their O&M obligations; (2) users actions against state control over their operations through the mobilisation of politicians; (3) a lack of political back-up for government agencies to maintain the colonial disciplinary O&M policy due to the strong-hold of opposition parties in the Save valley. The Nyanyadzi case cannot be seen in isolation. It is exemplary of what happened during the post-independence era in all Save valley smallholder irrigation schemes (the so-called old schemes).[9] The Musengezi case exemplifies irrigation management reform through a formally constituted turnover experiment that was started by Agritex in 1989. Four phases of operation and maintenance regimes are discussed based on who pays for the continued operation of the 5 smallholder schemes. To keep their schemes in operation the farmers display a remarkable capacity to secure new funds from different sources during each phase. Donors, government and politicians line up to chip in each time the schemes cease to function. The case study brings the various threads and changes in donor initiated turnover initiatives together.

Table 6.4 Nyanyadzi blocks: official data compared with *de facto* use of the scheme (1997)

Block	Water source	Official data (Agritex Chimanimani files, 1997)		Plot survey (Bolding, 1997)		
		Command area (ha)	Registered plot holder	Deceased plot holder	Actual plot holder	Plots leased
A	Odzi and Nyanyadzi	136	123	20%	192	7%
B	Odzi and Nyanyadzi	147	193	n.a.	n.a.	n.a.
C	Nyanyadzi	65	67	54%	158	5%
D	Odzi and Nyanyadzi	69	75	n.a.	n.a.	n.a.
Total		417	458			

Nyanyadzi Irrigation Scheme: Turnover by Default

> Nyanyadzi irrigation scheme, the economic lifeblood for 500 plot holders in Chimanimani, has been dry since last March after the Zimbabwe Electricity Supply Authority and the Department of Water cut off supplies for non-payment of power and water bills. In a recent interview, one of the plot holders, Mr King Dube, said the beneficiaries from the irrigation scheme were ready to meet the electricity and water bills 'if Government tells us we are now paying on our own'. It has been Government Policy in the past to subsidise farmers in the irrigation schemes by paying maintenance costs, water and electricity bills (*Manica Post*, 18 December 1998).

This newspaper article relates the story of Nyanyadzi irrigation scheme and various other pump-irrigated schemes along the Save valley being closed on account of Government not being able to own up to its financial obligations. The same article mentions the opening of a new irrigation scheme in Chimanimani district by the Minister of Lands and Agriculture, whilst announcing that 'Government now expected irrigation scheme farmers to meet their own costs'. No policy to this effect had been formally endorsed and neither were the irrigation plot holders informed of this decision. Below an analysis is given of the various factors that led to this financial turnover by default. For the characteristics and problems of Nyanyadzi irrigation scheme during the post-colonial era reference is made to Table 6.4 above and Box 6.2 below.

Box 6.2 Characteristics of Nyanyadzi irrigation scheme

> Nyanyadzi irrigation scheme was started in 1934 by ED Alvord, an American missionary, as one of the first African smallholder schemes constructed by the Government. The scheme expanded over time to grow to its present size of 414 hectares, split up in 4 different blocks (see Table 6.4). All blocks can get water from the Nyanyadzi river through the main canal, whilst only block A, B and D can get water that is pumped from Odzi river into the Night Storage Dam. Plot holders irrigate their border strips with the help of siphons. Crops are grown during 3 seasons each year. The main summer crops are maize and cotton. Cash crops like beans and tomatoes dominate the dry winter season, whilst during the ensuing hot season wheat is the main crop.
>
> The scheme suffers from frequent water scarcities, with on average one out of three years being water scarce. During the 1980s and first half of the 1990s this problem was exacerbated by a number of factors. Rainfall during this period was below average (Makarau, 1999) and as a consequence Nyanyadzi river water supply became highly unreliable. The problem was further compounded by the fact that the main canal is unlined and loosing up to 50% of its total abstraction from Nyanyadzi river on its way to the Night Storage Dam (Pearce and Armstrong, 1990). Ever since the war ended, more irrigation furrows have been constructed upstream of the Nyanyadzi intake. This has resulted in fierce competition over Nyanyadzi river water between the Nyanyadzi plot holders and their upstream counterparts (an estimated 100 irrigation furrows abstract water upstream (Bolding, 1999)). Furthermore water abstraction from the Odzi river has been unreliable due to frequent pump break-downs and diminishing budget allocations reducing the number of pumping hours (Bolding, 1996).
>
> Maintenance requirements of Nyanyadzi irrigation scheme are quite high, due to heavy siltation, which has become more severe over time. After independence a lot more land has been taken into cultivation upstream (even in the river bed) and as a consequence the weir and main canal intake are frequently silted. Also land along the main canal has been taken into cultivation, though this used to be forbidden during the colonial era. As a consequence about 13 large gullies have developed that discharge their load straight into the main canal. The sump of the pump house on the Odzi river is also frequently silted, causing pump break-downs. All this deposit has to be regularly scooped employing manual labour (Bolding, 1996).

A New Start with a Popularly Elected Irrigation Management Committee (1980-83) Nyanyadzi irrigation scheme had been a 'hot place' during the war of liberation

(1972-1980). The first political killing of a white man took place in July 1964, in close vicinity of Nyanyadzi. The attack had been facilitated by young plot holders from the scheme, organised in a local ZANU party branch. During the subsequent war, infiltrating freedom fighters allowed the scheme to continue functioning, since Nyanyadzi plot holders and the scheme's management staff secretly provided food and clothing in exchange for personal security. However, during the 1978/79 growing season, when control over Chimanimani district was virtually in the hands of the freedom fighters, Nyanyadzi plot holders had been instigated to refuse payment of water rates as an act of resistance to the colonial regime. The irrigation manager subsequently closed the scheme. The Lancaster House independence negotiations reigned in a new dawn by the end of 1979, and the white irrigation manager re-opened the scheme on condition that water rates would be paid after harvesting the crops. However, when the time of payment was due (April 1980), the kraal head committee on behalf of Nyanyadzi plot holders refused to pay up. The committee's stance was merely cashing in on promises of free government services after independence, made by freedom fighters during their nightly rallies. A promise that had been honoured by the newly elected Member of Parliament for Chimanimani district during his victory speech, when he suggested that water rate payments should be done away with. The out-going white District Commissioner, however, firmly stood behind the irrigation manager by suggesting to the new Minister of Local Government to cut off water supplies to the scheme in case the kraal head committee persisted in its refusal to pay. To resolve the ensuing standoff, a team of government officials from responsible departments visited the scheme in May 1980. Their investigations revealed that the plot holders had a number of grievances regarding enforcement of water rate payment during water scarcity years and exploitation by the marketing co-operative, which had been supervised by white colonial officers. Besides promising a rehabilitation of the canal infrastructure of the scheme and construction of a new pumping station, it was decided by the visiting officers that the existing kraal head committee should be done away with and be replaced by a popularly elected farmer committee.[10] Two farmers from each block were selected to sit on the committee, which consisted of a chairman, vice-chairman, secretary, vice-secretary, treasurer and 3 ordinary members. The duties of this Irrigation Management Committee were:

(a) To communicate problems and bring solutions
(b) To discuss problems with the irrigation manager
(c) To recommend individuals for plot allocation
(d) To disseminate information among plot holders.[11]

It was hoped that the newly promoted African irrigation manager and newly elected IMC would resolve any outstanding problems.

However, with persisting water shortages and local politicians calling for a new order based on free services by government, Nyanyadzi plot holders kept refusing to pay their contribution towards operation and maintenance of the scheme in the early 1980s. This was despite the fact that the farmers' contribution (at $35 per ha) only covered 23% of the required expenditure to keep the scheme operational. It was left to the local DERUDE staff to resolve the contradictory demands exerted on them. The new irrigation manager for Nyanyadzi and his African staff of extension assistants and water bailiffs (gatekeepers) were bent on reviving the old management style. Decisions on cropping programmes, water distribution schedules, maintenance activities and agricultural operations remained centralised in the hands of the Irrigation Manager. The IMC was instructed to draft by-laws that emphasize the maintenance of discipline and spell out appropriate punishments for violators of the rules. The IMC was only called to meetings to be told what to do by the irrigation manager.

Efforts to Revive the Old Management Style (1984-1992) In 1984 water charges were re-introduced in Nyanyadzi, this time called maintenance fees, at the rate of $145 per hectare (or 21% of the expense needed to keep the scheme running). Since payment was now done towards maintenance of the scheme and not towards water delivered, the recurring issue of wavering of water fees in years of water shortage could henceforth be avoided. Nyanyadzi scheme and its management were however confronted with three thorny issues that required appropriate action: competition for Nyanyadzi river water with upstream irrigators; illegal cultivation along the main canal; and continued deference of maintenance fee payments by a majority of plot holders. Government staff tried to resolve these problems by invoking the colonial management style.

During the dry winter season of 1984 Nyanyadzi scheme faced a water shortage. Top government officials attributed the shortage to illegal water abstractions upstream of the Nyanyadzi river intake. By October 1984 only a trickle of water reached the scheme and the Irrigation Manager (IM) decided to act. The IM organised a raid along the Nyanyadzi river destroying all irrigation furrows that abstracted water, which he claimed legitimately belonged to the scheme. To his own surprise, the IM was remanded in custody. He was made to understand that he was not to interfere with upstream furrows that had been issued water rights by the District Administrator.[12] After mediation by the local MP, the upstream irrigation furrows were rebuilt. This in turn led to bitter reactions from the IM who requested the 'top officials' to make up their minds 'whether to have the existing scheme or legalise the 80 hectares along the Nyanyadzi river.' More upstream raids were organised by successive IMs of Nyanyadzi irrigation scheme, often at the instigation of the plot holders. These raids (e.g. in 1987, 1988, 1991 and 1994) invariably resulted in the IM being faced with disciplinary action and virtually no improvement of water supply to the scheme (for more details see Bolding et al., 1996: 206-11; Bolding, 1999).

In May 1987, the IM directed his attention at illegal garden cultivation along the main canal in Block C. These gardens had been taken up by relatives of a former IMC member and kraal head for block C, causing a major siltation problem. The kraal head had acted on the understanding that the prohibition to cultivate along the main canal was no longer in force, since freedom fighters had fostered the idea that everybody would be free to settle and cultivate anywhere once independence was attained. The gardens were set ablaze and destroyed. During a general meeting in June 1987 the IM was publicly scolded for his actions. A local police officer declared his loyalty to the ruling party ZANU(PF) and then asked the affected people to step forward to be granted the right to continue cultivating their gardens.[13] Again the IM had been over-ruled by party directives.

Another pernicious problem was provided by the fact that the majority of Nyanyadzi plot holders continued to defer payment of maintenance fees. Whenever local government staff raised the issue and threatened eviction of defaulters, as had been customary during the colonial days, local leaders threatened to approach the Minister or Prime Minister to effect exemption. In 1987 provincial government staff approached the provincial ZANU(PF) leadership to find out what the party's position was on payment 'particularly in those schemes with large sums still outstanding'.[14] However, the party left it to the irrigation staff to find out what their authority was. When in April 1989 a new IMC is elected in Nyanyadzi, consisting exclusively of defaulting plot holders, the provincial agricultural officer decided to act. He personally explained to the new IMC office bearers that he expected them to maintain discipline amongst plot holders and encourage payment of maintenance fees. In his view the IMC was meant to 'assist and never replace Agritex.'[15] All plot holders were urged to pay their arrears or else face eviction from the scheme. The provincial Agritex officer also threatened to dissolve the sitting IMC if its members failed to pay up. In turn, angry IMC members told him to go home and never visit them again.

Rather than leaving it at that, the Provincial Agritex Officer mounted a province wide irrigation inspection (the so-called 'physical exercise'). Plot holders were requested to wait at field edge of their plot on the announced day of inspection with their identity papers and receipts of maintenance fee payments at hand. Those guilty of offences like arrears in payment, sub-division of plots amongst offspring, lease of plots or cultivation of illegal gardens, were to be served with eviction orders on the spot. The exercise resulted in resurgence in payment of fees and restoration of 'discipline' in all irrigation schemes, except Nyanyadzi, where 'comparatively the disorder ... is extreme.' Nyanyadzi also had the highest amount of outstanding fees (zim $ 202,521). Still, Agritex staff felt confident and pressed on, as the minutes of a staff meeting in July 1989 show:

> Action in form of either eviction (forcefully) or withholding water was called for, despite the mass movements from the plots that would ensue. Co-operation with district councils, governor, Provincial Administrator would be vital in this exercise to avoid back-firing on Agritex face if it carries out the task lonely.

Whilst the 'physical exercise' yielded good results in other schemes, the IM for Nyanyadzi had less favourable news to report in October 1989.

> [A]t Nyanyadzi the committee and farmers refused the physical exercise (...) The farmers were influenced by the committee (...) They uttered bad words against government. That the physical exercise is not government policy. That they are not going to vote in favour of the government in the coming election. (...) The whole staff must be removed from the scheme and remain communal. That they will make a follow up to the President.[16]

The Provincial Agritex Officer replied by sending eviction orders for all registered defaulters in Nyanyadzi. The director of Agritex, visiting Nyanyadzi in an attempt to pacify relations, was told to go home by Nyanyadzi plot holders. Meanwhile the IMC sent delegations to the provincial chairman of ZANU(PF) and the president's office. The ZANU(PF) politicians in the end decided to instruct the director of Agritex to cool down on operations in Nyanyadzi, thus forcing Agritex to repeal its actions.[17] The ruling party was eager not to loose support in an area that was known to be under the influence of two opposition parties.[18] The move frustrated the Provincial Agritex officer, who stressed to his staff that before talking of scheme rehabilitation or farmer training to facilitate irrigation management turn-over, farmers would have to prove that they could foot the bills.

In the years to come the Nyanyadzi plot holders, by mouth of their IMC, often mobilised local, provincial and national ZANU(PF) politicians to press for solutions to their problems. Eager not to fuel support for the opposition, politicians of the ruling party managed to press for exemption of fee payment during drought years and speedy construction of a new electrical pump station on the Odzi river to replace the failing diesel engines of the old pump station.[19]

The Final Attempt at Government Control by a New Irrigation Officer (1993-95)
Early 1993 sees the arrival of a new Irrigation Manager at Nyanyadzi irrigation scheme. Since he was newly trained as irrigation engineer he requested the Provincial Office for the irrigation policy on smallholder irrigation schemes. The Provincial office replied that:

> there is no gazetted policy (...) However there is quite substantial written work classified as *Policy papers*. These seem to be proposals more than anything else. They are very *valuable* and *operative* in most cases. If anything else they need *updating* and *formal Endorsement as Policy* (...). In conclusion it can be observed that there is no policy...

On management, it was observed that '[a]lthough the dominant figure should be the Irrigation Committees, there is a marked degree of paternalism by Government departments and Donor Agencies.'[20]

On the basis of this information the new IM decided to draft his own policy together with his Nyanyadzi office staff during two separate meetings in June 1993. The old line of command stretched from the IM to the agricultural supervisor to his 4 subordinate extension workers (one for each block) to the 6 water bailiffs down to the general hands. Farmers were organised in the IMC and in Block IMCs. However, the extension staff noted that discipline amongst farmers had gone down since the war: 'Staff said they cannot use the stick method which was used by the whites because the IM has no support from the top officials.' The new IM instructed his extension workers to work closely with the Block IMCs: farmers would no longer be allowed to visit the office unless they had clearance from their extension worker. In order to curb undisciplined behaviour by Block IMCs it was suggested to select 'workable farmers' and 'manageable committee members'. It was also decided to 'prevent the IMC to talk straight to Provincial *Chefs*,[21] because that resulted in local Agritex staff not being considered important. Furthermore it was noted that the IMC did not enforce byelaws itself, but left implementation at the discretion of the water bailiffs. The IM decided it was the task of the extension worker to enforce IMC decisions, if necessary.[22] In October 1993, the new IM decided to discipline the kraal head and IMC member for block C. The latter was called into the local Agritex office and told to stop issuing out more land along the main canal and to stop instigating block C plot holders to refuse paying maintenance fees.[23]

However, this effort at regaining control over the scheme by the local Agritex office was short-lived. One factor in particular compromised the position of Agritex. Ever since the government of Zimbabwe had embarked on an Economic Structural Adjustment Programme in 1991, the public service faced further budget cuts. These cuts resulted in less money becoming available for operational activities like canal maintenance. Furthermore many government posts were abolished and early retirement packages issued to superfluous staff. As a result the Agritex Nyanyadzi office had less general hands available to do the hard needed desilting of the pump sump of the new pumping station and cleaning of the main canal. In February 1994 both sources of water for Nyanyadzi irrigation scheme (Odzi pump station and Nyanyadzi main canal) were blocked with silt. Since the DWD was officially responsible for keeping the pump station operational, the new IM decided to direct his general hands at cleaning the blocked main canal from Nyanyadzi river. This decision produced disastrous results. Block A, B and D, relying heavily on Odzi river supplies, were only reconnected after three weeks. As a consequence the standing maize crop was lost. On 11 March 1994 a general meeting was held, where the District Administrator (DA) was asked to mediate between the angry farmers and Agritex. The new IM was scolded by farmers for not taking their interest at heart.[24] Equally so the IMC was scolded for not taking action. The DA observed that there had been a breakdown in communication lines. It was resolved that in future Agritex would keep in touch with the DWD Provincial head quarters and would inform the IMC in turn. The IMC would be tasked with calling emergency meetings with farmers to address any problems at hand (Bolding, 1996: 76-77).

As a result of these events and the continuing problems with the Odzi pump station in the following seasons, some form of cooperation between the IMC and Agritex staff ensued. During calamities joint meetings were held (for instance to discuss a new system of water rotations). In case Agritex needed more manpower to desilt the pump sump or main canal, the IMC would call general emergency meetings with the farmers to discuss the problem at hand together with Agritex (see Bolding, 1996).

Turn-over by Default: Devolution of Financial Obligations from 1996 Onwards
Towards the end of the 1990s two concurrent processes have been responsible for a *de facto* devolution of financial responsibilities from the Government to smallholder irrigators. The first process is that of the proposed privatisation of the DWD by the formation of the Zimbabwe National Water Authority (ZINWA). This authority will operate like a parastatal, sustaining itself by selling water from

government dams. In October 1995, the Provincial Water Engineer and the national newspaper the Herald announced that smallholder irrigation schemes were to pay for their water at the rate of the national blend price.[25] Agritex irrigation engineers held meetings and decided that these costs could not be borne by smallholder irrigators. However, in 1998 the ZINWA Bill was passed by Parliament.[26]

Secondly, the on going restructuring of the public service and increasing budget constraints under ESAP, resulted in a shortfall on the operational budget of the DWD. The electricity bills for the Nyanyadzi pumps could no longer be covered. The DWD budget allocation for the whole financial year (July 1995 – June 1996) only amounted to half the costs made during the first seven months. The shortfall (zim $ 187,980) was presented to Agritex. Agritex, however, could not bear these costs and suggested to DWD that the only way to pay the electricity company would be to recover the outstanding amount from the Nyanyadzi plot holders. However, the Provincial Water Engineer refused to come down to Nyanyadzi to explain the problem to the plot holders. Instead, he left Agritex personnel to do the job, since they had 'more experience with dealing with farmers'. In the end the IMC managed to collect zim $ 20,000 from plot holders in block A, B and D on the 4th of July 1996 to cover for the electricity bill from May to July 1996. The plot holders, however, contributed the money on the understanding that they would get it back: they thought it was only meant to help out DWD during the last months of the financial year. When it transpired that the electricity bills were to be paid by them at the end of each month, they refused to do so, arguing they could not bear the costs. The electricity company was then forced to close the pump station. Since then Nyanyadzi irrigation scheme has relied on water from Nyanyadzi river. Effectively this has split the scheme in two: block C plot holders who irrigate first, and block A, B and D who have to wait and hope some water has been left. Luckily 1996/97 and 1997/98 were above average rainfall seasons, allowing for some irrigation activities to continue. However, during the winter of 1998 the scheme was closed due to a lack of water in Nyanyadzi river.

The Politics of the Irrigation Management Committee The evidence provided above has pointed out that the IMC was very successful at resisting Agritex' efforts to impose the old management style. This was normally done through the mobilisation of local, district, provincial and ministerial politicians, who were afraid to affect their feeble electoral base in Nyanyadzi. Later on, the IMC started using the same mechanism to get government departments to bring about improvements to the scheme (i.e. construction of a new pump station). One may wonder what role the IMC was actually playing in operation and maintenance activities in the irrigation scheme, and what the future beholds for this formally constituted farmer organisation. Below a short enumeration of their different activities is given.

Ever since the demise of the Nyanyadzi cooperative society at the end of the 1980s, due to mishandling of funds, the IMC was tasked with negotiation of cash crop contracts (beans, tomatoes) with horticultural companies. The IMC, in close cooperation with the local Agritex office, signed the contracts, which specified the acreage to be grown, minimum price to be paid by the company and exclusive marketing rights of the company. However, individual farmers frequently breached the contracts by 'side-marketing'. The IMC, in most instances, failed to curb this practice, which in turn frequently resulted in the company not fulfilling its transport obligations. The only successful example of crop marketing by contract has occurred in block C, where an uncle of the local kraal head has been tasked to maintain contracts with a tomato canning company. This uncle organises the crop plan, signs the contract and acts as an intermediary during crop collection and pay-out. He receives a small commission on each pay-out for his work.

After the 1994 events at the pump house, the IMC was formally tasked with organising labour from all blocks in case of siltation problems at the main canal or pump sump. However, in most instances only few plot holders fulfilled their labour obligations. The kraal head for block C, however, did succeed in mobilising labour from his own block to desilt the main canal during two periods of two weeks each year. However, this resulted in an effective split of the scheme. During water shortage periods, or pump break-downs, when the whole scheme had to rely on the little water which was supplied by Nyanyadzi river, block C irrigators blocked the main canal, so as to satisfy their water needs first. Agritex and plot holders of other blocks did not accept this. But block C plot holders have continued to practise occasional blocking of the canal, arguing that they have the right to do so, since they also maintain the canal.

The IMC and block IMCs have never been able to enforce the bye-laws they drafted, due to a lack of authority. Bolding (1996: 78-79) describes a case of water theft. When the IMC chairman wanted to impose a fine, the perpetrator violently threatened the chairman. A meeting was called and the chairman tendered his resignation. Only after the Irrigation Manager had reconfirmed the fine and had demanded to see the culprit in his office, did the IMC chairman have the courage to continue in his post. Furthermore IMC meetings normally end in a brawl, with various people accusing each other of wrongdoings. The block IMCs and overall IMC have always provided the stage for different political factions to air their concerns and point fingers at their (political) opponents.

The IMC has never been fully involved in the actual water distribution activities in the scheme. Water distribution is done by the water bailiffs (gatekeepers), who are accountable to the local Agritex office. It is the bailiffs who translate the official list of plot holders, with at least 40% of the registered plot holders dead, into the real people on the ground. They know exactly which sons and widows have taken over the numbered plot of a deceased plot holder. They also know the intricacies of sub-leasing of plots, selling of plots and other deals that are made with regard to the use of a particular piece of irrigated land (see also Table 6.4). However, by reporting according to the official plot numbers and by using the names of the registered plot holders, the bailiffs keep the official Agritex reality alive.

Initially the IMC was not involved in emergency operations, like the repair and follow-up on the repairs of pumps. However, during winter 1995, which was characterised by frequent break-downs of the pumps at Odzi pump station, the local Agritex office discovered that the procedure to follow-up on pumps that had been taken for repairs by the Provincial DWD office was tedious. Instead of following the bureaucratic channels from their office, to the district office, to the provincial Agritex office, to the DWD provincial office, local Agritex staff started pressurising IMC members to go and see the Provincial Water Engineer themselves to enquire about the repairs of the pumps[27] (Bolding, 1996).

From the above the impression may be gained that local plot holders hardly undertake any operation and maintenance activities. However, this is not true. Rather, different local and urban-based organisations that consist of Nyanyadzi people have taken up tasks that foster the functioning of their scheme. Besides the involvement of the block C kraal head and his kin in desilting the main canal and organising crop contracts, there is the Nyanyadzi Advisory Committee On Development (NACOD) that actively lures donor organisations into the scheme. NACOD consists of wealthy sons and daughters of Nyanyadzi plot holders, who either run private businesses or work for NGOs in Harare and Mutare (provincial capital). After commissioning two feasibility studies in 1995 on the possible construction of a dam on the Nyanyadzi river (funded by an Irish donor agency), NACOD succeeded in securing zim $ 3 million for the lining of the main canal from Nyanyadzi river in 1997. The European Union funded lining was partially completed in 1998 and has reduced conveyance losses considerably. Most of the NACOD members originate from blocks A, B and D and are known in Nyanyadzi for their support to the opposition. Policy makers and local government agencies have hardly noticed these new, locally based initiatives. Yet, they do show the potential and capacity of local plot holders to fill the gaps in operation and maintenance, left by a retreating and cash strapped government. Furthermore, since strict government control over the scheme was relinquished a variety of livelihood strategies consisting of irrigated farming, dry land farming, labour migration, gardening and sub-leasing of plots, has emerged.

Musengezi Irrigation Schemes: Turnover by a Donor-Funded Experiment[28]

This section discusses a case of irrigation management reform through a formally constituted turnover experiment. This is a case study of five irrigation schemes in the Musengezi resettlement scheme in Zimbabwe. Each scheme abstracts water from the Mupfure River by means of a pump station. All schemes were designed by Agritex irrigation engineers, who during the late 1980s had clear preferences for sprinkler irrigation infrastructure over canal infrastructure, for reasons of its supposed higher water efficiency rates. Agritex and the Resettlement Officer selected the farmers from amongst all farmers in Musengezi resettlement area (Bourdillon and Madzudzo, 1994). Most notable are two farmers from Zvimba Communal area, who claim to be related to the President of Zimbabwe.[29] Firstly an outline of the funding and the establishment of the irrigation schemes is given. After this, four phases of operation are discussed distinguished on the basis of who pays for operation and maintenance.

Establishment of the Musengezi Irrigation Schemes The story of the Musengezi irrigation schemes is a story of an experiment in turnover without a clear plan of action on the part of its originators (Agritex). Hamilton Hills and Johanadale 1 irrigation schemes were established in 1989 financed through the 'National Farm Irrigation Fund' (NFIF) described above. The idea was that farmers would finance the purchase of infield irrigation equipment by borrowing from the NFIF administered by the Agricultural Finance Corporation (AFC), while government financed the main system irrigation and water delivery equipment. This was a distinct departure from the normal way of developing smallholder irrigation schemes in which smallholder farmers only contributed labour during construction (Makadho, 1990). In this experiment Agritex assumed that if farmers paid for the purchase of infield irrigation equipment, they would be more willing to take over the operation and maintenance of their irrigation schemes. However the idea of using the NFIF approach was abandoned unceremoniously in 1993 when the other three schemes (Johanadale 2, Shamrock 1 and Shamrock 2) were established with help of the 'Support to Smallholder Irrigation Project' (SSIP),[30] funded by DANIDA (Danish donor agency). Table 6.5 shows the irrigation schemes and the source of funding.

Table 6.5 Musengezi irrigation schemes size, number of plot holders and funding source

Name of scheme	Total area ha	Type of scheme	Number of plot holders	Plot Size/ farmer	Year commissioned	Funding
Johanadale 1	10.5	Drag-hose	21	0.5	1989	NFIF +SSIP
Johanadale 1 after rehabilitation	10.5	Drag-hose	7	1.5	1993	SSIP
Johanadale 2	28.5	Semi-portable	19	1.5	1994	SSIP
Shamrock 1	28.5	Semi-portable	19	1.5	1992	SSIP
Shamrock 2	33	Semi-portable	22	1.5	1992	SSIP
Hamilton Hills	8	Drag-hose	16	0.5	1989	NFIF +SSIP
Hamilton Hills after rehabilitation	8	Drag-hose	4	2	1993	SSIP

Source: Agritex Chegutu district files, 1994.

Two main developments necessitated the unceremonious departure from the NFIF approach. Firstly Johanadale 1 and Hamilton Hills collapsed in 1993 following the 1991-92 severe drought (the worst in living memory for Zimbabwe). In October 1992 the State President visited the irrigation schemes on his 'meet the people' tour of the country to assess the effects of the drought. In his wide ranging speech he indicated that the Government was committed to provide irrigation infrastructure to smallholder farmers free of charge under the 'one dam, one irrigation scheme per District per year programme'. This created some animosities amongst the Musengezi irrigators and Agritex, since for the establishment of their scheme, Agritex gave them only the NFIF option. The farmers therefore felt cheated by Agritex. The President further promised the Musengezi farmers a grinding mill and new electric motors for their pumps.[31] Secondly Agritex adopted the policy of full time irrigation in smallholder irrigation scheme development. The aim was to give smallholder farmers larger irrigated plots (from 0.5 ha to 1.5 ha) but with no rain-fed plots. The assumption was that with only the irrigated plot, the farmers would be more committed to irrigated crop production and thus improve production. Agritex therefore introduced a two-year period of farmer acclimatisation in the establishment of smallholder irrigation schemes. During this period government assisted the farmers with the payment of all operation and maintenance costs for the operation of their irrigation scheme. The SSIP funds were used to achieve this purpose.

When Hamilton hills and Johanadale 1 where rehabilitated in 1993, the above changes were taken into consideration. This resulted in DANIDA paying off the outstanding NFIF loan for the farmers. However, the decision to pay off the outstanding NFIF loan on behalf of the farmers brought some confusion among the farmers. Although the loan was given to the farmers as a group, on repayment AFC equally divided the loan amongst the farmers and deducted this by stop order from the individual farmer's sales to the Grain Marketing Board (GMB) and the Cotton Marketing Board (CMB). The farmers were at different stages of repayment. However DANIDA simply paid the lump sum outstanding loan to the AFC with no consideration to the payments by the individual farmers. This confused the farmers with those who had paid more crying foul and feeling cheated by those who had paid less. DANIDA promised the farmers new equipment in return, but the new equipment was never delivered.

Operation & Maintenance of the Irrigation Schemes Four phases can be distinguished based on how the operation and maintenance activities at the irrigation schemes were financed (see Table 6.6). The main operation and maintenance requirements for each scheme can be captured in four main categories. Firstly funds are required to pay for the electricity bills generated by operating the pumps. Secondly during each cropping season seeds, fertiliser and chemicals have to be financed in order to be able to implement the cropping programme. Thirdly labour to run the pumps and cultivate the land is required. And finally maintenance of the main system equipment, in particular pump repairs, in case of a breakdown, has to be financed and at field level worn out sprinklers and other parts have to be replaced.

Phase One: Part-Time Irrigators Finance their Own Scheme (1989 – 1993)
During phase one the farmers borrowed money to finance electricity bills and crop inputs from the AFC as individuals. The crop-input loan also covered inputs for the five hectare rain-fed plot. They also borrowed as group to finance the infield

equipment as discussed above. One condition for access to AFC by a smallholder farmer was that Agritex approved the farmer's cropping programme. The idea was that by so doing only viable cropping programmes were financed and thus loan repayment at the end of the growing season was guaranteed. However, due to poor monitoring, as soon as the AFC loan was secured farmers proceeded to grow crops they wanted (not necessarily those specified in the cropping programme). Important to note is that during this phase, the farmers were not given a loan for the maintenance of the pumps and irrigation equipment. As a result the farmers did not follow the routine maintenance specified for pumps and the irrigation equipment.

Table 6.6 Summary of the phases and source of funds for operation and maintenance

Phase	Period	Operation		Maintenance	
		Electricity	Crop inputs	Main system	Infield
One	1989/90	AFC	AFC	Farmers	Farmers
	1990/91	AFC	AFC	Farmers	Farmers
	1991/92	AFC	AFC	Nil	Nil
Two	1992/93	DANIDA	AFC	DANIDA	Farmers
	1993/94	DANIDA	AFC/Farmers	DANIDA	Farmers
	1994/95	DANIDA/ Farmers	AFC/Farmers	Nil	Farmers
Three	1995/96	MP / Credit contract with ZESA	Canpack/ Zimfreez	Farmers	Farmers
	1996/97	Bonduelle / Natbrew	Bonduelle / Natbrew	Farmers	Farmers
	1997/98	Credit contract with ZESA	Whole Sale Fruters / farmers	Whole Sale Fruters / farmers	Farmers
Four	1998/99	DDF	Farmers	DDF	DDF

Source: Zawe (1999).

The situation was further complicated when in 1992 the country experienced its worst drought in living memory. This affected repayment from the irrigated plot. Even if the farmers made money from the irrigated crop, some of the money was used to service the rain-fed loan. The farmers failed to fully service their debts and they became defaulters. The combined result was that after the 1992 drought, the AFC was not prepared to issue any loans to finance operation and maintenance of the schemes. The local Agritex office however came to the rescue of the farmers by luring the local DANIDA co-ordinator at Agritex head office into clearing the outstanding AFC loan for infield works. This was a recipe for disaster as farmers soon developed a habit of relying on other people's money.

Phase Two: a Danish Donor Agency Helps Out (1993 – 1995) During this phase DANIDA paid the electricity bills and the maintenance of the main system, while the farmers paid for the input costs and the maintenance of the infield equipment. Some farmers from Johanadale 1 and Hamilton Hills could not secure loans from AFC for crop inputs, since they had not serviced previous loans for crop inputs. AFC could only fund programs based on assured markets. But since farmers wanted to grow high value crops, which were not centrally marketed like maize and cotton, the AFC demanded that farmers sign crop production contracts with food processing companies like Olivine Industries and Canpack for the growing of Michigan-pea beans and French-fine beans to ensure that there was ready market for the produce. The AFC hoped that the farmers would service their debt once the crop was marketed. Three important developments occurred towards the end of this phase. Firstly, in April 1995 DANIDA unexpectedly decided not to extend the SSIP programme, leaving the farmers with no funds for the payment of electricity. Secondly, 1995 was a dry year and as a result there was shortage of water during the month of October 1995. Thirdly, AGRITEX decided it was time to leave the farmers to pay for all the operation and maintenance costs, as there were no more funds available to assist the farmers.[32] As a result most of the farmers failed to service their loans with AFC. The farmers had no choice but to look elsewhere for financial support. This paved the way for the entry of the local MP into centre stage.

Phase Three: Government Withdraws, Horticultural Companies Move In (1995 – 1998) The MP negotiated with the power authority ZESA for a power supply credit contract on behalf of the farmers. Since the farmer organisation was not legally constituted, ZESA agreed to the credit contract only when the MP agreed to guarantee the payment himself. At the end of the 1996-1997 season, farmers failed to pay the electricity bills. ZESA demanded payment from the MP who promptly paid zim $ 400 per farmer (totalling zim $ 28,400). This amount was paid around the same time that the MP received his pay-out from the War Victims Compensation Fund.[33]

For crop inputs the farmers depended on crop contracts with horticultural processing companies and own savings for grain and fibre crops like maize and cotton. For the maintenance of the irrigation system from the main system to the infield equipment, the farmers depended mainly on their own savings, except for one scheme (Johanadale 2) that was assisted once by the horticultural processing Company Whole Sale Fruters. Table 6.7 presents the details of the contracts. The contracts normally entailed free provision of seed and other inputs (fertiliser and chemicals), though in some occasions the farmers also managed to secure pre-financing of the ZESA pumping bill. However, the farmers never managed to reap a good crop and income due to various reasons. This forced them to look for a new company and new crop contract every season. The 1995-96 summer season contract with Canpack ended in a disappointment, due to late picking of the beans and strict grading, resulting in only one third of the crop being marketed in the end. The 1996 winter season contract with Zimfreez proved to be an outright racket. Zimfreez, a government owned company, failed to collect and pay for the crop, since it went broke. The farmers were then swindled of the whole crop by one of the company representatives, who formed his own briefcase company and took off with the crop proceeds without ever paying the farmers. The 1996-97 summer season contract with Bonduelle ended in controversy. The farmers over-committed themselves by borrowing too much money from the company. The company employed two field assistants to advise farmers on proper husbandry of the crop. However, at the end of the season the farmers were expected to pay these assistants. Moreover the company re-graded the crop proceeds at their factory resulting in a lower pay-out than the outstanding amount the farmers had borrowed. Finally, the summer 1997-98 contract with Whole Sale Fruters also ended in controversy. The company over-committed itself by repairing the pump set for Johanadale 2 scheme. It recovered its money from the sales of the rest of the irrigation schemes. Also, some farmers did not use the inputs provided by the company to grow baby corn, but instead used these for growing their grain crop.

Table 6.7 Summary of the crop contracts during phase 3

	Summer				Winter			
	Company	Crop	No. of farmers	Area/ farmer (ha)	Company	Crop	No. of farmers	Area/ farmer (ha)
1995-96	Canpack	Fine-beans	71	0.5	Zimfreez	Peas	71	0.5
1996-97	Bonduelle	Fine-beans	71	0.5	Natbrew	Barley	71	1
1997-98	WSF	Baby-corn	71	0.5	Nil	Nil	71	nil

However, by winter 1998 the farmers had exhausted all possibilities for entering crop contracts with new horticultural companies. They had to look for other ways to keep their schemes operational.

Phase Four: Politicians Seeking to Extend Their Political Mileage (1998 – 1999)
By 1998, due to a lack of repairs, the pumps in all five-irrigation schemes started to experience major break-downs requiring large sums of money. Also the power credit contract with ZESA collapsed when the farmers failed to repay the MP at the end of 1998. The result was that ZESA cut off the power supply in January 1999. The farmers then approached the provincial governor for support. However, the local MP and the Provincial Governor were two politicians fighting for control over the provincial leadership of the ruling party ZANU PF. The situation soon became a political battlefield in which the two were using their power and positions to influence government departments, financial institutions and other service organisations to assist the farmers in a bid to extend their political mileage.[34] Instead of a co-ordinated approach to the problem each of the two

politicians decided to look for assistance from different government departments and organisations to help the farmers. This resulted in further confusion among the farmers and government departments.

The MP decided not to use his money this time. But not wanting to be outwitted by the governor, he approached the District Agritex office for an inventory of the irrigation equipment needing replacement at the irrigation projects. He also asked Agritex to develop a cropping programme capable of paying off the costs of replacement. The Agritex officer provided the inventory and the cropping programme. Using the inventory, the local MP then approached the Commercial Bank of Zimbabwe (CBZ)[35] to negotiate for a loan on behalf of the farmers. Parallel to this the MP approached the District Development Fund (a donor funded quasi-government organisation tasked with rural development) to source for financial and material help for the resuscitation of the irrigation projects. The MP knew that DDF had many diesel-powered pumps imported from the USA lying idle at their provincial and national offices, which could be utilised by the irrigation projects. He also could make use of his relations with the Minister of Water Resources and Rural Development, who was the minister responsible for DDF. It could not be established during the research period under what program the pumps were imported. CBZ however rejected the proposal, citing it as being too ambitious. Some farmers were voicing concern as to the wisdom of borrowing from CBZ at even higher (commercial) interest rates when they failed to pay AFC at much lower rates. DDF however was swift. They brought the pumps and paid off all the outstanding ZESA bills. However the DDF diesel powered pumps were not of the required head. As a result DDF took all the pumps for service. However disaster resulted when the funds committed to the DDF operational budget were redirected to other government needs in October 1999. All the operations of DDF were halted and as result the pumps sent for service to an Irrigation Company in Chegutu could not be paid for.

Meanwhile the governor approached the Agritex provincial office and demanded that something be done to solve the plight of the farmers at the irrigation schemes. The governor was aware of the SISP Irrigation rehabilitation and irrigation management reform programme that the government was launching through Agritex, with funds supplied by IFAD and DANIDA (see section 6.4). The governor requested that the irrigation schemes be included in the SISP programme. As a result three of the 5 irrigation schemes were included. The other two however could not be included since the SISP programme works on a maximum of three projects per District. One of these however managed to sign a contract with a vegetable vendor at the main open market in Mbare in the capital city Harare. The vendor paid for the repairs of the pump, the crop inputs and the electricity bills. The farmers grew the vegetables and marketed to the vendor who recovered his money from the produce sold to him. The vendor provides the transport to carry the vegetables to market. The fifth scheme has ceased operating.

Turnover or Dependency Syndrome? The Musengezi case shows what can happen when irrigation management turnover is executed without a clear plan of action. In this case for example farmers that actually took the experiment with the NFIF loan facility seriously and serviced their loans for in-field equipment, were punished in the end by a donor agency that paid off the outstanding loan irrespective of being a defaulter or good creditor. The turnover experiment soon amounted to a game of finding the right 'saviour' to bail the farmers out on operation and maintenance costs when the acclimatisation period was unceremoniously ended. Horticultural companies and politicians took turns in trying to salvage the situation but to no avail. The farmers seem to have developed a collective survival strategy that is commonly known in Zimbabwe as the 'dependency syndrome'. The schemes with their heavy reliance on pumps were doomed to be financially unviable, and thus unsuitable for an experiment in turnover. The political career of the local MP has set the scene of what happened at the schemes since 1995. The reasons for his and the Governor's interest in keeping the schemes alive can only be speculated at. Is it that both politicians try to improve their personal relationship with President Mugabe over the heads of the Musengezi irrigators? Or have the Musengezi schemes become an 'island of salvation', a political showcase for the electorate of two fighting politicians?

Conclusions and Issues for Debate

In a conventional perspective public policy should be accompanied by a clear statement on objectives, means to achieve those objectives and clearly spelled out implementation plans that are known by all relevant people involved (be they government officials or farmers). From such a perspective it can be safely concluded that irrigation management turnover in Zimbabwe is characterised by quarter and half measures. What has become blatantly clear in the course of this paper, is that there is and has been no elaborated policy on irrigation management turnover in post-independence Zimbabwe. As a consequence there is also no need to talk of implementation or instrumental failure of the policy. The analysis could stop there and recommend that a coherent policy framework be designed, adopted and implemented.

But the Zimbabwe case is more interesting than this. The framework for policy analysis we sketched in section 2 made us look beyond the mere existence of a policy statement on smallholder irrigation. From that perspective it would be observed that a policy discourse has developed that stresses the need for turnover for reasons of supposed better performance under farmer management, and most importantly, for reasons of improved cost recovery on operation and maintenance expenditures required to keep the schemes running. The discourse is silent on rights and responsibilities of farmers. The main interest of government has always been limited to that of cost recovery. This policy discourse has been given shape through various policy papers (government initiated and donor initiated), research projects, donor initiated stakeholder workshops and donor funded pilot projects. However, this emergent policy discourse has never been cemented in official law, public policy statements endorsed by the cabinet or implementation guidelines for use by the relevant government departments.

This is not to say that government staff in the field has not frequently called for a clear-cut policy on smallholder irrigation. Despite this lack of public policy, turnover of smallholder irrigation systems in Zimbabwe has occurred. From the mid-1990s onwards the government could no longer live up to its operation and maintenance obligations and a process of turnover by default was triggered. A declining economy combined with an increasing impact of the economic structural adjustment programme, embarked on in 1991, forced the government to devolve its financial responsibilities to smallholder irrigators. This happened more or less over-night. As a result many pump operated smallholder irrigation schemes have been forced to close, since irrigation farmers have not been able, nor been prepared, to shoulder the high recurrent operation and maintenance costs.

Lacking a public policy prescription, it becomes important to look at policy practice. The question then becomes: what strategies have been applied by the various actors, involved in smallholder irrigation, to further their interests and shape public policy processes? This paper has tried to analyse policy practice by presenting two detailed case studies on smallholder irrigation management. From these case studies we can derive a number of observations on turnover policy in practice.

Local DERUDE (and subsequent Agritex) staff after independence was preoccupied with reviving the colonial management style which is characterised by strict centralisation of control over irrigation operations in the hands of the Irrigation Manager. This should come as no surprise, since they had been the subordinates of their colonial predecessors, quickly promoted for the sake of speeding-up the process of Africanisation of the public service. Their emphasis was on cost recovery and strict implementation of control regulations. The newly formed Irrigation Management Committees were seen as the long arm of the state: advisory in nature and mainly devised to enforce discipline amongst fellow farmers. The commitment to gradual turnover of management to farmers, reflected in the unofficial DERUDE policy paper, were re-interpreted by local staff on the ground. The explicit emphasis in the DERUDE document of three annually renewable permits, that had formed the corner stone of the colonial management policy, allowed government staff in irrigation schemes to view the policy as a continuation of the old line of command. However, the required policy back up for implementing strict measures on cost recovery and control over farming operations, was found to be lacking in independent Zimbabwe.

Post-independence politics in Zimbabwe have been characterised by a commitment to subsidising the smallholder farmer sector under the banner of a social spending programme inspired by a Marxist-socialist orientation. This should come as no surprise, since the whole war of liberation had been led under the motto to return the lands to their legitimate owners. Part of the promise of ZANU(PF) to its electoral base in the rural areas was to do away with restrictive measures that had been imposed on the African majority by the settler minority. The freedom fighters, operating in small-holder irrigation schemes, often emphasised during their nocturnal rallies that people should refuse to pay water rates: after independence they wouldn't be imposed anyway. There would be free education, free farming, free access to river water and free irrigation services. After independence these promises were not forgotten by freshly elected Members of Parliament, who kept electoral gain at heart especially in areas that were renowned for their electoral support to opposition parties (Save valley in particular). This led to a game of hide and seek, which was fully exploited by irrigation farmers. Whenever DERUDE or Agritex staff resorted to strict imposition of irrigation regulations, farmers quickly mobilised local and even national politicians to back their cause. This has led to an intense feeling of disillusion and disorientation within the government agencies responsible for smallholder irrigation.

Irrigation farmers, on the other hand, have exploited this political vacuum in the irrigation schemes. A process of re-appropriation of smallholder irrigation schemes has taken place. A variety of livelihood strategies, based partly on irrigated farming, partly on dry land cultivation, migrant labour, gardening, and sub-leasing of plots, have emerged. The Irrigation Management Committees have proven largely ineffectual in mobilising farmers or enforcing irrigation bye-laws. However, they have proven to be very effective platforms to mobilise irrigation farmers against strict government control over their operations. One of the persistent aims of the farmers has been to postpone financial devolution of operation and maintenance obligations. In the Musengezi schemes farmers have proven themselves very effective in finding outside funding for sustained operation (whether from donors, horticultural companies or ambitious politicians). Yet, one should not underestimate the capacity of local farmers to maintain and even improve their scheme, as has been shown for the Nyanyadzi case. What is remarkable, however, is that the officially constituted Irrigation Management Committee is not utilised for such initiatives. Rather it is kin-based networks of local and urban kind that have mobilised new and viable avenues for future operation of the scheme (i.e. canal cleaning, canal lining, sourcing of funds for construction of a dam). Unfortunately, policy makers and local government agencies have been blind to these initiatives, suffering from either a 'technical fix' (trying to foist an expensive pump set on the farmers) or a 'disciplinary control' orientation.

Donor agencies have played only a minor role in changing this policy reality on the ground. Whilst their intentions were often directed at increased farmer participation in operation and maintenance activities, their approach has often been piecemeal and lacking in persistence. The brief involvement of DANIDA in the turnover experiment in Musengezi is a good example. The biggest contribution of donor agencies probably lies in the development of an elaborate policy discourse on smallholder irrigation, as is reflected in a stream of policy papers, feasibility studies, irrigation strategies and workshop proceedings.

This brings us to another observation: we can distinguish two separate policy arenas that hardly ever link up. On the one hand there is the policy discourse arena, which produces new policy initiatives and pilot projects on a regular basis. Agritex and DWD head office staff have been heavily involved in this arena. Donor agencies have shown a preference for hiring directorate level staff to do consultancy assignments related to policy formulation or project formulation. Despite this involvement of top government officials, these initiatives have not resulted in concerted efforts by the Zimbabwe government to formulate a clear policy on smallholder irrigation, let alone a policy on turnover of irrigation management. On the other hand, there is the local arena on the ground, which is characterised by a lack of policy guidelines and a lack of political backing for government staff. Government staff have been left to their own devices in trying to operationalize some kind of irrigation policy. The few occasions where these two policy arenas link up are formed by donor-funded stakeholder workshops. These are often characterised by hefty confrontations between Agritex field staff and the Agritex directorate. Whilst field staff often calls for an irrigation policy to be effected, the directorate often claims such policy is already 'in the bag'. Irrigation farmers hardly play a role during such stakeholder consultations. Their views have been summarised in the consultancy report, based on rapid rural appraisals. During the actual stakeholder workshop, which is normally held at a conference resort in Harare, normally very few farmers attend.

That leaves us with one of the main questions raised by this paper: Why has there been no concerted effort at policy making with regard to smallholder irrigation? Here we can only speculate at some of the root causes. More research needs to be done to establish the exact reasons and motives. One contributing factor has been that despite the various changes in government departments involved in the smallholder irrigation sector, staff on the ground has remained largely of the same 'blood-group'. These were mainly former DEVAG staff, who had been trained and experienced in the old management style. New irrigation officers who have been trained abroad and been exposed to different ideas on farmer management, started infiltrating the department only during the late 1980s and 1990s. However, these are the same staff that has left government employment, looking for greener pastures in the private sector or donor circuit. Also, senior government officials from head offices in Harare might not be all that interested in coming up with a clear government policy on smallholder irrigation. Such a policy would preclude highly lucrative donor initiatives at formulating a new policy. That might also explain why the directorates of Agritex and DWD have failed to lobby for the necessary political support for their staff on the ground. On the other hand, the ruling party and central government have never been keen on losing control over the countryside. The resettlement policies of the 1980s are characterised by a high level of central government control over farming operations. Incidentally, the same annually renewable permits for cultivation, residence and depasturing livestock that were used in colonial irrigation settings, have been implemented in resettlement schemes. Another contributing factor is that irrigation farmers have not organised themselves to lobby for a clear government policy. Before the turnover by default was brought about due to acute fiscal problems, there was also no incentive for irrigation farmers to organise themselves. Since the financial devolution was effected in 1996, some irrigation farmers have organised themselves in federations in order to protest against the sudden devolution of financial responsibilities. These federations, e.g. the Save valley irrigation schemes plot holder federation and the Musengezi schemes federation, provide new alternatives for donor and government agencies to turn to in formulating irrigation policy.

We would like to conclude our paper on a high note. Despite the obvious lack of government policy on irrigation reform, the sector has proven highly dynamic in finding a way for the future. The financial bankruptcy of the government, which resulted in the sudden turnover process, has put the irrigation farmers in the driving seat. It seems to us that this is an opportune moment to go back to the drawing board. If the onus of policy development is put on the actual users of the irrigation facilities, viable policy initiatives are likely to emerge.

Notes

1 Of interest here is to observe that one of the authors and main researchers behind this study was the then director of Agritex, the government agency responsible for smallholder irrigation schemes in Zimbabwe.
2 Rhodesian officials of the Ministry of Internal Affairs often flirted with the idea to do away with plot holders totally and run African irrigation schemes as state ventures with hired African tenants. This model of tenant irrigation had proven very successful in other African colonies according to these officials (the Gezira in the Sudan and Mwea in Kenya). For a critique of the 'success' of this tenant scheme approach, see Barnett (1977) and Chambers (1969).
3 The key word here is 'discipline'. Rhodesian officials stressed the need for discipline as an essential ingredient in the execution of successful irrigation operations.
4 For a more elaborate discussion on how the white settler state 'invented' and 'manipulated' traditional authority structures, see Holleman (1969), Weinrich (1971) and Ranger (1983). During the community development drive of Rhodesia under the Smith regime, tribal authorities were re-created to serve as leaders of 'natural' communities.
5 Most probably the policy was formulated at directorate and provincial level by a mixture of officers that originated from DEVAG and CONEX (the white farmers extension service), since the document reflects a mixture of old ideas and new initiatives. Note that the director and deputy director of DERUDE later moved in to become successive directors in Agritex, in 1985 and 1988 respectively. These two were the prime movers behind the subsequent take over of smallholder irrigation schemes by Agritex in 1987.
6 During a training on irrigation management for Agritex staff in irrigation schemes in 1995, parts of the DERUDE policy document were used and presented as government policy. However, the part addressing turnover to farmers was left out of the curriculum, whilst the part on the three annual permits and eviction order form were discussed at length (field notes Bolding, February 1995).
7 During the liberation struggle many chiefs and headmen were regarded collaborators of the Smith regime and subjected to attack by freedom fighters (Ranger, 1985; Kriger 1988).
8 Development planning and decision making was done in District Development Committees and Provincial Development Committees that consisted of government bureaucrats and were almost void of elected representatives.
9 The old schemes in the Save valley in South-Eastern Zimbabwe were all constructed during the 1930s and 1940s as a means to combat famine, modernise African production methods, and relieve population pressures in other areas. The schemes are Mutema (started in 1932), Nyanyadzi (1934), Maranke (1937), Chakohwa (1937), Chibuwe (1944) and Devure (1945) (Roder, 1965).
10 This move towards democratically elected structures of local governance was one of the policy priorities of the new ZANU(PF) government. The new structure of local governance was intended to sideline the traditional leaders (Chiefs and headmen) who were accused of being collaborators of the Smith regime during the war (Ranger 1985,

Kriger 1988). Since most of the members of the new committee were drawn from the local ZANU(PF) party branch, it was also a move to facilitate control of the ruling party over an area that was known for its support to an opposition leader (Ndabaningi Sithole).

11 Report of meeting at Nyanyadzi irrigation scheme, dd 8 May 1980, PC Files.
12 The District Administrator, as the legitimate land holding authority in communal areas, has to second the application for a water right by a communal farmer.
13 Report by IM Sithole on farmers committee meeting with all plot holders in Nyanyadzi, dd 10 June 1987. Agritex Provincial Files.
14 Letter from acting Provincial Agricultural Officer to the acting Provincial chairman ZANU(PF), dd. 7 October 1987.
15 Report by Provincial Agricultural Officer Manicaland to the District Agricultural Officer for Chimanimani, dated 19 May 1989. Agritex Chimanimani District files.
16 Report by IM Sithole to the Provincial Agricultural Officer Manicaland, dated 18 October 1989. Agritex Provincial files.
17 Interviews with Agritex Chimanimani district office staff, Field notes Alex Bolding, 1995.
18 Nyanyadzi was renowned for its nationalist activities during the period of open African nationalist politics (1958-1964). During the liberation war most people remained loyal to Ndabaningi Sithole, the disposed president of ZANU (1975, when Mugabe took over the leadership). Many people in Nyanyadzi vote ZANU-NDONGA, the opposition party led by Sithole. The Zimbabwe Unity Movement (ZUM) party led by former ZANU(PF) secretary general Edgar Tekere also had a large following in Nyanyadzi running up to the 1990 general elections.
19 Letter from MP Chimanimani to the Chief Agricultural Extension Officer (CAEO) for Manicaland, dd 17 December 1991, requesting for update on commissioning of raw water pumps at Nyanyadzi. Letter from MP Chimanimani to CAEO Manicaland, dd 22 January 1992, demanding information on problems at various Save valley irrigation schemes and progress on Nyanyadzi pumps. Letter by Chimanimani District Party Commissioner to the DAEO Chimanimani, dd. 13 May 1992, requesting withdrawal of maintenance fees on account of drought.
20 Letter from the Principal Agricultural Extension Officer (Technical) for Manicaland, to Irrigation Manager Nyanyadzi, dd. 11 February 1993.
21 The word *Chef*, as used in Zimbabwe, signifies somebody of great importance. Superiors or politicians are popularly addressed to as *Chefs*.
22 Report by Irrigation Manager Nyanyadzi on staff meetings on 17 and 23 June 1993, Agritex Chimanimani district files.
23 Letter from Irrigation Manager Nyanyadzi, to the District Officer Chimanimani, dd 23 October 1993, Agritex Chimanimani district files.
24 The IM had already received an anonymous letter by Nyanyadzi farmers requesting his removal from the scheme. Later some women stripped naked in front of him singing that they would make a coffin for him. Report on Nyanyadzi staff meeting, dd 28 July 1994, Agritex Chimanimani District Files.
25 The national blend price was at the time zim $47 per mega litre (1,000 m3). By 1997 it was increased to zim $185 per ML. Nyanyadzi irrigators require on average 18 ML per year per hectare for three cropping seasons.
26 At the close of the period that this paper covers (November 1999) ZINWA had not yet become operational.
27 Partially this action was the effect of mounting frustrations of the Irrigation Officer regarding the non-fulfilment of obligations by DWD. He was fed-up with being blamed

by plot holders for responsibilities that were formally not those of his office. In a letter to his superior, the Irrigation Officer refers to the repeated failure by the DWD to supply manpower to desilt the pump sumps: 'This office (Agritex Nyanyadzi) is being seen as the most devil which has ever existed in the history of the scheme, whilst nothing ill is being spoken of the very people who are responsible (meaning DWD)'. Letter dd 20 June 1994, Agritex Chimanimani District files.
28 The information presented in this section is based on MSc research work done by Conrade Zawe during 1999.
29 Zvimba communal area is the home area of President Mugabe.
30 The DANIDA 'Support to Smallholder Irrigation Project' provided manpower, funds to rehabilitate small-scale irrigation projects and equipment for the newly (1992) established irrigation construction units within Agritex.
31 This created further confusion amongst the farmers. They never saw the grinding mill. It is alleged to have been misappropriated by an Agritex staff member.
32 This was announced by the local Agritex officer during a meeting at the start of the 1995-96 growing season.
33 The War Victims Compensation Fund was a state administered facility that catered for mental and physical damages incurred during the liberation struggle. The MP was a former freedom fighter. In 1997 a scandal broke out surrounding the actual assessment of victims. Some Ministers and other high-ranking officials were found to have claimed over 100% disability.
34 It seems that the two Musengezi farmers who claimed to be related to the President played an important role in triggering this political tug-of-war between the Governor and the MP.
35 It must be noted here that CBZ is a government managed bank and the local MP was trying to use his influence to secure a loan from the bank on behalf of the farmers without collateral security.

References

Alexander, J. (1994), 'State, peasantry and resettlement in Zimbabwe', *Review of African Political Economy*, Vol. 21(61), pp. 325-45.
Alvord, E.D. (1958), *Development of native agriculture and land tenure in Southern Rhodesia*, Unpublished manuscript, Salisbury.
Anderson, James E. (1997), *Public policymaking: An introduction*, Houghton Miflin Company, New York.
Barnett, T. (1977), *The Gezira scheme. An illusion of development*, Frank Cass, London.
Beinart, W. (1984), 'Soil erosion, conservationism and ideas about development: a southern African exploration, 1900-1960', *Journal of Southern African Studies*, Vol. 11(1), pp. 52-83.
Bolding, A. (1996), 'Wielding water in unwilling works. Negotiated management of water scarcity in Nyanyadzi irrigation scheme, winter 1995', in E. Manzungu and P. van der Zaag (eds), *The practice of smallholder irrigation. Case studies from Zimbabwe*, University of Zimbabwe Publications, Harare, pp. 69-101.
Bolding, A. (1999), 'Caught in the catchment. Past, present and future of Nyanyadzi water management', in E. Manzungu, A. Senzanje, and P. van der Zaag (eds), *Water for agriculture in Zimbabwe. Policy and management options for the smallholder sector*, University of Zimbabwe Publications, Harare, pp. 123-52.

Bolding, A. (2003), 'Alvord and the demonstration concept. Origins and consequences of the agricultural demonstrator scheme, 1920-44', in: A. Bolding, J.K. Mutimba, and P. van der Zaag (eds), *Agricultural intervention in Zimbabwe; new perspectives on extension*, University of Zimbabwe Publications, Harare, pp. 24-52.

Bolding, A., Manzungu, E. and van der Zaag, P. (1996), 'Farmer managed irrigation furrows in the Eastern Highlands, some preliminary observations', in E. Manzungu and P. Van der Zaag (eds), *The practice of smallholder irrigation; Case studies from Zimbabwe*, University of Zimbabwe Publications, Harare, pp. 191-218.

Bolding, A., Manzungu, E. and van der Zaag, P. (1999), 'A realistic approach to water reform in Zimbabwe', in E. Manzungu, A. Senzanje, and P. van der Zaag (eds), *Water for agriculture in Zimbabwe. Policy and management options for the smallholder sector*, University of Zimbabwe Publications, Harare, pp. 225-53.

Bourdillon, M.F.C. and Madzudzo, E. (1994), 'Small-scale irrigation schemes in Zimbabwe', Unpublished manuscript, DANIDA, Harare.

Chabayanzara, E. (1994), 'Smallholder irrigation development: impact on productivity, food production, income and employment', in S.A. Breth (ed), *Issues in African development 2*, Winrock International, pp. 185-98.

Chambers, R. (1969), *Settlement schemes in tropical Africa. A study of organisations and development*, Routledge and Kegan Paul, London.

Chavunduka Commission (1982), *Report of the commission of inquiry into the agricultural industry*, Government printers, Harare.

Chitsiko, R.J. (1988), 'Irrigation for agricultural extension workers', Agritex training branch, Harare.

Chitsiko, R.J. (1995), 'Irrigation development in Zimbabwe – constraints and opportunities', Paper presented at ZFU workshop water development for diversification within the smallholder sector, Harare.

DERUDE (1983), *Policy paper on small scale irrigation schemes*, Department of Rural Development, Ministry of Lands, Resettlement and Rural Development, Harare.

Drinkwater, M. (1989), 'Technical development and peasant impoverishment: land use policy in Zimbabwe's Midlands Province', *Journal of Southern African Studies*, Vol. 15(2), pp. 285-305.

Drinkwater, M. (1991), *The state and agrarian change in Zimbabwe's communal areas*, MacMillan, London.

FAO (1990), *Irrigation subsector review and development strategy. Technical report and Annexe*, Food and Agriculture Organisation, Harare.

FAO (1994), *Irrigation policy and strategy*, Republic of Zimbabwe and Food and Agriculture Organisation, Harare.

GKW Consult (1985), *Rehabilitation and development of small scale irrigation schemes in communal land. Volume II: sector review. Final report April 1985*, GKW Consult, Mannheim, Hunting Technical Services Ltd, London, and Brian Colquhoun, Hugh O'Donnel and Partners, Harare.

Government of Zimbabwe (1981), *Growth with equity: an economic policy statement*, Government printers, Harare.

Grindle, M.S., and Thomas, J.W. (1991), *Public choices and policy change: The political economy of reform in developing countries*, The John Hopkins University Press, Baltimore.

Holleman, J.F. (1969), *Chief, council and commissioner. Some problems of government in Rhodesia*, Oxford University Press, London.

Howlett, M. and Ramesh, M. (1995), *Studying public policy: Policy cycles and policy subsystems*, Oxford University Press, Oxford.

Hughes, A.J.B. (1974), *Development in Rhodesian Tribal Areas, an overview*, Tribal Areas of Rhodesia Research Foundation, Salisbury.
Hunt, A.F. (1958), *Manicaland irrigation schemes, economic investigations*, Department of Native Economics and Marketing, Salisbury.
Kerr, Donna H. (1976), 'The logic of 'policy' and successful policies', *Policy Sciences*, Vol. 7, pp. 351-63.
Kriger, N. (1988), 'The Zimbabwean war of liberation: struggles within the struggle', *Journal of Southern African Studies*, Vol. 14(2), pp. 304-22.
Long, N. and van der Ploeg, J.D. (1989), 'Demythologizing planned intervention: an actor perspective', *Sociologia Ruralis*, Vol. 29 (3-4), pp. 226-49.
Makadho, J.M. (1990), *The design of farmer managed irrigation systems: experiences from Zimbabwe*, ODI/IIMI Irrigation Management Network Paper 90/1d.
Makadho, J.M. (1994), *An analysis of water management performance in smallholder irrigation schemesin Zimbabwe*, Unpublished PhD thesis, University of Zimbabwe, Harare.
Makarau, A. (1999), 'Zimbabwe's climate: past, present and future', in E. Manzungu, A. Senzanje, and P. van der Zaag (eds), *Water for agriculture in Zimbabwe. Policy and management options for the smallholder sector*, University of Zimbabwe, Harare, pp. 3-16.
Makombe, G., Makadho, J.M. and Sampath, R.K. (1998), 'An analysis of the water management performance of small holder irrigation schemes in Zimbabwe', *Irrigation and Drainage Systems*, Vol. 12, pp. 253-63.
Manica Post, 18 December 1998, 'Nyanyadzi, Mutema irrigation schemes dry up.' Regional newspaper, Manicaland Province.
Manzungu, E. (1995), 'Engineering or domineering? The politics of water control in Mutambara irrigation scheme', *Zambezia*, Vol. 22(2), pp. 115-136.
Manzungu, E. (1998), 'The development of farmer participation methodologies for the transfer of authority and responsibility to farmers at government-run smallholder irriagtion projects in Zimbabwe', Main report Farmesa, Mimeo, Harare.
Manzungu, E. and Van der Zaag, P. (1996), 'Continuity and controversy in smallholder irrigation', in E. Manzungu and P. Van der Zaag (eds), *The practice of smallholder irrigation; Case studies from Zimbabwe*, University of Zimbabwe, Harare, pp. 1-28.
Meinzen-Dick, R. (1993), 'Objectives of irrigation development in Zimbabwe.', Paper presented at the University of Zimbabwe/Agritex/IFPRI workshop on 'Irrigation performance in Zimbabwe', Juliasdale.
Mollinga, P.P. (1998), 'On the waterfront. Water distribution, technology and agrarian change in a South Indian canal irrigation system', Unpublished PhD thesis, Wageningen.
Mupawose, R.M. (1984), 'Irrigation in Zimbabwe. A broad overview', African regional symposium on smallholder irrigation, Hydraulics Research, Wallingford, and University of Zimbabwe, Harare.
Pazvakavambwa, S.C. (1984), 'Management of small holder irrigation systems in Zimbabwe: the committees approach', in M.J. Blackie (ed), *African regional symposium on small holder irrigation*, 5-7 September 1984, Hydraulics Research, Wallingford/University of Zimbabwe, Harare, pp. 421-26.
Peacock, T. (1995), ' Financial and economic aspects of smallholder irrigation in Zimbabwe and prospects for further development', Paper presented at ZFU workshop 'Water development for diversification within the smallholder farming sector', May 1995, Harare.

Pearce, G.R. and Armstrong, A.S.B. (1990), *Small irrigation scheme design, Nyanyadzi, Zimbabwe: summary report of studies on field-level water use and water distribution*, Report OD 98, Hydraulics Research, Wallingford.

Phimister, I. (1988), *An economic and social history of Zimbabwe: 1890-1948. Capitcal accumulation and class struggle*, Longman, London.

Ranger, T. (1983), 'The invention of tradition in colonial Africa', in E. Hobsbawm and T. Ranger (eds), *The invention of tradition*, Cambridge University Press, Cambrdige, pp. 211-62.

Ranger, T. (1985), *Peasant consciousness and guerilla war in Zimbabwe*, James Currey, London.

Republic of Zimbabwe (1997), 'Smallholder irrigation support programme (SISP)', formulation report, October 1997, Harare.

Riddell Commission (1981), *Report of the commission of inquiry into incomes, prices and conditions of service*, Government printers, Harare.

Reynolds, N. (1969), *A socio-economic study of an African development scheme*, PhD thesis, University of Cape Town, Cape Town.

Roder, W. (1965), *The Sabi valley irrigation projects*, Department of Geography Research Paper No. 99, University of Chicago Press, Chicago.

Roe, E. (1995), 'More than the politics of decentralization: local government reform, district development and public administration in Zimbabwe', *World Development*, Vol. 23(5), pp. 833-43.

Rukuni, M. (1985), 'State of the art of irrigation research in Zimbabwe, a historical bibliography', *Zimbabwe Agricultural Journal*, Vol. 82(3), pp. 85-90.

Rukuni, M. and Makadho, J. (1994), 'Irrigation development', in M. Rukuni and C.K. Eicher (eds), *Zimbabwe's agricultural revolution*, University of Zimbabwe Publications, Harare, pp. 127-38.

Weinrich, A.K.H. (1971), *Chiefs and councils in Rhodesia. Transition from patriarchal to bureaucratic power*, Heinemann, London.

Weinrich, A.K.H. (1975), *African farmers in Rhodesia*, Oxford University Press, London.

Chapter 7

Irrigation Policy Reforms in Pakistan: Who's Getting the Process Right?

Edward J. van der Velde and Jamshed Tirmizi

The achievement of British Indian engineers in the late nineteenth and early twentieth centuries was not so much to eliminate silt within the canal system, but rather to define a new, 'scientific' way of looking at the problem that would justify excluding local 'communities' from a role in the administration of major canal channels.[1]

The principal objective of Pakistan's irrigation reforms is to reverse the deteriorating performance of its irrigation systems and the consequent stagnating productivity of irrigated agriculture in the Indus Basin. This is to be accomplished primarily through two institutional policy initiatives. One is the restructuring and decentralization of the provincial irrigation departments (PIDs) into autonomous Provincial Irrigation and Drainage Authorities (PIDAs) and Area Water Boards (AWB). The other is to include local irrigation communities in the management of the irrigation system by devolving operational and maintenance as well as fiscal responsibilities for secondary canal channels to them through independent, self-sustaining Farmer Organizations (FOs). It is an ironic twist in the history of Indus Basin irrigation development that so much effort having been expended to align irrigation operations and maintenance (O&M) with the apparent efficiency requirements of engineering science by excluding local communities from active roles and participation therein, the farmers are now to be actively encouraged to formally organize themselves in order to participate in irrigation system management!

Hindsight is 20/20: Institutional Keys in the Policy Reforms Context

Geography and the *de facto* division of water rights following sub-continent partition and the independence of Pakistan in 1947 necessitated several major physical engineering interventions that were intended to assure the productivity and sustainability of the country's Indus Basin irrigation systems. Although less evident at the time, each of these interventions also was accompanied by important institutional developments that changed, weakened or constrained the heretofore dominant position of the Irrigation Department in the irrigation affairs of the Indus Basin.

WAPDA and the Indus Basin Plan

The first post-Independence crisis in Pakistan irrigation was the replacement of the water supplied by the eastern Punjab rivers, the Sutlej and the Ravi, that now were controlled by India. The most obvious course of action was an engineering solution: construct several large, trans-basin canals to transfer water from Indus tributaries in western Punjab to replace the flows lost to the eastern canal systems. Far less clear, however, was how the enormous capital requirements for such a major engineering undertaking were to be met.

A new multinational donor, The World Bank, itself seeking a major development finance capital undertaking to further confirm its reputation, soon became the primary lender for the remarkable design and construction activities of the Indus Basin Plan (IBP). Arguably the civil works resulting from the IBP were the salvation of the national political economy of nascent Pakistan from the 1950s through the 1960s, and into the 1970s. They remain the essential underpinning of the Indus Basin canal systems and the irrigated agriculture economy based upon them.

Less noticed in the process was the associated establishment of a new national institution, the ostensibly independent Water and Power Development Authority of Pakistan (WAPDA), as the primary agency for coordinating the design, development, construction and initial operation of these physical facilities. WAPDA subsequently emerged as the pre-eminent national agency for the execution of all new water sector projects in Pakistan, including measures to control waterlogging and salinity. No longer was the West Pakistan (Punjab) Irrigation Department the premier irrigation institution in the country, although a decade or more would pass before that fact would become more widely recognized.

From the outset, many of Pakistan's most promising and ambitious bureaucrats, civil engineers and other professionals were attracted to WAPDA for their careers. An unintended consequence of that development was the beginning of the slow decline in the quality and professionalism of the officer cadre that comprised the West Pakistan Irrigation Department and the provincial irrigation departments into which it was subsequently devolved.

SCARP Projects and Public Tubewell Development

By the early 1960s the extent and serious consequences of widespread waterlogging and salinity in the command areas of the Indus Basin irrigation system were beginning to be recognized. Although by no means phenomena new to Pakistan's irrigation farmers or system administrators, the apparent rapid spread of both waterlogging and salinity now were viewed as primary threats to irrigated agriculture in the Indus Basin as well as to the massive financial investments that were being made in its physical infrastructure.

Once again, a technological engineering solution appeared to be at hand, if only the huge capital requirements could be met. And, as it had in the previous decade, The World Bank, was forthcoming, providing the necessary loans so that the first Salinity Control and Reclamation Project (SCARP) could be initiated.[2] This program of deep tubewell development which was intended to reduce both high water tables and to provide farmers (in good quality groundwater areas) extra irrigation water supplies was implemented first in Punjab Province and subsequently throughout the country by WAPDA.

However, although WAPDA was responsible for the design, construction, initial operation and subsequent monitoring and research of these deep tubewell projects, it was the PIDs that assumed responsibility for the continuing operation and maintenance requirements of the SCARP tubewells. This had several unforeseen (and certainly unintended) institutional consequences for these agencies over the next two decades.

Whereas the irrigation department formerly was the exclusive professional preserve of civil engineers, it now had to be shared with a new officer cadre of mechanical engineers. The former remained dominant in the operation and administration of the canal systems as well as in the bureaucratic ranks of the department, while the latter were a minority concentrated in separate administrative field units, almost never penetrating the more important and senior departmental administrative hierarchy. This administrative, almost caste-like separation of officers (and their respective support staffs) typically meant that at field level in SCARP project areas, canal and tubewell operations were rarely if ever coordinated for effective conjunctive use of irrigation water.

In SCARP project areas, ID officers and non-professional field staff now had visible and active roles and responsibilities at the watercourse level because virtually all of the public tubewells pumping useable quality groundwater were directly connected with the *sarkari khal* (the sanctioned watercourse) to supplement less plentiful surface water supplies. This contrasted markedly to and did not connect with the traditional (and continuing) situation in the canal administration divisions where the activities and responsibilities of canal officers and supporting field staff were largely confined to above the watercourse outlet.[3] Compounding this contrast was the early and rapid unionization of SCARP tubewell operators, a development which soon placed them beyond the traditional confines and discipline of departmental rules and regulations whilst injecting an entirely new dynamic in personnel relationships between irrigation officers and the field staff they supervised.

It was the financial consequences of operational and maintenance responsibility for the SCARP tubewells, however, which impacted most severely upon the provincial irrigation departments, particularly in Punjab. There the annual costs of operating and sustaining more than 10,000 deep public tubewells (80% of which were in fresh groundwater areas) and their supporting infrastructure were enormous, not least because tubewell life expectancies proved to be less than half that estimated when the SCARP projects were first conceived. By the early 1980s, the rising annual O&M costs of SCARP tubewells exceeded one-half the budget of the Punjab Irrigation Department even though hundreds of tubewells were either inoperable or barely functioning. Concomitantly, funds for canal system O&M were increasingly constrained and the operational effects of deferred canal maintenance were already discernable to experienced canal officers. Clearly the financial burden of operating and maintaining public tubewells was rapidly becoming unsupportable.

On-Farm Water Management and Water User Associations

Field research carried out by American and Pakistani researchers in central Punjab canal commands from the mid-1960s to mid-1970s revealed that the dilapidated and poorly maintained water distribution infrastructure below the canal turnout, the system of farmer managed watercourses and lateral field ditches, was responsible for a significant proportion of overall system water losses. This was especially the case in canal command areas where SCARP tubewells had been sited at the watercourse heads, but for which no provision for watercourse remodelling had been made in SCARP projects. The resulting water losses, reported to be 40% or more, contributed to the growing problems of waterlogging and salinity and were a primary cause for shortages in irrigation supplies for tail end farmers.

Beginning in 1976 a five-year USAID-assisted pilot On-Farm Water Management (OFWM) Project was implemented in Pakistan to demonstrate that a focused program of watercourse improvements and agricultural extension activities could change these conditions, thereby increasing the productivity of irrigated agriculture. The package of physical improvements included lining a portion of the *sarkari khal* to reduce seepage losses, hard turnout structures (*pucca nakkas*) at farmer irrigation points to eliminate embankment cutting and filling as well as ponding at the intersections of lateral field ditches, and precision land levelling. Farmers were to participate in project implementation through an organizational component that established a Water User Association (WUA) comprising all the shareholders in the watercourse command. Once the physical improvements to the watercourse had been completed, the WUA was to remain as the primary contact point for continuing agricultural extension activities.[4]

The pilot OFWM Project was initially proposed to the provincial irrigation departments for implementation, but when that proposal was rejected, notably by the Punjab ID, On-Farm Water Management Directorates were established for that purpose in the provincial agriculture departments (PAD). Although not evident at the time, that decision would accelerate the PID's eroding influence and control over water distribution at the secondary/tertiary irrigation system interface.

From 1981 onward, a nation-wide series of OFWM Projects were funded by the World Bank and the Asian Development Bank as well as through bilateral assistance programs. While provincial ID O&M budgets were increasingly strapped for funds through the 1980s and into the 1990s, vigorous and well-funded On-Farm Water Management Directorates emerged in the provinces, especially in Punjab, and attempted to challenge the provincial ID's primacy in irrigation matters, as WAPDA already had succeeded in doing at the national level.

No less important was the provision to form representative organizations of water users, and the adoption of legislation in all four provinces intended to facilitate that process by providing WUAs a legal basis.[5] Although it is now clear that the watercourse level WUAs were largely transitory, typically sustained only as long as necessary to complete the watercourse improvements, their appearance and encouragement marked a significant shift in irrigation policy. Organizations of farmer water users were now seen as important for maintaining, even increasing, the productivity of the Indus Basin irrigation systems as well as capable of sharing some of the financial burdens for their physical improvement and maintenance.[6]

The Proposed Irrigation Reforms: Getting There

While it is generally understood that the current irrigation reform effort in Pakistan began with the circulation of the World Bank prepared document, *Pakistan Irrigation and Drainage: Issues and Options*,[7] hindsight suggests that the trend in this direction was already fairly well established by the early 1990s. In that respect, the pilot SCARP Transition Project, begun in 1987 and also World Bank financed, was important. It was the beginning of a major institutional initiative to divest the PIDs of the enormous financial costs of operating and maintaining the public sector SCARP tubewells in good quality groundwater areas.[8] It also initiated a clear shift in Pakistan's irrigation policy toward encouraging a wider and more active privatization of groundwater development and use.[9] In turn, that raised the issues of water markets for more widespread discussion, and in no province was this more evident than in Punjab.

Translating Issues and Options into Action: An Extended Minuet

Clearly, however, it was the World Bank's *Issues and Options* paper which sharply focused policy making attention and debate upon the future of the Indus Basin irrigation systems. It argued that the many problems that beset these systems "waterlogging and salinity, low efficiency in water delivery and use, inequitable water distribution, over-exploitation of good quality groundwater, and insufficient cost recovery" were essentially a consequence of government policies which treated irrigation water as a public good. Instead, it was urged that irrigation water 'should be commercialized and later privatized', facilitated by legislation that established enforceable property rights to water. It outlined a strategy for the phased implementation of such changes in irrigation policy that included:

> enabling legislation on water rights, quality, and markets; pilot projects at canal command level and setting up public utilities; and organizing and strengthening Provincial Water Authorities, and strengthening federal water agencies.[10]

The *Issues and Options* paper confidently predicted that the resulting new system of irrigation and drainage would both end the present crisis in the Indus Basin canal systems and bring about sustainable benefits for farmers, including greater farm output and income. It would guarantee a reliable supply of the minimum amount of water determined by the basic right, provide opportunities for farmer organizations to plan ahead for purchases of additional water, and reduce water losses as well as water charges (the latter because part of system O&M would be done more efficiently by the farmers). It appeared to be, and subsequently has proven to be, too good to be true.

Within less than a year, scarcely sufficient time for debate upon such fundamental changes in existing irrigation policy to become reasonably engaged, full scale feasibility planning was initiated for the first proposed pilot canal command project, the Second Irrigation and Drainage, Sukh Beas/Lower Bari Doab Canal Project. It was anticipated that this project would be one of perhaps several pilot projects to be implemented in Punjab under the planned National Drainage Programme (NDP), the intended vehicle for transforming irrigation and drainage policy in Pakistan.

The resulting feasibility study proposed major remodelling of the physical irrigation and drainage system of the Lower Bari Doab Canal (LBDC) system in the context of extensive institutional changes. Those changes included a greatly enhanced role for organized farmer participation at all levels of the system O&M as well as cost recovery to sustain system requirements. The institutional reforms required for farmer organization and participation would emerge through appropriate legislation that would be followed by the establishment of a provincial commission to implement the reforms. The provincial commission then would enable a system (LBDC)-based water utility with powers and objectives secured through the reform legislation.[11]

Four levels of farmer management organization were proposed: WCAs (Watercourse Associations), hydraulic-based Unions of WCAs, Federations of Unions, and finally a single system Confederation that eventually would assume responsibility for operating the entire LBDC irrigation and drainage infrastructure. Key elements in this institutional strategy included granting land-owning farmers specific rights to water, farmer participation in hydraulic unit remodeling, transparency in water distribution within the hydraulic unit, and granting the Union of WCAs a long-term concession to operate, maintain and develop its irrigation and drainage infrastructure.

Meanwhile, in August, 1995, the Government of Pakistan had communicated to the World Bank a strategy for decentralizing and transferring management of the Indus Basin irrigation and drainage systems. That strategy envisaged transforming the PIDs into autonomous Provincial Irrigation and Drainage Authorities (PIDA). Management of irrigation and drainage at the system level would be decentralized from the PIDA to newly created Area Water Boards (AWB). And, at the distributary or minor canal level, farmers would be encouraged to engage in an increased management role through Water Users Formations (WUFs).[12] Because the number of WUFs that would be formed would be very large – and the prevailing uncertainty whether or not farmers would be up to the task – a pilot approach would be used to form WUFs.

Subsequently, when the details of the feasibility study for the LBDC finally emerged in 1995 along with news of a draft of potential enabling legislation being discussed between the Government of Pakistan and the multi-lateral donors for the National Drainage Programme, a firestorm of criticism resulted. It centred around the issues of privatizing the canal system and separating land from water so that the latter could be sold or traded as a commodity. The Punjab Irrigation Department flatly rejected the LBDC feasibility study, and the other provincial IDs joined it in its opposition to the institutional reforms.

The then President of Pakistan, himself also the leader of a powerful landowning clan in the Dera Ghazi Khan district of Punjab, acknowledged that Pakistan had been asked by the multilateral donor agencies to 'privatize the canal system.' However, he said that could only be done slowly 'and on the basis of scientific experiments because we don't just want to destroy the system.'[13] Soon thereafter, he retreated from this lukewarm endorsement and forcefully announced that there would be no privatization of Pakistan's irrigation system.

That did not, however, mollify the critics of the proposed irrigation reforms as the details of the model draft reform legislation being discussed became more widely known. By early 1996, there was a nearly continuous barrage of press articles virtually always critical of the proposals, and it was evident by this time that opposition to the reforms was now wide-spread and deeply rooted among the national farmer organizations dominated by large and influential landowners, provincial and national politicians, the officer cadres of the provincial irrigation departments, and professional engineering societies.[14]

There is little doubt that the building criticism, allegations and rumors were intended to raise fear for the future of existing water supplies among the majority small farmer community, to confuse them on the issues and sway them into opposition to the proposed reforms. In that context, hyperbole and fiction were used effectively, whether by the PIDs' non-professional operations and revenue field staff or the English and vernacular press.[15]

By and large, the misinformation/disinformation campaign was successful. Certainly it had galvanized members of and spokespersons for both the Federal and Provincial Governments to speak out forcefully in opposition to any privatization of the canal system. Moreover, it had motivated a *de facto* (and incongruous) alliance between small farmers and organizations such as the Kissan Board of Pakistan (KBP) and FAP that consistently represented the interests of the largest land owners. These unnatural allies now stood together in vigorous denunciation of a perceived conspiracy to open up Pakistan's most valuable resource to exploitation by international capital and certain to impoverish its farming community!

However, this was not the only arena where the proposed reforms were encountering major difficulties. Significant differences had emerged between the National Drainage Programme donors and GOP negotiators over the latest draft (March, 1996) of enabling legislation that had been prepared by the GOP and was to give final shape to the irrigation reforms. Among the former many expressed the view that the GOP was trying to avoid any substantial reforms to the status quo, whereas on the Pakistani side, the feeling was that this most recent draft went as far in reforming the irrigation sector as it was possible to go.

Specifically, donor critics viewed the draft legislation as too narrow, focused mainly upon just one institution, the PIDA, that was to replace the PIDs. Their position was that the draft legislation should reflect the transformation of the entire irrigation sector, providing a framework for the decentralization of management to independent system-based Area Water Boards and secondary canal-based Farmer Organisations, the allocation of water rights, and the establishment of water markets. They were critical that all the laws pertaining to the irrigation sector, notably the Canal and Drainage Act of 1873, had been left in place, instead of a comprehensive set of new irrigation laws having been drafted. Also criticized was the restriction of water charges to recover only O&M costs, rather than providing for investment or capital cost recovery as well.

Clearly the positions of the two groups remained very much at odds despite several drafts of enabling legislation having been prepared and extensively discussed by teams of representatives for both sides over the previous year. But neither the National Drainage Programme donors nor the GOP were as rigid in their respective reform agendas as perhaps might have been expected. Among the donors, the issues of cost recovery and new participatory institutions apparently were more important than any others, while on the Pakistani side any appearance of 'privatization' was an anathema and there was a pragmatic appreciation for the amount of foreign exchange at stake in the proposed $785 million financing for the National Drainage Programme.

Hence, there was some optimism that the differences finally would be resolved when a new round of discussions was set for late April, 1996 at the World Bank in Washington, D. C. with the objective of producing an acceptable informal draft of enabling legislation that subsequently could be circulated among and cleared by the respective Chief Ministers of the provinces. Quite unexpectedly those meetings were abruptly terminated by the rapid emergence of major differences between the parties in both substance and style. Consequently by May, 1996 progress on finalizing the proposed reform enabling legislation as well as negotiations on loans for the National Drainage Programme slowed to a crawl, if not completely stopped, despite the fact that all the provinces had agreed in August, 1995 to form Provincial Irrigation and Drainage Authorities to replace the PIDs by July 1, 1996.[16]

In late July, 1996, the Government of Pakistan signalled its intention to proceed with at least the PIDA component of the reforms when at a high level meeting of the President, the Prime Minister and the four provincial Chief Ministers it was decided to assign the task of preparing a consensus draft of legislation to establish the Authority in all the provinces to the Federal Minister of Water and Power.[17] That and a similar subsequent effort by a committee of 'technocrats' made little progress in securing consensus on draft reform legislation over the ensuing months, perhaps not least because of the continuing efforts of the larger farmer organizations to frustrate the process, although national political events in October-November were likely just as important.[18]

During the latter half of 1996, however, the multi-lateral donors also had set a firm deadline for action on the reform proposals. They had made it very clear to the GOP that if formal agreement to establish PIDAs in all four provinces was not reached by the third week of December, there would be no loan agreements for the National Drainage Programme, and future funding for Pakistan's water sector was likely to be jeopardized as well.[19] Apparently unwilling to take the chance of once more testing donor resolve, the GOP – an interim or caretaker government at this point – secured the necessary provincial agreements, and PIDA Ordinances were finally notified in early 1997. These Ordinances were subsequently replaced by formal PIDA Acts passed by the provincial assemblies and notified by mid-year.

The Example of the Punjab PIDA Act

The Punjab Irrigation and Drainage Authority Act, 1997, was formally notified on July 2, 1997, having been passed five days previously by the Provincial Assembly. In its details, it differs only very slightly or not at all from similar acts passed by the provincial assemblies of Sindh, NWFP and Balochistan. Specifically, it provides for the establishment of a Punjab Irrigation and Authority as a body corporate, consisting of a number of members to be notified by the Government, of which at least six will be farmers and the number of non-farmer members will not exceed the number of farmer members. The Chairman of the Authority is the (Provincial) Minister for Irrigation and Power. The work of the Authority is to be carried out by the Board of Management that it appoints, with the prior approval of Government, consisting of a Managing Director and three General Managers.

The comprehensive powers and duties of the Authority are detailed in thirty-two specific articles that cover virtually all aspects of irrigation and drainage in the Province. Included here is the duty to exercise all the powers under the Canal and Drainage Act, 1873, and the Soil Reclamation Act, 1952, as well as all other laws in force related to irrigation and drainage in Punjab. Several other paragraphs clarify related aspects of the Authority.

By contrast, virtually nothing is specified concerning the powers and functions of the new institutions that are to be created for the decentralization of the provincial irrigation and drainage system, the Area Water Boards and Farmers' Organizations. The Act provides that these institutions may be established by Government, who will also 'assign to them such functions as it may deem fit.' Their establishment was to occur within a year of the creation of the Authority and was to be done on a pilot program basis, 'in a phased manner in accordance with the relevant Bye-Laws and Regulations framed by the Authority.'

Given the heavy emphasis that the National Drainage Programme donors had placed upon the creation of an appropriate legal and regulatory framework for the institutional reform of Pakistan's irrigation and drainage sector, the final results clearly fell far short of expectations. They were not the strong foundation for the entire decentralization and irrigation and drainage management transfer process that had been sought. The record of institutional reform implementation subsequent to the passage of the PIDA Acts appears to confirm this conclusion.

Is Anyone Getting the Process Right?

Over the past six years that the irrigation policy reform process has been actively underway in Pakistan, there have been three principal stakeholder groups, *viz.* the Government, primarily the provincial irrigation departments, of which the Punjab

Irrigation Department has been the most active, and arguably is the most important, agency; the multi-lateral donor organizations, with the World Bank in the leading role; and the farmers, broadly differentiated into two groups: the numerous, but largely powerless small farmers, and the far smaller group of influential and large landowners. Each stakeholder claims to be a supporter of the proposed reforms, particularly the key component, a substantive role for farmers in the O&M of the canal system and responsibility for mobilizing the required resources for it – in other words participatory irrigation management or PIM.[20] In that case, why have the reforms scarcely progressed beyond the establishment of the provincial PIDAs, and why does PIM remain effectively stalled in Pakistan? A contextual examination of stakeholder roles and actions in the reform process provides useful insight.

The Farmers

It is evident that the proposed irrigation reforms were not initiated by the farmers, whether the anonymous and typical small landowner or the influential and wealthy large farmer. From their perspective, the administrative and legal structure under which Indus Basin canal systems have operated has been in existence and apparently immutable for several generations, long before their lifetimes. Hence it is unrealistic to expect that farmers in Punjab or elsewhere in Pakistan have anything like a clear, well-conceived plan for assuming control of secondary canals, or how the Irrigation Department should be re-structured. In fact, few, if any, farmers had prior familiarity with or commitment to the concept of local control and management of the irrigation system outside of the watercourse command.

That does not mean, however, that Pakistan's farmers did not and do not have clear views and strong opinions on the subject of irrigation policy reforms, especially whenever the proposals were or are clearly described and accurately explained to them. After all, Indus Basin farmers have long been acutely aware of many causes for and the consequences of inadequate, unreliable and inequitable water deliveries. They have developed their own strategies for dealing with those conditions, as well as the general water scarcity principle that was designed into the canal system. No less are they conscious of the applicable rules and regulations and they have learnt how to manage relationships with the hierarchy of field staff and officers that administer them. Most farmers also appear to understand the canal system sufficiently well to appreciate that they are involved in essentially a zero sum game in which any gain they may achieve is at the expense of someone else and *vice versa*.[21]

Consistently lacking throughout the development of the irrigation policy reforms, however, has been a systematic farmer consultation process, one in which a broad and representative cross-section of Punjab (and other Pakistani) farmers openly participated.[22] Neither has there been any organized effort to regularly and accurately inform farmers through various media and by other means about the content of the proposed irrigation reforms or their purpose and likely consequences for the farmers.

Thus the average Pakistani farmer has been particularly susceptible to and bombarded by unsubstantiated rumour and purposeful misinformation about the irrigation reform proposals from a variety of sources – the vernacular press, functionaries of the PIDs, the associations and lobby groups of large farmers, local influentials and politicians, *et alii*. This condition, of course, contributed enormously to the environment of widespread farmer opposition to the initial reform proposals – perceived as de-linking water rights from land ownership and privatizing whole systems for the benefit of multinational corporations – that was encountered throughout the Punjab in 1995-96.

That farmer response has been interpreted by many as reaffirming a more general concern that Pakistan's socio-cultural and political environment will preclude successful system-level PIM, notably management turnover of distributary canals to FOs. The poor record of sustainability of watercourse level WUAs also has bolstered those concerns. Such views, however, do not accurately reflect the changing realities in rural Pakistan that are well illustrated, for example, in Punjab where informal organizations of water users do function effectively at the level of the watercourse community.

The watercourse community is comprised of shareholders – all landowning and permanent tenant farmers – who participate in the *warabandi*, the roster of irrigation turns in the watercourse command.[23] Although all resident kinship-based groups – *quoms, biradries* – are present in the informal watercourse community, the factionalism and competition among them that still typifies so much of rural life is now largely kept separate from irrigation operations and related activities. That is a consequence of the increasingly widespread recognition that irrigation water is too scarce, too valuable and too essential for productive agriculture and economic well being to be wasted, for its flow to the fields of the community to be disrupted by factionalism or *dharabandi*.[24] It is this growing separation of local community irrigation activities from other rural socio-cultural and politically based conflict that provides a reasonable opportunity and basis for optimism that farmer organization and PIM can be successful at higher levels in the irrigation system.[25]

Farmer Organization Principles This opportunity, however, is unlikely to be translated into the substantive forms of farmer participation in irrigation management envisaged in the institutional reforms unless several basic principles and pre-conditions, confirmed by farmer organization experience elsewhere, are met.[26] Those principles include the following: (1) the basis of the farmers' distributary organization is the watercourse irrigation community (2) the distributary Farmer Organisation is hydraulically-based (3) Farmer Organisation members are landowners with a confirmed right to canal water (4) the objective in canal water distribution is equity, a fair and practical equality, and (5) Farmer Organisation members are proportionately liable for O&M costs and other water charges.

These farmer organization principles, however, will be effective only in an appropriate environmental setting where at least two additional prerequisites have been met. Government must provide a suitable legal framework for the establishment of Farmer Organisations and to enable them to function effectively in discharging their responsibilities for distributary canal O&M, resource mobilization and dispute resolution. Most farmers in Punjab and elsewhere in Pakistan are well aware that without such a legal basis, the Farmer Organisations cannot be successful. In the absence of an appropriate legal cover, a Farmer Organisation would be unable to enforce rules, penalize water theft and vandalism, enter into contracts for water delivery or to secure specialized legal services, or to mobilize revenues from its membership to meet O&M costs and related water charges; indeed, the entire basis of the Farmer Organization would be in doubt.[27]

From the irrigation operations perspective of the Farmer Organization, farmers also require the important assurance that water supply by main system management, the PIDA/AWB, to the head of their canal will be at least its currently authorized or sanctioned discharge. A contractual agreement of this condition detailing the responsibilities of both parties is likely to be required, including the identification of specific environmental circumstances that would cause an adjustment of water deliveries. All water-related transactions both between main system management and the Farmer Organization and within the secondary canal command of Farmer Organization operation will have to be transparent and subject to easy mutual verification by the farmers. This prerequisite is a consequence of widespread farmer awareness of the multiple ways canal water deliveries can be and are tampered or 'adjusted' in exchange for long term, seasonal and short term 'considerations' throughout the Indus Basin systems, where, in fact, such interventions to appropriate canal water have become commonplace in the past decade.

Farmers and FO Pilot Projects Beginning in 1995, nine pilot projects have been initiated by two agencies to establish secondary canal-based Farmer Organizations in Punjab, Sindh and NWFP (Table 7.1).[28] Although they have varied in the specifics of implementation, the generally shared objectives of these pilot projects have been to (1) empirically devise and 'test' an organizational model for a secondary canal Farmer Organization (2) to gain experience in identifying and solving problems in the Farmer Organization development process, and (3) to establish farmer – irrigation agency joint management activities for secondary canal O&M.

The ordinary farmer's response to these pilot institution building activities has been generally positive and cooperative, reflecting both the overall consultative and explanatory mode adopted by the implementing agencies in the pilot projects, and their own genuine concerns for the continuing deterioration of conditions in the irrigation environment. Although there has been no evidence of any pre-existing commitment by the average farmer to the concept of control of the irrigation system by the local irrigation community, farmers in the pilot projects typically have grasped its advantages after detailed explanation and discussions of the irrigation reforms.[29]

As a result, substantial progress has been made in each of the pilot Farmer Organization projects with respect to the first two objectives. In each pilot project command area, a formal Farmer Organization structure is now in place, comprising a representative general body, an executive committee and several officers.

Positions in the latter two categories commonly have been filled by irrigation community consensus rather than through formal (and more divisive) elections. Depending upon the size of the canal command, a two or three tier Farmer Organization model has been used in every pilot project.[30] The process followed in establishing the Farmer Organizations also has been very similar in the pilot projects, typically with the implementing agency organizing and directing broad participation. Both farmers and the implementing agencies have gained significant experience in identifying and working through to solution a range of Farmer Organization development problems.[31]

Table 7.1 Distributary farmer organization pilot projects in Pakistan, 1998[32]

Distributary or Minor	System	Location	Design Q (cfs)	CCA (acres)	No. Outlets	No. Shareholders
Pabbi Minor	Warsak Gravity	NWFP	3.5	913	6	438
Surizai Minor	Warsak Gravity	NWFP	11	2792	10	684
Bahaderwah Minor	Eastern Sadqia	Punjab	82	19267	52	2207
Bhukan	Eastern Sadqia	Punjab	12	3027	8	322
Hakra 4R	Eastern Sadqia	Punjab	193	43400	123	4690
Innuana	LCC	Punjab	22	6900	15	1690
Bareji	Nara	Sindh	34	14300	24	197
Dhoro Naro Minor	Rohri	Sindh	52	13382	25	421
Heran	Nara	Sindh	58	15410	31	539

Although all of the pilot projects have had active implementation support from the water management directorates of the provincial agricultural departments, none have been implemented with the cooperation and assistance of the respective provincial irrigation departments (or PIDAs, following their establishment in mid-1997). This latter condition, more than any other single factor, has precluded any establishment of meaningful PIM activities in each pilot canal command, all of which remain firmly under formal irrigation agency control. No pilot Farmer Organization has become involved in and shared responsibility for the primary activities of distributary canal O&M, nor have they engaged in the mobilization of resources on a sustained basis for that purpose, despite clear evidence of farmer willingness in every pilot project to initiate PIM activities.[33] Consequently and unfortunately there has been no real progress to date in achieving the third objective of these pilot projects.

That does not mean, however, that serious efforts have not been made to enlist the participation of the PIDAs (and previously the PIDs) in these pilot projects, to enable the Farmer Organizations to undertake distributary canal O&M and resource mobilization activities. In the Hakra 4R pilot project, for example, the Farmer Organization formally approached the PIDA to enter into a joint management agreement for canal and drainage operations and maintenance activities. Unfortunately, the resulting draft Memorandum of Understanding proposed by the PIDA to govern such organized farmer participation was so limited and distinctly one-sided in the allocation of an operational role, authority and responsibility to the Farmer Organization that the farmers had little choice but to reject it.[34]

In the case of the Sindh pilot projects, the experience has been somewhat different, at least for a brief period. In the later part of 1997, joint management agreements actually were concluded between the SIDA and each Farmer Organization as well as the pilot project-implementing agency. They were in effect for about two weeks before the Government of Sindh unilaterally and indefinitely suspended them. During that two-week period, the distributary Farmer Organizations jointly participated in certain canal operations activities with field level agency staff, albeit with mixed results. Although the first (and only) two week period of PIM experience can not possibly provide any reliable basis for judging the future success or failure of the irrigation reforms, events in the Sindh pilot projects were indicative of some of the problems likely to occur.

In all three pilot projects, farmers were initially so exuberant about finally having a significant role in canal operations that it became a practically difficult matter for the Farmer Organization to prioritize, focus and manage group activities. In one canal command, the necessity for jointly monitoring canal discharge with agency staff was quickly lost in farmer enthusiasm to modify watercourse outlets to allocate a new, somewhat larger and apparently assured water supply to be provided under the joint management agreement. Unfortunately the resizing of watercourse outlets was begun in the head reach of the canal by the Farmer Organization, rather than in the more chronically water short tail reach. Thus when the agency field staff that operated the distributary head gate also began to manipulate discharge into the canal, in the absence of sustained Farmer Organization monitoring and verification, the resulting water deliveries to tail reach farmers were significantly less than they had been before the implementation of joint management operations. The resulting tail-ender versus head-ender acrimony severely damaged the credibility of the Farmer Organization leadership, created doubt about the impartiality of the agency assisting Farmer Organization development, and threatened the viability of the Farmer Organization concept itself.

That very brief experience highlighted some key issues, from the farmers' perspective, that will require the development of appropriate monitoring and intervention methodologies if the transition to participatory irrigation management is going to be successful. In the water scarce environment of the Indus Basin irrigation systems, it is very likely that some farmer members will attempt to manipulate or control the Farmer Organization in order to capture more than their fair share of water at the expense of other farmer members. This is not a new insight, but that ordinary farmers also may be tempted by the possibilities here – as the large and influential farmers always have been assumed to be – perhaps is.[35]

The Sindh experience as well as the continuing strong opposition to the irrigation reforms and PIM in the provincial irrigation agencies, particularly among field level operations and revenue staff, confirms that farmers will have to vigorously and continuously monitor all canal water deliveries to ensure that sanctioned or agreed supply targets are met.[36] Mutually verifiable and easy to understand mechanisms for doing so will be required at all key control points. Creating that farmer monitoring capability should be a high priority for agencies and institutions that facilitate and assist Farmer Organization development. As a corollary, the rapid development of the Farmer Organization's capacity to understand and manage the oversight of distributary canal O&M as well as water fee assessment and collection is far more deserving of emphasis than has been generally evident in the pilot Farmer Organization projects.

The current pilot projects also have focused attention on who can best facilitate and assist Farmer Organization development and for how long as well as at what level of intensity should development support be provided. There can be little doubt that the participation of the provincial water management directorates of Punjab and NWFP in the implementation of five pilot Farmer Organizations has been viewed by the respective provincial irrigation agencies as an unwelcome intrusion by OFWM into their own legally defined area of irrigation system responsibilities.[37]

The bureaucratic conflict thereby created aside, it also is clear from this pilot Farmer Organization experience that the provincial OFWM agencies have little capacity to facilitate the establishment of representative and sustainable Farmer Organizations as envisioned in the irrigation reforms. Available personnel are inadequately or inappropriately trained for the task, and well established agency objectives have resulted in a rapid shift in institutional priorities to establishing formal WUAs, delivering traditional OFWM services and working to arbitrary numerical targets in the pilot Farmer Organizations they have assisted.[38] In that context, it is difficult to be sanguine about the institutional objectives of the World Bank-assisted OFWM IV Project wherein as many as 200 Farmer Organizations are proposed to be established by the Water Management Directorates of the provincial agricultural departments to further 'test' the irrigation reforms and PIM.

The experience of the International Water Management Institute (IWMI, formerly IIMI), Pakistan, in these pilot Farmer Organization projects has been significantly more encouraging. Farmer members of the organizations that they have assisted and supported appear to be more confident and better prepared to deal with formal institutional issues, as well as some of the simpler functions of canal O&M. Whether or not that will translate into effective and sustainable farmer participation in irrigation management over the longer run also remains unknown because distributary canal turnover or joint management – excepting the brief period in Sindh Province – has not occurred in these pilot Farmer Organization projects either. IWMI has also pioneered the application of a form of farmer-to-farmer extension to facilitate Farmer Organization development in its pilot projects by using locally recruited and trained Social Organization Volunteers.

It is far too early to reach a confident conclusion regarding which organization has been most successful in facilitating Farmer Organization development and PIM, not least because none of the pilot Farmer Organizations have been engaged for any reasonable length of time in the irrigation management activities for which they have been established. It is clear from the pilot Farmer Organization projects, however, that Indus Basin farmers will need substantial assistance and sustained support from well-prepared and committed organization facilitators – as well as a responsive institutional framework in which they can function – if the irrigation reforms and PIM are going to be successful. It also is readily apparent that Pakistan's present institutional resources for such a venture are very limited and already under considerable pressure from existing project activities. The sheer scale of canal system size and farmer numbers should be sufficient to make this a priority concern. But to date, almost no program or policy planning attention has been focused upon how such needs can best be met, not even for the imminent expansion of PIM activities under the National Drainage Programme and the OFWM IV Project.

The Multi-Lateral Donors

It is clear from the record that the emergence of the current reform proposals for Pakistan's irrigation and drainage sector was primarily at the initiative of the two major multi-national donors, the World Bank and the Asian Development Bank. That was not particularly surprising given the extent to which these two international organizations had been involved in providing development finance assistance to water sector activities in the country over the previous two decades in particular, as well as the key role the World Bank had played in virtually all important institutional developments in Indus Basin irrigation since the late 1950s. From the donor agency perspective, a genuine concern had emerged by the early 1990s that the trend of rising O&M costs for Pakistan's irrigation and drainage systems coupled with the capital requirements for new and improved system infrastructure simply was not sustainable, especially when measured against overall system performance.[39]

In their diagnosis of the causes of irrigation sector problems, a range of institutional issues were identified, of which the most prominent were the dominance of irrigation by inefficient and administratively focused public sector agencies, weak or non-existent farmer organizations and the consequent lack of farmer participation in system O&M, and the absence of formal water markets. The irrigation reforms subsequently proposed by the World Bank and ABD were specifically designed to address those key institutional issues. In and of itself, the institutional focus of the reforms represents a marked change from all other proposals over the previous three decades made by these multi-lateral agencies to solve Pakistan's irrigation problems insofar as a technological and/or an engineering solution was not the exclusive or primary objective. In short, for the first time, it was the 'software' of Pakistan's irrigation environment that was proposed to be changed as much as, if not more than, its 'hardware' environment.

This paradigm shift in the identity and specification of the principal culprit for irrigation problems in Pakistan was certainly not welcomed or embraced by the indigenous irrigation establishment, notably the provincial irrigation departments, nor those interests who were (and still are) disproportionally benefiting from the existing *status quo*, specifically the large landowners and feudals as well as the engineering profession. They were quick to appreciate that a substantially changed institutional environment, such as was now being proposed, was certain to transform both control over and the pattern of the distribution of irrigation benefits to their disadvantage, unless somehow they were able to affect, manage or direct the reform process. The resulting opposition to the irrigation reforms and the various counter-proposals that subsequently emerged in the ensuring years from these groups reflects such a concerted effort.

This was a development that should not have surprised the multi-lateral donors either in its sources or in their resolve, nor should they have been unprepared for it. After all, similar responses already had occurred in a number of countries that were much further along in the irrigation policy reform process, *e.g.* The Philippines, Sri Lanka, Mexico. Unfortunately, however, that does appear to have been the case. There is no evidence, even now, that the multi-lateral donors had a coherent response strategy to counter the domestic opponents of the irrigation reforms, other than perhaps the implicit or explicit threat to reduce or terminate financial support to Pakistan's irrigation sector.[40]

They did not, for example, do much in attempting to foster the development of a strong indigenous advocacy for the irrigation reform proposals among political, administrative and/or community leaders in the provinces.[41] Nor apparently did the multi-lateral donors formulate, participate in or otherwise encourage an appropriate public relations campaign to counter the widespread disinformation and misinformation spread in the media about the reforms and their objectives, especially in 1995-96 when opposition to the reforms was at a public peak.

It is not clear why the donor agencies were lax in this regard, although their very nature does require these institutions to exercise a certain probity in promoting policies that could be construed as unduly interfering with national sovereignty. Whatever the reasons, the consequences of this approach were not positive for indigenizing the institutional reforms.

Even now, more than five years into the process, there is still no evidence of prominent political, agriculturist, or community leadership support in Pakistan for the irrigation reforms, much less an active group or lobby that advocates their effective implementation on merit. This contrasts markedly with the range of politically influential and organized institutional opposition to the reforms. For better or for worse, the view remains widespread and largely unchallenged in the country that the irrigation reforms were a foreign imposition that were accepted only because the multi-lateral donors 'held the nation's feet to the fire', as it were, at a time of national financial crisis.

In the longer run, however, the consequences of this apparent public relations lapse may not prove to be the most serious difficulty for institutionalizing irrigation policy reform and genuine PIM in Pakistan. Much more important is likely to be the multi-lateral donor acquiescence to an acknowledged weak and markedly inferior legal framework for irrigation reform implementation, especially compared to that which was initially being advocated. For this, the World Bank, as the principal co-financier of the National Drainage Programme, would appear to share the largest proportion of responsibility.

From the start, it was evident to the multi-lateral donor proponents of irrigation policy reform and PIM in Pakistan that a strong legal basis for such structural changes was essential for its success. At the core of their concern was the Canal and Drainage Act of 1873, a particularly powerful piece of colonial legislation, which vested virtually all meaningful control of the irrigation and drainage system in government institutions, specifically the provincial irrigation departments. The early model drafts of irrigation reform legislation that were prepared through multi-lateral donor initiative recognized the centrality of replacing the CDA with an alternative legal framework for accomplishing the reform objectives, just as certainly as local opposition to the reforms targeted the retention of that Act to delaying, if not completely defeating them.

However, the structure of the process whereby the multi-lateral donors facilitate and manage development project formulation, planning, and financing is something of an Achilles heel when it comes to protracted negotiations over policy details that, in the last analysis, will require final preparation and passage of enabling legislation through a domestic legislative process over which even government control can be tenuous. Maintaining the project cycle, the costs of continuing delays or failure in meeting lending targets, the financial exigencies of the borrower, agreed international commitments, and project development, approval and implementation in other social and economic sectors are just a few of the multiple considerations that can and do affect the complex of interactions between the multi-lateral donors and the borrowing country for any one project.

That appears to have been the case with respect to securing a strong legal framework for implementing irrigation reform and PIM in Pakistan. The success of the opposition to the reform in weakening that legal framework was evident first in the PIDA Ordinances promulgated in early 1997 and subsequently in the language of the PIDA Acts passed by the provincial assemblies that replaced the Ordinances. Faced with the prospect of either working within the new law or further delaying the National Drainage Programme while again trying to negotiate the adoption of stronger reform legislation, the multi-lateral donors opted for the former course of action.

Undoubtedly the worsening economic and foreign exchange situation facing the Government of Pakistan was also a factor in the multi-lateral donor decision to accept the PIDA Acts as 'the best that could be obtained under the circumstances' and to 'work with them' during the implementation of the National Drainage Programme.[42] So, too, was a general fatigue with these issues that almost inevitably emerged following the more than four years of (often intense and sometimes acrimonious) debate, discussion and negotiation that preceded the passage of the PIDA Acts. Nevertheless, while these and other reasons for donor acceptance of a weak legal framework for the reforms are perfectly understandable, they do nothing to change that earlier assessment of a strong legal basis for PIM and new irrigation institutions as a prerequisite for success.

The unfortunate consequence of multi-lateral donor compromise on this objective, even if only temporarily, is likely to be a protracted period of legal challenges to and administrative delays in implementing any serious or meaningful form of PIM. Almost certainly that will work to undermine farmer willingness to participate in the new reform institutions and hinder the development of confidence in their capacity to effectively manage canal system O&M and resource mobilization for that purpose.

The Irrigation Agencies and Government

The Provincial Irrigation Departments and, since the passage of the PIDA Acts, the Provincial Irrigation and Drainage Authorities that are intended to subsume them, have been the most important agencies of government in shaping the irrigation reforms as they have emerged in the National Drainage Programme. That is primarily because constitutionally, irrigation – as well as agriculture – is the responsibility of the provincial government in Pakistan and because transforming the PIDs was (and remains) a primary objective of the reforms.[43] Among the PIDs, the actions of the Punjab Irrigation Department probably have been the most influential and important. It's professional staff was and is the largest and most active among the PIDs, and it is administratively responsible for more Indus Basin canal systems and command area than any of the others. Arguably as the Punjab ID responds to PIM and implements the irrigation reforms, so will the IDs in the other provinces.

From the start, both professional and non-professional staff of the PIDs was nearly unanimous in vigorously opposing the irrigation reforms and PIM for several practical reasons. Those reasons, however, tended to be masked by a disingenuous line of argument couched in the polemics of farmer impoverishment and privatization. They argued that the large bureaucratic infrastructure of the PID is essential because of the factional and dysfunctional nature of rural society in Pakistan. In such a socio-economic environment, an alternate management structure that required organized farmer cooperation and shared responsibility to operate and maintain a complex system of thousands of secondary canals, as well as to mobilize the resources needed for that purpose, was doomed to fail.[44] Unsophisticated farmers and their organizations would be easy prey in the private market environment thereby created for the unscrupulous entrepreneurs led by the large and influential farmers who inevitably would capture control of Pakistan's most scarce and valuable natural resource ... water.

The far more likely reality of PID opposition was that the reforms openly threatened the job security of thousands of non-professional staff for whose skills the floundering Pakistan economy offers few if any employment alternatives, let alone more attractive ones. PID officers were equally quick to recognize that, for many, their own professional futures were similarly limited; meanwhile, no longer would the PIDs be the exclusive preserve of the engineering fraternity.[45]

Not least, the reforms and PIM were certain to greatly reduce, if not completely eliminate, the manifold and lucrative opportunities to extract illegal rents as well as the power and influence exercised by irrigation staff through the existing institutional structure of irrigation. Although many PID functionaries were willing to concede that by the 1990s departmental discipline had deteriorated to levels that adversely impacted upon canal system O&M, few were willing to acknowledge any institutional responsibility for that condition. Instead, they were quick to place it all in the larger context of declining societal conditions wherein influential farmers, politicians and corrupt law enforcing agencies had usurped their 'powers' and continuously interfered with the internal operations of the PID.

Despite the passage of the PIDA Acts and the PIDs having been absorbed into Provincial Irrigation and Drainage Authorities, this opposition continues essentially unabated, although expressed now in much less confrontational and more sophisticated ways. That is at least in part because neither government officers nor supporting field and administrative staff can publicly oppose what is now official government policy and the law. Here, the response of the Punjab ID provides useful insight to how the process of irrigation reform and PIM implementation has been progressing.

The Punjab ID/PIDA Even as the Managing Director of the Punjab Irrigation and Drainage Authority (and Secretary, Punjab ID) reiterates official commitment to implementing the irrigation reforms and the concept of PIM as set forth in the PIDA Act of 1997, others in the agency's professional cadre smoothly express their continuing doubts and apparently reasonable concerns for doing so. These include the difficulties of organizing the farmers in Punjab, their lack of technical expertise and the absence of professionalism among them, the risks of organizational manipulation by influential farmers and politicians, and the absence of sustainability in existing farmer organizations, such as the watercourse level WUAs.[46] For these reasons, they argue that the development of Farmer Organizations and the implementation of PIM should go forward only on a gradual and restricted pilot basis as provided for in the PIDA Act under the National Drainage Programme.

And, so far, PIDA progress in implementing the irrigation reforms and PIM has been very slow. Although provisions in the PIDA Act (and National Drainage Programme agreements) called for the establishment of pilot Farmer Organizations and a pilot Area Water Board within a year of its passage, more than two years have passed since the Act's approval by the Punjab Assembly. Neither pilot Farmer Organization development projects nor a pilot Area Water Board has been initiated through the National Drainage Programme, although a canal system locale for this work has been identified. Discussions also have been held with representatives of the existing (unofficial) pilot Farmer Organizations concerning their formal recognition or sanction by the PIDA, but as previously noted farmers have been unwilling to accept the severe constraints the PIDA proposed to place on PIM.

Thus the PIDA/ID commitment to irrigation reform implementation and PIM remains largely abstract in its most important and sensitive areas of concern. The prevailing institutional attitude toward PIM is top-down and directive, as well as patronizing, with an emphasis on a cautious, gradual approach to action in developing the pilot Farmer Organizations. This is evidenced in such PIDA/ID phraseology as: 'the pilot projects must be designed cautiously'; 'try to build on the existing community structure gradually'; 'setting up . . . appropriate financial controls and accountability mechanisms'; 'develop a well-defined implementation mechanism.'[47] It is as though the PIDA/ID is beginning the process with a clean slate, oblivious to either the potential of the on-going Punjab pilot Farmer Organization projects and what can be learned from them, or the lessons of PIM experience in other countries of South Asia. Under these circumstances, it is difficult, therefore, to avoid the conclusion that the Punjab PIDA/ID is following a *de facto* policy of avoidance and delay when it comes to the key irrigation reforms of creating autonomous Area Water Boards, facilitating Farmer Organization development and implementing PIM.

That does not mean, however, that the PIDA/ID has not been aware of the potential of the irrigation reforms, specifically the PIDA Act, for consolidating its own position with respect to irrigation affairs in Punjab. In its most significant institutional initiative to date, the PID proposed to the Punjab Government in early 1998 the formal relocation of the Water Management Directorate from the Agriculture Department into the newly established PIDA. The Water Management Directorate would become an additional circle-level administrative organization in the ID headed by its own Chief Engineer, the present Director General.

The rationale for this proposal was adroitly based upon the language of the PIDA Act, specifically the Preamble. Therein specific reference is made to establishing the PIDA for the purpose of implementing 'the strategy of the Government of Punjab for streamlining the Irrigation and Drainage System' by 'replacing the existing administrative set up and procedures with more responsive, efficient and transparent arrangements', in order to achieve more economical and effective O&M of the irrigation and drainage in the Province, and to 'introduce participation of beneficiaries in the operation and management' of the irrigation and drainage system.

The Water Management Directorate and the PAD in general were caught completely off-guard by the PID's proposal, which quickly received the endorsement of several key officials in the Punjab Government. Nor was it easy for them to refute the apparent logic of the proposal given that a major objective of the entire irrigation reforms package was to eliminate bureaucratic inefficiency and create a more coherent management structure for irrigation system O&M. In theory, such an institutional consolidation appears to make good policy sense, as perhaps it did when it was first proposed to the PID in the 1970s when the On-Farm Water Management programme was being developed. As a practical matter now, however, the creation of an even more powerful irrigation bureaucracy certainly would make the process of devolving O&M responsibilities to representative farmer institutions and establishing PIM even more difficult than it already is. Following a period of extended bureaucratic in-fighting, a ministerial decision was taken to place the PID's proposal in abeyance 'neither formally accepted nor rejected by Government' for an indefinite period.

Overall, there continues to be little hard evidence that the PIDA/ID is following-up on its commitment to the irrigation reforms and PIM either as an institutional priority or with a coherent programme of action, whether through the National Drainage Programme or outside of it. Admittedly economic and political conditions throughout much of 1999 in Pakistan have not been encouraging for and supportive of such a bureaucratic initiative. Nevertheless, even before the most recent governmental upheaval, it was clear that the PIDA/ID remained a reluctant and an uncooperative participant in its own institutional transformation. It had not even begun the process of seriously planning its role in the implementation of the most critical of the pilot irrigation reform activities 'creating the Area Water Boards and the Farmer Organizations' much less concerning itself with whether or not it was getting the process right.

The Legal Framework for the Reforms The most important task of the Provincial Governments was to provide an appropriate legal framework to support the irrigation reforms policy and PIM. Experience in other countries already has demonstrated just how critical the supporting legal environment is to establishing effective and sustainable farmer organizations for PIM. In fact, if there is not a suitable legal basis for establishing the Farmer Organization and the exercise of the legal powers required for it to engage in canal O&M, resource mobilization and dispute resolution, the Farmer Organization will be unable to function for the purposes for which it was established. Farmers will quickly perceive the futility of such a situation and cease to actively participate in the organization.

In the early drafts of 'model' reform legislation prepared for Government consideration, a very strong legal framework was proposed to support the devolution of responsibility for secondary canal O&M and resource mobilization to newly formed and representative farmer organizations, and the establishment and operation of quasi-independent institutions for main system management such as the Area Water Boards and the PIDA. It was recognized that the Canal and Drainage Act of 1873, which has the basis for provincial administration of the Indus Basin irrigation systems for more than a century, was essentially incompatible with the irrigation reform objectives and would have to be replaced by Government. In the subsequent policy debate over the irrigation reforms that culminated with the passage of the PIDA Acts by the four provincial assemblies, that position was effectively compromised. Again, the Punjab case illustrates the current situation accurately.

The official view of the Punjab Government is that the PIDA Act confirms its determination to establish a framework of new participatory institutions for the self-sustaining operational management of the irrigation and drainage systems of the Province. However, a careful review of that legislation confirms that what the Act does for the Punjab Irrigation and Drainage Authority – specifying its constitution and composition, its management, and its powers and duties – it does not do for the proposed Farmer Organization and Area Water Board. The Act does not clearly spell out the basis of formation, legal status, powers and duties of the Farmer Organization that are to be created, and neither are these essential elements specified for the proposed Area Water Board. Importantly, it does not provide a clear corporate personality for either institution, leaving them without a secure basis to engage in the activities of resource mobilization that will be essential for them to become self-sustaining. Instead, the Act weakly prescribes that among the powers and responsibilities of PIDA, it is 'to formulate, adopt and implement policies aimed at promoting, formation, growth and development of Area Water Boards, Farmers Organizations.'[48]

Apparently the process of establishing FOs and Area Water Boards, whenever initiated by the PIDA, will be implemented through Rules notified by Government and/or Regulations notified by the Authority. However, the Act also left virtually intact all previous existing and relevant irrigation law, in particular the comprehensive and powerful Canal and Drainage Act of 1873. Thus, an approach to implementing Farmer Organizations and Area Water Boards that relies upon Rules and Regulations is much less certain, if it does not completely fail, and likely to result in weak and unsuccessful participatory institutions that are highly vulnerable to legal challenge.

Legal specialists have noted that the PIDA Act may be the first legislation of its kind in Punjab that vests the powers and functions of various laws in the organization that it creates (the PIDA) while simultaneously allowing those laws to remain in force unchanged. One consequence of this unusual situation is that it can not be successfully remedied through the rule and regulation making process. That is because laws established through rule and regulation are subordinate and can not prevail over the superior law of legislative Acts.

Under the circumstances, it is not surprising that the PIDA Act so far has not been an effective instrument for implementing the irrigation reforms and PIM in Punjab. The PIDA/ID appears to be fully aware of its limitations. For example, it points to legal uncertainties as the primary reason for its inability to identify anything more than a very limited role for FO participation in canal management and resource mobilization when negotiating with the unofficial pilot Farmer Organization projects for their formal recognition.[49] Because the CDA has been maintained completely in tact in the PIDA Act, the PIDA/ID so far has been able to use it to fully retain the very powers and responsibilities that the irrigation reform proposals intended would be devolved to the Farmer Organization.

The alternative for Government, if indeed it is seriously committed to the irrigation reforms and PIM, is to provide a more suitable and unambiguous legal framework for the establishment and operations of the Farmer Organizations and the Area Water Boards. This could be accomplished either through the replacement of the existing PIDA Act by a new Act, or by selectively amending the present Act. At this time, however, Government does not appear to be disposed to adopt either strategy; instead it is content to maintain the *status quo* of legal ambiguity.

That does not augur well for the future of the irrigation reforms and PIM in Pakistan. It significantly raises the odds against a successful pilot program of new, representative institutions for farmer participation in irrigation management because the PIDA/ID retains its full legal powers and remains in a position of top-down control over their development and operation in the critical pilot testing phase. It is difficult to be confident that the largest bureaucratic organization in each of Pakistan's provincial governments will be able to facilitate and ensure the success of those very institutions that are intended to replace it in controlling the distribution of and access to the country's most valuable and largest quantum of irrigation water.

Last Observations

In the post-Independence decades in Pakistan, several major multilateral donor-assisted water sector interventions facilitated the continuation of what David Gilmartin has described as the efficient, engineering-based 'scientific' management (administration) of irrigation. Those primarily engineering interventions created the physical fabric of the late 20th century Indus Basin irrigation system in which was embedded a hierarchical socio-economic system of landed gentry and feudals largely unchanged from the pre-Independence period, and that through post-Independence politics emerged to dominate and control the irrigation system for its primary benefit.

However, in the present decade, a set of proposed irrigation policy reforms threatened to upset that dominance by devolving primary responsibility for secondary canal O&M to representative Farmer Organizations as well as transforming costly, inefficient and engineering-dominated irrigation bureaucracies into quasi-independent irrigation authorities modelled to function along the lines of private water utilities. The reforms are to be implemented in the context of the National Drainage Programme, a massive $750+ million co-financed and World Bank-led water sector project.

The National Drainage Programme is intended to initiate the process of sustaining Indus Basin irrigated agriculture, as envisioned in the Government of Pakistan's Water Sector Investment Plan. In the National Drainage Programme, the key institutional changes and irrigation reforms were planned to begin first. Only afterward would the major funding for additional engineering technology fixes be forthcoming to repair the severely decayed physical fabric of the irrigation system and to create new drainage infrastructure to facilitate system sustainability. Unfortunately, that script has not been followed, and the reverse already has occurred.

Among the key groups of participants in the irrigation reforms and PIM in Pakistan, it appears at this juncture that it is mainly the ordinary, average farmer that has made the most promising and effective progress in the implementation process. That progress is evidenced primarily in the 'unofficial' pilot Farmer Organization development projects in Punjab, Sindh and the NWFP. Admittedly it has not been achieved without difficulty as, for example, the brief experience of the Sindh pilot Farmer Organization projects with actual joint management of secondary canals in late 1997 illustrated. However, in the extended learning process of participatory organization development, mistakes are part of the process; far more important is whether or not learning occurs from them. It remains unclear when officially recognized or sanctioned pilot Farmer Organization development activities will begin as intended under the National Drainage Programme. Even less certain is the time frame for implementing such PIM activities as canal O&M and resource mobilization. It is clear, however, that it is not farmer unwillingness or reluctance to participate that is obstructing and delaying the process.

Disappointingly, Government in Pakistan has not yet seriously followed through on its apparent commitment to create an enabling and supporting environment for the irrigation reforms, in particular for the new farmer-based institutions. Perhaps the most important component of that environment is a strong and unambiguous legal framework to support the reforms, and that does not yet exist. Without such an environment, it is doubtful that the new institutional vehicles for farmer participation in irrigation management can be successful. Government certainly has not gotten the process right thus far, although it also must be acknowledged that over the last half of 1999 it has been engaged by several more pressing issues than the irrigation reforms.

Curiously, the multilateral donors, so prominent in the initial formulation of the irrigation reform policy agenda for Pakistan have retreated into relative quiescence. The reasons for this are not clear, but they may be related to the often mentioned condition of 'donor fatigue' and preoccupation with other elements in the continuing project cycle. This would suggest that multilateral donors have a limited capacity to sustain support and pressure for institutional change over an extended period. There is little reason to doubt that the worsening state of the Pakistan economy over the past year has been a factor in their relaxation of a hard line on the institutional preconditions for moving forward with implementing other National Drainage Programme activities. Unfortunately that also likely has cost the multilateral donors important leverage to move Government to become more proactive in creating a supporting environment for the irrigation reforms.

Although progress in implementing the irrigation reforms and PIM in Pakistan has been limited to date and there are legitimate reasons to be sceptical about the likelihood of its success in the foreseeable future, the fact remains that the changes in irrigation institutions that have occurred so far could hardly have been anticipated a decade ago. Because our attention span is short and our impatience with the slow rate of institutional change great, it is easy to lose our perspective on what is only the first stage in what will be a lengthy learning process if farmer participation in irrigation management is to be successful. As the participating institutions learn to be effective, mistakes will be common and the required resource inputs high relative to results. Getting the process right is no easy task.

Notes

1 Gilmartin (1994:1137, emphasis added).
2 The World Bank also remained the primary lender to Pakistan for its subsequent SCARP Projects.
3 The principal exception here had always been the Irrigation Department's revenue officers and field staff.
4 Painter et al. (1982:7-9). Farmer participation in subsequent OFWM Projects also has included a progressively increasing financial component comprising both labor and cash.
5 Although it has yet to occur in NWFP and Sindh Provinces, the WUA Ordinances

permitted the formation of farmer organizations – essentially federations of WUAs – at the distributary canal level. However, this was not the case in the Punjab WUA Ordinance, which limited the farmer organizations to the watercourse level. See Federal Water Management Cell (1987: 15-59).

6   Kerry J. Byrnes (1992) provides an extensive review of the WUA experience in World Bank-assisted OFWM Projects in Pakistan.

7   The World Bank (1993). That report was based upon the findings of two World Bank missions to Pakistan in 1992.

8   That effort has continued through the Second SCARP Transition Project and the ongoing Punjab Private Sector Groundwater Development Project.

9   As a policy matter, private groundwater development had been banned in all SCARP project commands, although by the mid-1980s, it was apparent that this policy was no longer being seriously enforced by the PIDs.

10 The World Bank (1993:iv).

11 WAPDA (1995).

12 A consistent and generally accepted usage of organizational terminology for these new farmer organizations has not yet emerged in Pakistan, although WUA (Water User Association) is used everywhere to refer to the watercourse-based farmers' organization. The use of WUF has been particularly confusing because in the National Drainage Programme and other irrigation reform documentation it has been the acronym used for Water User Formation, a generic organization of farmers above the watercourse level. However, in the pilot organizing work of various agencies and in some research reports, WUF stands for Water User Federation, a grouping of WUAs at the canal command level. Other agencies have used WUO (Water User Organization) either as a generic term for WUA and/or WUF, or to refer to an intermediate grouping of WUAs. Eventually a consistently used terminology will emerge, but until then, researchers and others will have to pay careful attention to contextual definitions. In this paper, FO (Farmer Organization) and WUF will always mean the distributary (secondary) canal level organization of farmers; WUA will always mean the watercourse level organization of farmers.

13 *Dawn*, December 6, 1995. However, in a communication to the GOP in early November, 1995, the World Bank already had formally clarified that it did not advocate privatization of the Indus Basin irrigation and drainage systems, but rather it was seeking decentralization and increasing participation of farmers in the management of those systems.

14 Spokespersons and members of the first two groups were often one and the same, *e.g.* the leader of the Farmers Association of Pakistan (FAP) also was State Minister for Parliamentary Affairs in the Federal Government and Chairman of the Institutional Reform Group, as well as a very large landowner. The same was frequently true in the case of the latter two groups, e.g. the President of the Pakistan Engineering Congress (PEC) also was a long-time Special Advisor to the Secretary, Punjab Irrigation Department. Other cross-linkages between these groups were common as well, *e.g.* a large and influential landowner, formerly an Executive Engineer of the Punjab Irrigation Department, was Vice-President of PEC, and also the owner of a successful engineering consulting firm that worked primarily in the irrigation sector.

15 Examples of all abound. By mid-1996, it was practically impossible to discuss the subject of irrigation with Punjab farmers in the field without encountering stories about how government was proposing to sell distributary canals to foreigners who would then collect rents and water charges from shareholding farmers in those commands, or sell their water rights to others if they did not pay. The commonly

mentioned source of the story was the canal *patwari*. In an April, 1996 article in the English language newspaper, *The Muslim*, under the headline 'Privatization of canal system to be disastrous for economy' spokesmen for FAP and PEC were quoted as saying: 'What was being proposed was not even genuine privatization. The plan is to sell irrigation channels to big landlords under the umbrella of this newly created Provincial Irrigation and Drainage Authority.' 'It is a diabolical scheme against the rural masses and our agricultural economy. The proposed law denies water rights to poor farmers by changing entitlement of irrigated lands to water by making canal irrigation water freely and independently tradable to land owners with money and under their own authority.' 'Under the guise of better water distribution what is being planned will prove to be disastrous for small landowners. Big landlords are being given a handle to oust small landowners or farmers from the land by cutting their water supply.' (The Muslim, April 9, 1996). In the Urdu daily, *The Daily Nawa-e-waqt*, under the headline, 'World Bank control over Punjab water, Management of two canals handed over,' it was reported that 'The Punjab Government has delivered to the World Bank management of two canals of Sahiwal and Faisalabad . . . .' and that '. . . World Bank control over Pakistan's water resources will increase our agricultural and national economic problems . . . .' (*The Daily Nawa-e-waqt* July 10, 1996).

16 Both the World Bank and the Asian Development Bank had attached conditionalities to their proposed loan agreements for the National Drainage Programme requiring formal provincial government agreement to pass legislation which would enable the irrigation reforms *before* the loans could be finalized.

17 Reported in *Dawn*, July 22, 1996.

18 Specifically the removal of Benazir Bhutto as Prime Minister and the fall of the second PPP led government. One interesting proposal from the large farmer associations was an offer '. . . to pay a flat rate as water charges in order to generate resources . . . rather than depending on (the) IMF and ADB. Still, the vested interests did not agree with us,' a spokesman was quoted as saying. (*The Nation*, December 6, 1996).

19 The GOP had been informed that the deadline for expiry of ADB's loan commitment for the National Drainage Programme was December 19th (*The Nation*, December 6, 1996). Without ADB co-financing, the World Bank had indicated to the GOP that it would be unable to go forward with its own financing of the National Drainage Programme.

20 With the caveat that there is scarcely any public record of the views of the small farmer diaspora, primarily because they have had no real representation in this process and practically no one has bothered to ascertain their opinions on the various elements that comprise it.

21 Tirmizi et al. (1998:3-4).

22 A major irony of PIM in Pakistan is that from the perspective of the average farmer, it has been participatory generally after the fact. To the extent that any participation or consultation occurred, it was dominated by the large, influential farmers – some of whom were also members of the national or provincial assemblies – and the lobby groups they control.

23 The warabandi specifies each shareholders time and length of turn for directing the flow of canal water in the watercourse to his fields from his *nakka*, the specific location in the watercourse at which the shareholder takes his irrigation turn.

24 As the rural population has grown and the number of shareholders in the watercourse *warabandi* increased, coupled with the increased awareness that access to canal water is essentially a zero sum game, this understanding has been driven home more and

more powerfully. Both an informal organization and informal institutional arrangements work best in separating *warabandi* from *dharabandi*, and this is likely a major reason for the lack of interest and effort in sustaining WUAs.

25 The pattern of increasing separation of *dharabandi* and factionalism from irrigation operations was first recognized in 1992 and documented by Jamshed Tirmizi in the Punjab during the final project evaluation of the USAID/World Bank-assisted Command Water Management Project. It was subsequently confirmed by him and further documented in the command of the Lower Bari Doab Canal system (in a study of the Sociology of the Watercourse and Irrigation Communities of the LBDC, as part of the feasibility study for the Second Irrigation and Drainage, Sukh Beas/Lower Bari Doab Canal Project in 1994); in southern Punjab in the Abbasia, Desert and Muzaffar Garh canal commands (during the evaluation of the ADB-assisted Second Punjab On-Farm Water Management Project and during project preparation for the proposed third ADB-assisted Punjab On-Farm Water Management Project); and in the central Punjab canal commands of the Lower Chenab Canal and the Pakpattan Canal systems (during project feasibility studies for the ADB-assisted Punjab Irrigation Management Project, now the Punjab Farmer-Managed Irrigation Project).

26 See the discussion in Ostrom (1992:67-79).

27 This is based in part upon intensive PRA focus group meetings conducted by the authors in June, 1996, with farmers in 10 different watercourse commands divided evenly between Dijkot Distributary (LCC system) and 12-L Distributary (Pakpattan system), Punjab; the meetings were open to all watercourse shareholders and designed to obtain farmer responses to specific FO/PIM proposals. More than 150 farmers participated in the PRA focus group meetings. In these meetings, farmers consistently stated that without a strong legal basis a distributary FO could not function and PIM would not be possible. Specifically, farmers cited the inability of any FO to assess and collect a water rate or service fee without appropriate legal powers; similarly they insisted a FO would require suitable powers to manage the sensitive issues presently falling under Sections 20 and 68 of the Canal and Drainage Act.

A follow-up farmer consultation among 17 (additional) watercourse communities in the same two distributary canal commands was completed in 1998. It reconfirmed the strong farmer fear that without powers similar to those of the PID under the Canal and Drainage Act, any irrigation farmers' organization responsible for distributary management will be ineffective (Tirmizi, et al., 1998:6-8). IIMI social organization teams reported similar concerns voiced by water users during their organizational development activities in three pilot FO projects in Sindh (Bandaragoda and Memon, 1997:43).

28 All of the pilot projects pre-date the National Drainage Programme by at least two years and have been implemented in the context of other irrigation sector projects, all of them with World Bank (and for a few, later with OECD) assistance. Thus they are separate from the pilot FO development and PIM activities planned to occur under the National Drainage Programme, although it is clear that in most cases the implementing agencies had anticipated or hoped that when the National Drainage Programme became operational, these pilot projects would be subsumed under it. As of late 1999 that had yet to formally occur in any case. Five of the pilot projects have been implemented directly or indirectly through provincial water management directorates; the remaining four pilots have been implemented by the Pakistan Division of the International Irrigation Management Institute (IIMI), now the International Water Management Institute (IWMI). Not included in this discussion is an interesting pilot project in Sindh Province being implemented by Oxfam Sindh in the Rahuki Canal

Irrigation Policy Reforms in Pakistan 235

command the objectives of which are not so explicitly (and narrowly) FO development and PIM. Documentation for this activity has not been readily available, however, see Khoso, et al. (1996) for a brief description of it.

29 See, for example, the discussion in Bandaragoda and Memon (1997:21-7), and Bandaragoda et al. (1997:35-65).

30 Individual WUAs form the base level; for large canals, an intermediate organization is formed from WUA representatives within specific hydraulically-defined command areas; and at the top for the entire canal command is the FO formed around representation from either the base or intermediate level.

31 The extent to which that experience has been institutionalized and processed for later transfer and application in an expanded program of FO development as is envisaged under the National Drainage Programme and the OFWM IV Project, however, remains to be seen.

32 Vander Velde, *INPIM Newsletter* 1998.

33 In several pilot projects, the FO has mobilized local resources for a major canal desilting and maintenance activity in which much of the general membership has participated. However, in most cases these have been one time events. Although positively indicative of FO resource mobilization potential, such episodic activities are not a true test of FO capabilities to mobilize O&M resources on a sustained basis.

34 Specific draft terms of reference in the MOU maintained the full, complete and final authority of the Executive Engineer and Sub-Divisional Officer, as Canal Officers, in *all* matters of canal operations. Any operational role or responsibility proposed to be granted the FO could be overridden at the discretion of the responsible Canal Officer. The FO was assigned tasks for which completion would incur significant financial expenses, but for which no mechanism was provided for the FO to recover its costs, whereas the Divisional Canal Officer would be able to recover any costs or charges from the farmers resulting from any failure of the FO to perform its tasks. See 'Draft Memorandum of Understanding' and 'Draft Notification Assigning the Responsibility of Collection of Canal Dues to Finance Secretaries of Water Users Association on 4-R/Hakra Distributary System' (undated, but confirmed as 1998).

35 The common condition of outlet tampering in most Indus Basin canal commands provides an additional reason for anticipating such occurrences. For every 'sale' of a tampered watercourse turnout by PID field staff there have been willing buyers, and they are not always the large and influential landowners. That farmer understanding they are involved in a zero sum game *vis-à-vis* their canal water supply is increasingly widespread provides no assurance that attempts to manipulate the environment for individual or group benefit will not occur.

36 It appears likely that pilot FO projects under the National Drainage Programme, and the OFWM IV Project as well, will be implemented with irrigation agency field staff remaining present in the canal commands selected for these activities. The pilot projects are intended to 'test' the viability of the PIM concept, with FO participation in canal O&M and resource mobilization activities expected to proceed in a 'phased manner' over an extended period of time. It is difficult to be sanguine about the success of such an approach. Given the intensity of their opposition to the reforms, it would be naive to expect that irrigation agency personnel will not take advantage of opportunities to manipulate water supply and other O&M events in ways intended to cripple if not cause the FO pilots and PIM to fail outright. A vigorous FO oversight capacity and program of monitoring may provide a positive counter to such efforts.

37 The experience of the two pilot FO projects in NWFP illustrates the problem. There, the FIDA has continued the NWFP Irrigation Department's policy of ignoring,

delaying and rebuffing every initiative of the NWFP Water Management Directorate to formally recognize the FO's legal right to participate in secondary canal O&M and resource mobilization activities, even though NWFP's OFWM Ordinance specifically legitimizes secondary canal water user organizations.

38   Again, the NWFP pilot projects illustrate the problem. The social organizers assisting FO development there have been ordered by their superiors in the Water Management Directorate to secure 100% farmer compliance with watercourse improvement in the pilot canal commands, ignoring both WUA decisions to the contrary as well as local soil conditions which preclude the economic viability of watercourse improvement. In the Punjab pilot FO projects of OFWM, much more emphasis has been put upon watercourse improvement in the canal command and farmer use of such OFWM services as precision land leveling by the OFWM support teams than on developing FO O&M capabilities.

39   By any of several widely used measures – e.g. the productivity of irrigated agriculture, return on O&M investment, water delivery efficiency, equity in water distribution, O&M cost recovery – irrigation system performance in Pakistan is low. Several of these were discussed succinctly in the World Bank's 'Issues and Options' report. The World Bank, 1993:2-15. By at least one performance measure, however, Pakistan's irrigation does reasonably well, *viz.* productivity per unit of irrigation water. When their crop water use efficiency is compared to Indian and California experiment station data, the top 10% of Punjab wheat farmers apparently are on a par (M. Akhtar Bhatti et al., 1991). Although this is not the place to address the topic, such findings raise important questions concerning the choice of appropriate irrigation performance measures.

40   Arguably it was the warning that ADB co-financing of the National Drainage Programme would no longer be possible and future support to the irrigation sector jeopardized, if government's formal commitment to enact appropriate reform legislation was not forthcoming by the third week of December, 1996, that was instrumental in that decision, culminating in the promulgation of the PIDA Ordinances early in 1997.

41   To be fair the multi-lateral donors did sponsor two activities that may have been related to such an objective. Several 'study' trips or tours to such countries as Spain, Morocco, The Philippines, and Mexico where various irrigation institutions and policies similar to the reform proposals are operative were arranged for a significant number of key senior federal and provincial government water sector bureaucrats. One objective or rationale for those trips may well have been to build consensus at a certain level of government for the reform proposals. However, this technique has so long been a feature of irrigation sector development projects in Pakistan that it has become corrupted into a 'perk' of those positions, typically available to whomsoever occupies them during project preparation and/or implementation. It is rare for anyone other than government officers or 'officials' to participate in such 'study' tours, and since they are widely acknowledged to be primarily opportunities for international travel, shopping and visiting friends/relatives residing abroad, it is doubtful that much significant learning results from them.

A corollary activity was the organization of two PIM seminars or workshops in Pakistan in 1995 and 1996 by the World Bank's EDI unit. Invited international specialists in the field and representatives of relevant national and provincial agencies as well as research organizations in Pakistan working in the irrigation sector met for several days to discuss papers and exchange views on the proposed institutional reforms, especially the farmer participation component. Although the dialogue was

42 undoubtedly useful for the participants, there is no evidence to confirm that its impact spread much beyond those who were immediately involved.

42 Just as that situation was a significant factor in the GOP's decision to formally commit to adopting reform legislation in late 1996 rather than risk the possible loss of significant external financing.

43 Of course the PIDs were not the only governmental organizations in Pakistan that have been involved in the process of developing the National Drainage Programme and the irrigation policy reforms contained therein. Various units of the Federal Government's Ministry of Water and Power (e.g. the Office of the Chief Engineering Advisor and the Water Wing of WAPDA) and the Ministry of Finance (e.g. the Planning and Development Division) also were involved at different stages of the process, as is the normal condition in Pakistan for the formulation and approval of any major, internationally co-financed, multi-year water sector development programme comprising a set of complex, inter-linked projects. Counterpart organizations at the Provincial Government level (e.g. the Planning and Development Departments, the Finance Departments) also were active participants at various stages in the evolution of the National Drainage Programme. Throughout, however, the PIDs were central in the process both as actors and objective focus.

44 Many PID officers were quick to offer as 'proof' of this argument the evidence that with all of its professional and technical competency as well as the 'powers' at its command provided by the Canal and Drainage Act, the PID itself was no longer able to operate and maintain the system as designed and intended!

45 Unlike the case in other government departments in Pakistan, from the earliest days of British establishment of the Punjab Irrigation Department (and the PIDs which ultimately emerged from it) all key administrative positions, to and including the position of the Secretary, have been occupied by an officer cadre trained as engineers and promoted from within the organization. Although there were exceptions to this tradition, they were relatively few and far between. The irrigation reforms threaten to permanently change this situation by opening PIDA positions to recruitment from a wide range of educational or professional backgrounds and applicable institutional management experience, including the private sector.

46 Asrar-ul-Haq (1998:126-28). These and other reasons now commonly advanced by PID spokespersons to justify the continuing delays in reform and PIM implementation are largely self-serving and/or have little foundation. Indeed, available empirical evidence is often contrary. Irrigation-focused farmer organizations are not difficult to establish, in Punjab (or elsewhere in Pakistan, for that matter) as the several unofficial pilot FO projects (and two decades of OFWM experience) have confirmed. It is doubtful that many (if any) Farmer Organizations would undertake direct control of canal O&M and resource mobilization as farmers are well aware that their own expertise (and interest) lies in farming. Moreover, abundant (and more cost-effective) technical expertise and professionalism in irrigation management is available for hire in the market place. Arguably organized farmers are better able to manage relations with their influential neighbors and local politicians (whom they already know well) than are the employees of a government agency penetrated by patronage and corruption. In any case, there is no rationale for holding farmer organizations to a higher standard here than other institutions of Pakistani society. There also is much evidence to confirm that farmer organizations in the Punjab are 'sustainable' so long as there are good reasons for their continued existence, but those reasons are determined by farmer members, not outsiders. That WUAs have had a very good rate of repayment of loans advanced to them for watercourse improvement under

successive OFWM Projects, whether or not the WUA remains in formal operation, has been conveniently ignored.

47 Asrar-ul-Haq (1998:131). The author is a Superintending Engineer in the Punjab Irrigation Department, and currently assistant to the Secretary, PID and Managing Director, PIDA for institutional matters.
48 The Punjab Irrigation and Drainage Authority Act, 1997, Section 5, Paragraph 24.
49 This is evident as well in the implementation of projects in which the PIDA/ID is directly involved. For example, in the Punjab Private Sector Groundwater Development Project, an attempt in late 1998 to form Farmer Organizations for the limited purpose of farmer participation in minor canal lining and rehabilitation floundered because of the PIDA/ID's inability to ensure the Farmer Organizations would have legal corporate status under the existing law. Without such status, the FO could not legally collect monies from its membership or hold funds in a bank account in its name; neither could it enter into a binding contractual agreement for even the limited participatory purposes of the Project.

References

Anwar, Arif A. (1996), *Provincial Irrigation and Drainage Authority: The Future of the Irrigation Department?*, Monograph privately published by the author, Peshawar.

Asrar-ul-Haq (1998), *Case study of the Punjab Irrigation Department*, Report no. C-12, International Irrigation Management Institute, Lahore.

Bandaragoda, D. J. and Yameen Memon (1997), *Moving Towards Participatory Irrigation Management*, International Irrigation Management Institute, Lahore.

Bandaragoda, Don Jayatissa, Mehmood ul Hassan, Zafar Iqbal Mirza, Muhammad Asghar Cheema and Waheed uz Zaman (1997), *Organizing Water Users for Distributary Management; Preliminary Results from a Pilot Study in the Hakra 4-R Distributary of the Eastern Sadiqia Canal System or Pakistan's Punjab Province*, International Irrigation Management Institute, Lahore.

Bhatti, M. Akhtar, F. E. Schulze and G. Levine (1991), 'Yield measures of irrigation performance in Pakistan', *Irrigation and Drainage Systems*, 5:183-90.

Byrnes, Kerry J. (1992), *Water Users Associations in World Bank-Assisted Irrigation Projects in Pakistan*, World Bank Technical Paper no. 173, The World Bank, Washington, D.C.

Federal Water Management Cell, Ministry of Food, Agriculture and Co-operatives, Government of Pakistan (1987), *Co-operatives and Water Users' Associations and the Ordinances for the Provinces*, Islamabad.

Gilmartin, David (1994), 'Scientific empire and imperial science: Colonialism and irrigation technology in the Indus Basin', *The Journal of Asian Studies* 53(4):1127-49.

Khoso, Abdul Hakeem, Fauzia Rauf and Aijaz Nizamani (1996), *Report on Water and Population Dynamics in Rahuki Canal Area, Hyderabad, Pakistan*, IUCN, Karachi.

Ostrom, Elinor (1992), *Crafting Institutions for Self-Governing Irrigation Systems*, ICS Press, San Francisco.

Painter, James E., E. Baldwin and Sandra Malone (1982), *The On-Farm Water Management Project in Pakistan*, Project Impact Evaluation no. 35, U.S. Agency for International Development, Washington, D.C.

Tirmizi, Jamshed (1998), *Facilitating Capacity Building and Participation Activities Project* (Stakeholder consultation for receptiveness in organizational and physical change on

Punjab irrigation canals), TA No. 5592-REG, Asian Development Bank, Seer Pvt. Ltd., Lahore.
Vander Velde, Edward J. (1998a), *Progress in participatory irrigation management in Pakistan: a report on pilot projects in farmer organization at the secondary canal level*, Consultant's Report, Asian Development Bank, Manila.
Vander Velde, Edward J. (1998b), 'PIM pilot projects in Pakistan', *INPIM Newsletter*, no. 8 (December).
Vander Velde, Edward J., Jamshed Tirmizi and Robert Yoder (1997), 'Achieving PIM in Pakistan', *INPIM Newsletter*, no. 7 (December).
Water and Power Development Authority (WAPDA), Government of Pakistan (1995), *Second Irrigation and Drainage, Sukh Beas/Lower Bari Doab Canal Project (Feasibility Study). Final Report (4 vols)*, National Engineering Services Pakistan, Lahore.
The World Bank (1993), *Pakistan Irrigation and Drainage: Issues and Options*, Report no. 11884-PAK, Washington, D.C.
The World Bank (1997), *Pakistan National Drainage Program Project*, Staff Appraisal Report, Report no. 15310-PAK, Washington, D.C.

Pakistan Daily Newspapers:
The Daily Nawa-e-waqt
Dawn
The Muslim
The Nation

Chapter 8

Capture and Transformation: Participatory Irrigation Management in Andhra Pradesh, India[1]

Peter P. Mollinga, R. Doraiswamy and Kim Engbersen

Introduction

Since 1996-97 the State of Andhra Pradesh in south India is witnessing the implementation of a dynamic reform process in the canal irrigation sector, aiming to introduce PIM (Participatory Irrigation Management) in the irrigation systems of the State. In the South Asian context it is a rather special reform effort, as it receives strong political support of the State government – no lack of the infamous 'political will' – and is implemented by a dynamic group of committed reform managers. It is also the first large-scale effort at delegation of substantial water management powers to water users in the South Asian canal irrigation sector. The process has attracted a lot of national and international attention, and reference is now made to the 'Andhra model'.

This paper reports on fieldwork conducted in two secondary canals (called distributaries in India) in the Tungabhadra Right Bank Low Level Canal irrigation system.[2] The paper investigates what has happened to the irrigation reform policy 'on the ground'. The argument moves at two levels. The first is a description of the impact of the implementation of the reform package by looking at:

- The constitution and composition of the Water Users Associations (WUAs) and Distributary Committees (DCs) that have been formed;
- The conduct of the canal system rehabilitation works;
- The impact of the introduction of Participatory Irrigation Management on water distribution practices;
- The impact of Participatory Irrigation Management on area irrigated;
- Change in the relationship between farmers/water users and the Irrigation Department.

At the second level the impact data is used for an analysis of the way local interest groups have captured and transformed the reform policy aiming at the introduction of Participatory Irrigation Management. Instead of measuring success or failure in terms of achievements against stated objectives, the analysis of policy transformation looks at the process that has been induced by the policy reform initiative, and the constraints and opportunities for further reform inherent to that process.

Below we begin with an outline of the 'politics of policy' approach that underlies the paper's analysis of policy transformation (section 2). The paper then briefly describes the characteristics of the Andhra Pradesh irrigation reform experiment (section 3), after which the impact of the reform programme is presented (section 4). Section 5 analyses the capture and transformation of PIM policy 'on the ground'. Section 6 contains the conclusions of the paper.

The Politics of Policy Implementation

Irrigation management as a form of water control has three dimensions: 1) a technical dimension related to the regulation of physical processes, notably the flow of water, 2) an organizational dimension related to the regulation of human behaviour in day-to-day irrigation practices, and 3) a socio-economic and political dimension referring to the wider societal conditions of possibility for particular management practices to take place. These three dimensions are intimately related, and policies that seek to achieve changes in irrigation management therefore have to address all three simultaneously (Bolding, Mollinga and van Straaten, 1995; Mollinga, 2003).

In regions like South India where irrigation water is a scarce resource and livelihoods directly depend on it, water control is contested. This contestation is an inherently political process when politics is understood as the set of activities through which the balances of power that shape resource use are (re-)negotiated. The political contest around water use takes place at different levels (Mollinga, 2001). The first is the level of the everyday politics,[3] the day-to-day struggle over irrigation management. The second is the level of the politics of policy,[4] the social process in which policy formulation and implementation are contested. The third level is official state politics and inter-state politics. For the latter the term hydropolitics is commonly used.[5] The fourth level finally is the newly emerging level of the global politics of water. The recently held World Water Forum (The Hague, March 2000), aiming at the development of a world water vision and framework for action, is an example of this.

This paper focuses on the politics of irrigation policy implementation. Analysis of the politics of policy is a counterpoint to dominant models of linear or rational planning.

> According to [the latter] view, a proposed reform gets on the agenda for government action, a decision is made on the proposal, and the new policy or institutional arrangement is implemented, either successfully or unsuccessfully (Thomas and Grindle 1990, p.1164).

Implementation is seen as basically a technical task, to be undertaken by a strong enough institution with professional managers. When reforms do not materialize as envisaged, institutions need to be strengthened, managers trained, and enforcement monitored better.

Warwick calls this approach the 'machine model' of policy implementation, derived from classic administrative theory (1982, pp.40-3).

> This theory views implementation as a quasi-mechanical exercise in which organizational units and individual implementers form a delivery system and program clients become receptacles for the services delivered (ibid., p.40).
> It assumes that a clearly formulated plan backed by legitimate authority contains the essential ingredients for its own implementation (ibid., p.179).

Implementation in this view requires hierarchical authority, trained staff and close supervision. In practice, so it is argued, implementers' behaviour is always to some extent diluted, but the machine metaphor is strong both at the level of policy discourse and policy implementation.

In South Asia policy making and implementation is characterized by strongly legalistic and administrative approaches, with a top-down nature. This is also true of the Andhra Pradesh irrigation policy reform exercise. It exhibits all three characteristics: strong authority, that is support from the highest political level; large investment in the training of policy implementers, both government officials and farmer office bearers; and an intensive monitoring system, including both statistical data collection and direct personal monitoring of field staff by high level bureaucrats.

The contrasting approach emphasizes a 'policy as process' perspective, while the linear planning framework can be characterized as a 'policy as prescription' approach (Mackintosh, 1992). A process-oriented framework starts from the observation that the outcomes of policy implementation are highly variable. Implementation is an ongoing, complex and interactive process of decision-making by the different interest groups involved: governments, managers, and 'beneficiaries'.[6] Policy implementation is an example of strategic action in which a government agenda becomes articulated with local interests, the policy content is renegotiated and transformed, and particular intended and unintended outcomes are produced. Analysis of such processes focuses on the interests and motivations of relevant actors, their strategies, the resources (financial/economic, material, social, cultural, political et cetera) they mobilize, the structure of the implementation process, and the outcomes produced.

Warwick calls this the 'transactional model' of policy implementation. He summarizes the seven assumptions on which the transactional view of policy implementation is based as follows (Warwick, 1982, pp.181-84).

1. Policy is important in establishing the parameters and directions of action, but in never determines the exact course of implementation.
2. Formal organization structures are significant but not deterministic in their impact.
3. The programme's environment is a critical locus for transactions affecting

implementation. 'The essence of implementation in the transactional view lies in the coping with environmental diversity, uncertainty, and hostility' (ibid., p.182). Environments are multiple, shifting, and difficult to predict.
4. Judged by its impact on implementation, the process of policy formulation and programme design can be as important as the product.
5. Implementer discretion is universal and inevitable.
6. In human services programmes, clients have a potent influence on the outcomes of implementation.
7. Implementation is inherently dynamic.

The case study will illustrate all of these assumptions about policy implementation in practice. However, it will also become clear that the design for the implementation of the Andhra Pradesh irrigation reform policy cannot be fully adequately characterized with the 'machine' metaphor. Though exhibiting some of its characteristics it also has consciously integrated 'transactional' elements.

Map 8.1 Andhra Pradesh and research site in India

The Andhra Pradesh Irrigation Reform Policy for PIM

Andhra Pradesh is a south Indian state with a population of 66.5 million people in 1991, and a surface area of 275,000 km². The state has three distinct regions: coastal Andhra, Telengana and Rayalaseema. The first is the eastern, low-lying flat part of the state: the deltas formed by the Godavari and Krishna rivers. It has a long history of agricultural intensification related to irrigation development. This was accelerated in the 19th century by the British colonizers, when they started to construct more weirs and irrigation canals (Atchi Reddy, 1990; Rao, 1985; Upadhya, 1988). Paddy cultivation is the core of the agricultural economy in this region.

The latter two regions, Telengana and Rayalaseema, are the upland parts of the state. They are hot and drought prone areas, with low and erratic rainfall (ranging around 600 mm per year), and a hilly and often rocky landscape. Large-scale canal irrigation development in this region dates from the 20th century, and mainly took place after Indian Independence in 1947.[7] The peculiarity of the canal irrigation systems in these two regions, as in most of upland south India, is that most of them have been designed as so called protective irrigation systems. The design of the systems envisaged thin spread of water over large areas for supplementary irrigation of primarily 'light' or 'irrigated dry' crops. These are crops like sorghum, millet, oilseeds and cotton. Only small areas of paddy and sugarcane were allowed. The water allocation system put in place after Independence to effectuate this is called the system of localization. It is a form of land use planning that prescribes per cadastral unit which crops farmers are allowed to grow and not to grow using irrigation water, and in which season (for detailed discussion of the concepts of protection and localization see Mollinga 1998). In the Tungabhadra Right Bank Low Level Canal that is the subject of this paper, the objective of spreading water has resulted in a design that gives villages pockets of localized area, with long stretches of canal and officially non-irrigable area in between (like melons on a vine).

The protective irrigation systems of south India provide classical cases of unequal water distribution in the head-tail pattern. Instead of thin spread of water the actual pattern is concentration of water use in upstream reaches of canals, much more paddy cultivation as envisaged as a result, and deprivation of tail-end areas. In the Tungabhadra Right Bank Low Level Canal there is also a lot of 'unauthorized irrigation', that is irrigation in the non-localized areas along the canals leading to the localized pockets near the villages. (For documentation of water distribution in protective systems see for example Wade, 1988, Ramamurthy, 1995; Mollinga, 2003).

Performance problems in the large-scale canal irrigation systems have been acknowledged since the mid-1960s (see GOI/PC/PEO, 1965). In the early 1980s the Irrigation Utilisation Committee constituted by the government of Andhra Pradesh produced a report that was highly critical of canal system performance and management practices (GOAP, 1982). In the 1980s and 1990s there were several efforts, by both government and NGOs to improve the working of the systems through farmer participation in management. These efforts concentrated on farmer organization at the lower levels of the system through outlet committees. They remained isolated pilot projects and were not sustainable because the issue of main system management (and Irrigation Department performance) remained unaddressed: local organization alone cannot solve erratic supply and distributional issues at system level (Wade and Chambers, 1980). By the mid-1990s the situation in Andhra Pradesh regarding canal irrigation was very similar to that in many other states: decaying canal systems, very low recovery rates of water charges and unequal water distribution, leading to a general perception that the systems perform far below potential.

In 1996 a new irrigation reform exercise was started as part of the overall economic and institutional modernization programme of the Chief Minister of the state, Chandrababu Naidu. The genesis of the irrigation and other reform programmes started by the Chief Minister has to be explained from a combination of general political factors (reproduction of political support through directly reaching out to local leaders/organization, circumventing the tiered system of elected representative bodies), socio-economic and financial factors (budgetary constraints, stagnating economic development), and sector specific problems that needed to be addressed (like poor production and financial performance in the irrigation sector). However, detailed discussion of the emergence of irrigation reform policy is outside the scope of this paper (see Narasimha Reddy, 1999 for more discussion). Whatever the overall political economy of the Andhra Pradesh reforms, it provided a window of opportunity for advocates of irrigation reform, which was used with great vigour by its managers, and gladly supported by outside funding agencies, notably the World Bank.[8]

The Andhra Pradesh irrigation reform policy seeks to address all three dimensions of irrigation management as a form of water control: technical, organizational and socio-political. The major focus of the programme is the organizational component. It involves a move away from government management through the constitution of a three-tier management system controlled by water users. Water Users Associations (WUAs) and Distributary Committees (DCs) have been formed at minor (tertiary) canal and distributary (secondary) canal level. Project Committees will be formed – according to the policy – at system (primary canal) level.

This organizational intervention is accompanied by a technical intervention and a socio-political change. The rehabilitation of the physical infrastructure is the technical intervention. The formulation of a new law that defines the powers delegated to the new institutions, and the overall procedures and conditions within which they can operate, is the most important element of the change of the socio-political environment. This can thus be called a comprehensive approach. Earlier efforts at achieving irrigation management improvement in South Asian canal irrigation were also characterized by the twin activities of physical works for rehabilitation and the establishment of Water Users Associations (usually limited to the local/tertiary level of the irrigation systems), but lacked the necessary legal and other general policy changes necessary for making the changes enduring.

The specific characteristics of the Andhra Pradesh approach to irrigation reform have been discussed in detail in several publications (see for example

Oblitas and Peter, 1999; Raju, 1999) and we will only highlight some of the main elements as they exist on the ground.[9] In 1997 the Andhra Pradesh State Legislature adopted the Andhra Pradesh Farmers Management of Irrigation Systems Act.[10] In the same year the water rates were increased by a factor of three, and statewide elections were held for the establishment of Water Users Associations (June) and Distributary Committees (November).[11] Water Users Associations are based on hydraulic units and composed of 4-10 Territorial Constituencies (TCs), each having one representative on the Water Users Association Managing Committee. Each Water Users Association has a president, elected by the TC members from among them in case of 'unanimous' nomination.[12] Clusters of minimum 2 Water Users Associations form a Distributary Committee (eight Water Users Associations in the case studied). The Water Users Association presidents form the Distributary Committee Managing Committee and they choose a Distributary Committee president among them. Till July 2003 Project Committees had not been formed.

The Water Users Associations and Distributary Committees have legal personality, and their own bank account. In the first years after implementation of the Act, the State government has given a fixed amount per acre for maintenance and rehabilitation work on the canal infrastructure, directly to the bank accounts of Water Users Associations and Distributary Committees.[13] The first year this was mainly used for canal clearing and desilting, the second year for repairs of structures. The Water Users Associations and Distributary Committees can prioritize works to be undertaken, and they in principle control the funds, and decide who executes the works. The Irrigation Department gives technical advice and makes estimates. These maintenance/rehabilitation activities have been the main focus of activity of the Water Users Associations and Distributary Committees. This shift in control over maintenance/rehabilitation budgets from Irrigation Department to farmers is a qualitative shift in the structure of irrigation management. Water Users Associations and Distributary Committees have also been empowered to organize water distribution themselves. A Government Order was issued that put the *laskars*, the irrigation field staff executing water distribution activities, under the control of the Water Users Associations. 3500 *laskars* in the State have opposed this order in the courts, and the legal battle is ongoing. The transfer of water distribution responsibilities has thus not been effectuated so far.

The implementation of the Act, that is the introduction of PIM, has been accompanied by massive awareness campaigns, including training of office bearers of Water Users Associations and Distributary Committees and issuing of newsletters and other written material. There is also regular monitoring by the State government of activities and progress, for example through the regular organization of videoconferences of the Irrigation Secretary with Irrigation Department staff.

Impact of PIM Introduction in the Tungabhadra Right Bank Low Level Canal

This section discusses the impact of the implementation of the government PIM policy in two distributary (secondary canals) in the Yemmiganur sub-Division of the Tungabhadra Right Bank Low Level Canal. This irrigation system has a projected irrigated area (command area) of approximately 100,800 hectares (40,800 in the upstream State of Karnataka and 60,000 hectares in the downstream State of Andhra Pradesh). It is designed for protective irrigation. As discussed above, this means that the design of the command area consists of long secondary and sub-secondary canals, running through non-localized areas, that is areas not planned to be irrigated, to bring water to patches or blocks of localized land near villages, that are meant to be irrigated. Canal lengths of 5 km to irrigate, say, 200 hectares are very common.

The two distributaries studied are located in the tail end section of the Tungabhadra Right Bank Low Level Canal (they take off around km 300 of the 324 km long main canal.[14] Some characteristics of the two canals are summarized in Table 8.1.

Table 8.1 Characteristics of distributaries studied

Distributary	A	B
Length (km)	16.7	19
Localized command area (ha)	4880	3260
Design discharge (m³/sec)	1.9	1.3
Number of WUAs	5	3

The two distributaries are adjacent distributaries and have one joint Distributary Committee, of which one of the Water Users Association presidents is the president.

This section focuses on 5 elements central to the present phase of the implementation.[15]

- The constitution and composition of the Water Users Associations and Distributary Committees that have been formed;
- The conduct of the canal system rehabilitation works;
- The impact of the introduction of Participatory Irrigation Management on water distribution practices;
- The impact of Participatory Irrigation Management on area irrigated;
- Changes in the relationship between farmers/water users and the Irrigation Department.

The Constitution and Composition of WUA and DC Management Boards

What is noticeable about the elections of Water Users Association TC members is that they were not always held. The government put a financial incentive of Rs.15,000 (US$ 300 approximately) on the proposal/election of consensus candidates. This incentive applied when all members of the Water Users Association Managing Committee were elected in this way. In distributary A all Water Users Association TC members and presidents were elected unanimously, that is without a vote. In distributary B all elections were contested. The intention behind the provision of a financial incentive for uncontested nomination may have been reduction of the organizational burden of holding statewide elections, and the avoidance of conflicts in local communities. However, the arrangement had some unintended consequences, which are discussed below.

Table 8.2 WUA characteristics in the study area

WUA	Distributary A					Distributary B		
	No.1	No.2	No.3	No.4	No.5	No.1	No.2	No.3
No. TC Members	7	5	5	7	9	7	7	7
TDP	6	3	4	2	6	4	2	2
Congress	1	1	1	2	-	-	-	-
Unknown /no party/ other	-	1	-	3	3	3	5	5
President								
Party	TDP	TDP, moved to Cong	TDP	Cong	None	TDP	TDP	Cong
Caste	Reddy	Reddy	Brahmin	Reddy	Kamma	Reddy	Brahmin	Balija
Land-holding[a]	28L 0.8W	14D	12W	3.2W 0.8D	24L 8W	20W 1.6D	10W	1.6W
Other activities[b]	B	P, C	D	B, P	None	P, C	B, C	P, C

a) in hectares; L= lift irrigation (well or river); W=localised area for wet crops (paddy); D=localised area for dry crops (sorghum, millet, oilseeds, cotton, etc.)
b) B=businessman; P=politician; C=contractor; D=doctor

Table 8.2 gives some characteristics of the Water Users Associations studied. Looking at the social status of the Water Users Association Managing Committee members it can be observed that most of the presidents belong to the higher/dominant caste groups (Reddys and Brahmins). TC members and presidents tend to be larger landholders also. In distributary A 85% of the Managing Committee members of the Water Users Associations had landholdings larger than 10 acres (4 ha). The presidents generally had much larger holdings, and many of them had additional activities as businessman or politician. A total of 4 of the 8 presidents, including the Distributary Committee president, undertook rehabilitation contracting work in their own WUA/DC jurisdiction (and elsewhere/other works). The membership of the Managing Committee was closely linked with party politics. In the local office of the ruling party (TDP = Telugu Desam Party) lists were kept of which TC members and presidents were party members, in the same way as these were kept for other local representative bodies. We assume that the same is done in the main opposition party office, the Congress Party. Though generally parties influenced candidature of TC members, there are also cases were proposed/elected members only became active in the party after this.

A low level of interaction between leadership and members characterizes the functioning of the Water Users Associations. The meetings held by Water Users Associations are reported in the Water Users Associations records, which are kept at the Irrigation Department office. In Distributary A each of the Water Users Associations held three to five meetings in the period January-August 1999. Most of these were about the rehabilitation works to be undertaken in the May-June period. In all five Water Users Associations one of the meetings was of the Water Users Association President only with the Irrigation Department. In the other meetings the recorded average attendance was 59 people (total 14 meetings). Our guesstimate is that this number is about 25-30% of the total membership.[16] However, we have reason to assume that the records paint a too optimistic picture. We have observed that in several cases the president basically worked on his own, without even consulting the TC members. In 1999 the government gave instructions to hold general body meetings, as these did not occur naturally. Cases have been reported to us where signatures for attendance were collected the day after the meeting.[17] Several presidents and TC members expressed to us that they had no clear idea who the members exactly were. These disappointing phenomena are partly the result of the consensus-focused process of appointment/election of office bearers. Had elections been held more generally, it would also have become clearer who the members actually were.[18]

Canal System Rehabilitation Works

The cleaning and repair works undertaken in the first two years of the PIM programme have no doubt improved the technical state of the system considerably.[19] There simply was an enormous amount of work to catch up, given that the Irrigation Department maintenance budgets were mainly spent on staff and establishment costs.

In the study area, the Irrigation Department staff has done the design of the rehabilitation works. For this they have used the standard designs and design criteria that they also used earlier. The orientation is to 'bring the system back to the design state'. Farmers generally also feel they do not have the expertise to make these designs, or even suggestions for designs. We have come across only very few instances where there seems to have been local discussion among water users on the design characteristics of the structures. One exception is a tail end Water Users Association covering a set of spread-out pockets/blocks for irrigation with a complex rotation system. In this case widths of openings in division structures were modelled in proportion to the area irrigated. Generally speaking however, there are no signs of a participatory technology development process.

Also in the execution of the works there was little farmer/water user involvement in the study area. The Irrigation Department Work Inspector was practically in charge of day-to-day construction works, and at each work a *laskar* was posted. In four cases the presidents themselves were the contractors for the execution of the works in their jurisdiction.[20] In other cases the contractors were outsiders (not members of the Water Users Association). Labour was also employed from outside the Water Users Association (that is farmers/water users were not the construction workers).

The quality of the work thus fully depends on the skills of the labourers and the expertise and motivation of the (Irrigation Department) supervisors. Field observations have shown a series of examples of poor quality construction. An example is construction with uncleaned, muddy stones. If this is done cement will not hold, and quickly these works will crack and partly collapse. There is no working system for quality control. Water Users Association members, with few exceptions, are not actively monitoring, and external monitoring through a special inspection force is also not adequate. The chances of getting caught are small, also because these inspectors do not always get full cooperation of the concerned Irrigation Department staff (through not making available transport facilities for example). Nevertheless, many farmers feel that the quality of work has improved, and that more cement per unit of sand is used.

The presidents, who, as indicated above, are sometimes contractors, execute the works without much consultation with their membership. The main contact they have is with the Irrigation Department officers. In most cases the accounts and other files of the Water Users Associations are kept in the Irrigation Department office. The Irrigation Department staff is strongly involved in the financial management of the Water Users Association budget. The funds are still channelled through the Irrigation Department administratively, the ID has to technically approve the design, and it draws up the final bill and releases the outstanding balance.[21]

Given this situation it is not surprising that the 15% anticipated contribution by farmers is not forthcoming, neither in cash, nor in labour. Farmers suspect there is leakage of funds to both Irrigation Department staff and presidents/contractors,[22] and therefore see no reason to contribute. Even when this is not the case, it is very difficult for presidents and TC members to collect money from farmers. Suspicion of misappropriation will always be there, which makes presidents less than keen to collect, as the social costs may be high. For many presidents the Water Users Association leadership role is part of a larger agenda of local leadership and building political careers. They are therefore interested in the accumulation of social capital, to which the sensitive task of collecting money may not contribute. Those presidents who do not interact with the membership at all, have no basis to collect money or demand labour contributions. The central issue thus is accountability of the leadership.

Water Distribution Practices

The peculiar design of the Tungabhadra Right Bank Low Level Canal, with its melons-on-a-vine pattern of localized blocks of legally irrigable land, has already been referred to above. This design has induced widespread illegal tapping of water from the 'idle' parts of the canals. Illegal outlets have been constructed, but more common is siphoning of water, and pumping. The pumping is sometimes direct pumping from the canal, but more commonly from wells close to the canal, largely fed by the seepage from the earthen canals. (but these farmers of course argue that they do not take water from the canal illegally).[23] This illegality has been institutionalized through payments that the concerned farmers make to the Irrigation Department officials for allowing this. The usual payment is Rs.500 per acre for a full crop, and Rs.100 for a single wetting (1 hectare = 2.5 acres).[24]

The new PIM policy declares the *ayacutdars*, the farmers with land in the localized areas as the rightful irrigators and Water Users Association members. Thus emerges an issue of water rights and entitlements. The formally illegal irrigators have a practical entitlement to irrigation water, are in a location where that entitlement can be denied only with great difficulty, and in a number of areas are supporters of the sitting TDP member of parliament (the MLA = Member of the Legislative Assembly). Tail end farmers consistently complained that the 'whole water distribution here is in the hands of the MLA'. We have found no indications that Water Users Associations are trying to address this issue on a systematic scale.[25] Water Users Associations have so far not undertaken activities in the field of water distribution, other than already existed (for example existing rotation schedules are continued).

Area and Intensity of Irrigation

Substantial increases in area irrigated as a result of PIM implementation has been reported for different parts of the State (see for example Oblitas and Peter, 1999; Raju, 1999). This is not an unlikely course of events: clearing, desilting and repair of canals and structures increases the efficiency of water use, and allows water to flow further down the canals. In the study area this phenomenon cannot be observed. Farmers and officials both stated that there is no increase in irrigated area over the past years, and this is also suggested by the statistics of cropping patterns (see Table 8.3).

Both farmers and Irrigation Department officials feel that factors like the rainfall pattern and availability of water in the main canal are the dominant factors in the determination of irrigated area. Farmers reported small expansions of irrigated area in particular sub-secondary canals as a result of canal repair (no overflow of canal banks, cleared cross section, *et cetera*). Some farmers also reported that they managed to give their crops three wettings now instead of two. These examples are all from the head and middle reach of the distributaries. There is qualitative evidence for the existence of a mechanism that in this case any gains in water use efficiency are absorbed in the head reaches (Jairath, 1999) also makes this point). When structures are repaired and canals cleaned, head reach sub-secondary canals may be able to draw more water, which is absorbed by more intensive irrigation in the (illegal) head reaches of these canals.

Table 8.3 Irrigated area of different crops (hectares)

	Distributary A		Distributary B		Sub-Division	
Year/season	*kharif*	*rabi*	*kharif*	*rabi*	*kharif*	*rabi*
1993-94	1405	3470	837	1681	4856	6475
1994-95[a]	-	3470	-	1710	-	8367
1995-96	1420	3473	837	1432	4813	8292
1996-97	1420	3477	834	1514	4822	8423
®1997-98	1420	3477	624	1223	4562	6965
®1998-99	1420	3477	822	1301	4894	7814
®1999-2000			796	1077	4576	6311[b]

a) Canals closed in *kharif* for repairs.
b) In this year registered non-*ayacut* irrigated was 846 hectares, mainly groundnut, while in the four years before it ranged from 38 to 187 hectares. Perhaps registration has improved, or there was a shift from *ayacut* to non-*ayacut* land. The reliability of the statistics on non-*ayacut* land is doubtful.
® Years reforms were implemented.

It can be hypothesized that in areas with full paddy irrigation, like the delta regions of the State, canal improvements do more quickly lead to area increase: the water that becomes available is as it were pushed out at the tail, because there are no possibilities for absorption in the head reaches. It can also be hypothesized that in the delta areas canal maintenance, particularly desiltation, is the primary issue rather than overall water availability (which is rather generous in that area). This would mean that the focus on physical rehabilitation works is particularly appropriate to the delta areas, and therefore the reform programme may have been particularly well received in this region. These issues need further research.[26]

Changing Relations between Farmers and Irrigation Department

For the Irrigation Department officials working in the canal command areas bureaucratic life has become more uncertain. Particularly the transfer of control of maintenance/rehabilitation works to farmers may imply a serious reduction of control over the system, and loss of additional income. However, because the

Water Users Associations in this area have been predominantly captured by rural elite presidents, regularly in the contracting business themselves, and operating rather individually, or in a small circle of supporters, the engineers are able to maintain much of their earlier control. There seems to have been a shift in who the contractors are who execute the works, but the set of relationships surrounding it have not fundamentally changed. Also the structure of social relationships in water distribution, with strong influence of politicians, seems to be unaltered. The Irrigation Department engineers are under considerable pressure from the top to work in a different manner, but hardly so from the bottom. The philosophy of the Act that Irrigation Department staff provides services to the Water Users Associations and Distributary Committees, on request as it were, has not (yet) materialized.

Capture and Transformation

We have no reason to doubt the sincerity and hard work with which the irrigation policy reform managers are pursuing their project. They try to incorporate the newest insights from (inter)national debates on irrigation policy reform in their approach, adapted to and within the constraints provided by the Andhra Pradesh situation.[27] The reform managers are probably taking a top-down approach to irrigation reform as far as it can go in the Indian context.[28]

However, the programme seems not to be sufficiently equipped to cope with on the ground realities such as those occurring in the Tungabhadra Right Bank Low Level Canal. In private discussions, senior policy actors in Andhra Pradesh have argued that the Tungabhadra system should not be used as a standard for evaluation because it is the most difficult situation to address, given the design characteristics of the system, and the socio-political relations in this region. There may be some truth in this. Nevertheless we feel that the processes and issues discussed below are likely to carry general relevance, though the degree and intensity of their occurrence may differ from place to place.

An overall interpretation of the reform programme so far could be that in the first few years the programme has realised the relatively easy gains related to much needed technical maintenance and rehabilitation. In many systems this was the first time in many years that serious investment in the physical system took place. If the new system for collection of water rates becomes a reality, a permanent resource base for continuing this may be created. After these first years the reform programme is now starting to face new issues, of a more difficult kind: those related to re-distribution of irrigation water, how to organize water rates collection by farmers, how to constitute the Project Committees, and how to transform the Irrigation Department.

In the study the rural elite has captured region the reform programme. This elite exhibits the usual pattern of the combination of economic and political power. A basic contradiction in the reform programme is that the political support base of both the ruling TDP and the opposing Congress relies on these rural elite members through the vote banks that they control. Taking up the more difficult issues of redistribution and resource mobilization is likely to create considerable tension within the parties.

The capture of the programme by the rural economic and political elite could occur because of several reasons. First, the programme was carried out with great speed in 1997 after acceptance of the Act in parliament. Despite publicity campaigns many ordinary farmers were hardly aware of the existence of the policy, and certainly not of its possible significance. Even during our fieldwork in 1999 we talked to many farmers who were unaware of the PIM initiative and the existence of a Water User Association in their area. The rural elite with its better contacts and access to information, is likely to have heard about, and realised earlier the possible importance of the programme.[29] Still, at the local level even the political parties were not fully prepared and aware, and overtaken by the speed of implementation. All agree that the next elections of the Managing Committees, in 2002, will be far more contested that the first one.

A second reason for the capture of the programme by the rural elite are the modalities of the maintenance and rehabilitation works. These works constitute an interesting business opportunity, and because of the financing structure of the works, affluent people are required to pre-finance the execution of the works. The Irrigation Department also willingly cooperated in this development, as it provided an opportunity to continue existing relationships with a partly changed group of contractors. In a sense the situation for the Irrigation Department improved, because more funds for physical works have become available.

A third reason is that unless special efforts are made there is no reason why the Water Users Association should be treated differently by local interest groups as any other local representative body: a stepping stone for political careers, a site for the accumulation of social and political capital, and an institution for the distribution of resources to supporters. The irrigation policy reform programme invested much time and effort in the formulation of the APFMIS Act, to prevent such pitfalls.[30] However, no Act can enforce democratic principles when local groups do not also actively claim these. The reform programme has not undertaken activities so far to enable the emergence of such claims.

In the study area the irrigation reform policy has been transformed into something that is controlled by the rural economic and political elite, and that allows the Irrigation Department to more or less continue business as usual (with an added possibility to escape from unpleasant tasks by stating that they are now the responsibility of the Water Users Associations and Distributary Committees). The continued pressure from the government policy reform managers keeps the Irrigation Department officials on their toes. In the absence of this the changes in mode of operation would in fact be minimal.

No sophisticated strategies were required to effect this transformation of the policy intentions, and the capture was contested only to a very limited extent. The struggle, at a personal level, and in terms of political parties, was more over which section of the rural elite would capture the reform policy. Given existing social, economic and political relations the observed pattern is the pattern that one would expect to occur, unless additional efforts would have been made.

Nevertheless, the outcomes are not altogether negative. The canal system has been technically improved and some efficiency gains are evident, even when in this study area these were absorbed in the already privileged head and middle reaches, due to the specific design features of the irrigation system. And, in principle a framework for more democratic forms of water management is in place. In this respect the next elections for the Water Users Association Managing Committees are crucial. They may provide an opportunity for establishing improved accountability and transparency at this level, but this will depend on how they are prepared. If no special efforts are made there is little reason to expect that the outcome will be qualitatively different from what it is now, at least in the study region.

Also in a more pragmatic perspective – 'forget about the redistribution for the moment, let them take up management responsibilities' – there is reason for doubt. The touchy issues of water distribution and water rates collection are not likely to be taken up by the Water Users Association Managing Committees with great enthusiasm in this region. Whether this will undermine the new system of water rates collection is a matter of speculation. What is also striking in this respect is that in the irrigation reform implementation no contractual arrangements regarding water delivery from the Irrigation Department to Water Users Associations and Distributary Committees have been introduced so far. The experience in lift irrigation makes clear that farmers are willing to pay considerable amounts of money for water supply (there are examples of 25% of the crop value) as long as the supply is secure. Though in large surface canal systems the arrangements for contractual delivery with liability on both sides are more complex than in lift irrigation, examples in other Indian States show that it is far from impossible.[31] Another step to be considered would be the introduction of financial autonomy of the Irrigation Department concerned, which might further induce the process of re-negotiation of resource use.

Conclusion

This paper has shown how, in the Tungabhadra Right Bank Low Level Canal study region, the local economic and political elite has captured the Andhra Pradesh irrigation reform policy. It has made available funds for system improvement, which are probably used more efficiently as they would otherwise have been. The design and construction activities are however not participatory in nature, but controlled by a small group, in cooperation with the Irrigation Department. Apart from this the situation has remained pretty much as it was. In the case study area no extension of irrigated area has taken place (though for reasons discussed above this is different in other systems), water distribution has not been taken up by Water Users Associations and Distributary Committees, the power of Water Users Associations to collect fees/water rates has remained unutilized, and head-tail issues have not been addressed.

The capture of the reform policy by vested interests took place without much contest. The approach chosen by the government for policy implementation was a top-down one. This more or less automatically resulted in the described pattern. One of the reasons to adopt a large-scale, State-wide top-down approach was the dissatisfaction with NGO induced reform in other States, which never developed beyond limited numbers of local examples of successful Water Users Associations (the 'scaling up' problem). This may be appreciated, and indeed a legal and policy framework is now in place in Andhra Pradesh that in principle allows far-reaching reforms in all areas. Ironically however, the reform programme now runs into the same problems that many of the NGO initiatives seek to address: devolution or decentralization of resource control to users is a highly complex, social and political issue, which requires special mechanisms to go beyond reinforcing the unequal and undemocratic status quo.

This can be illustrated by looking at the seven assumptions that, according to Warwick, underlie the 'transactional' approach to policy implementation.

The Andhra Pradesh case study provides ample evidence of Warwick's first point, policy is important but not fully determining action. There is evidence that there is regional, inter-system and local diversity in the way the policy was implemented/received. In some regions the emphasis on physical rehabilitation works may have been more appropriate than in others. In some systems expansion of irrigated area took place, while in others it did not, in relation to design factors, relative water scarcity and other factors. In some localities there were elections for Water Users Association Managing Committees, in others there weren't.[32] The policy managers have not been unresponsive to this diversity and need to adapt. Many amendments to the Act have been accepted, and there has been a lot of interaction with farmer leaders on implementation problems. The reform managers have thus also acknowledged Warwick's seventh point, that policy implementation is inherently dynamic, and his third point, the importance of the environment. Apart from the active way in which the reform managers have guided and shaped implementation, the introduction of new elements like water rates collection by Water Users Associations, in a consciously stepwise manner (see chapter VII of Oblitas and Peter, 1999) further illustrates this acknowledgement.

Still, we would like to argue that the policy implementation process is not sufficiently transactional in a conscious way, to be able to address the so-called 'second generation' problems now emerging. Warwick's fourth point states that the formulation process may have as important an impact on implementation as the product of that formulation process. In the Andhra Pradesh irrigation reform a lot of energy was invested in policy formulation (and building all-political party support for the policy), but water users were at best consulted. They did not play an active role in policy design. Also the Irrigation Department staff was only partially involved and co-opted. International debate on PIM brought in through consultants seems to have been more decisive. We would like to suggest that this has had an important impact on the reform process because it has left unquestioned, or has insufficiently questioned, some of its 'machine' approach features.

A first 'machine' feature is the functional view of Water Users Associations. These have been conceptualized as non-political bodies, with territorial constituencies, and no reservation for particular social groups. The assumption seems to be that legal force can bring about the emergence of such functional and democratic institutions. No further action has been taken to shape processes within the Water Users Associations apart from the organization of the elections, and training (to office bearers) explaining the features of the Act.

Another limitation following from the lack of involvement of users and Irrigation Department staff in policy design is the sector, irrigation system management focus of the Act. There is little recognition of multiple-use situations with regard to irrigation water, and the Act seems to be unaware of the possible consequences of consolidating the water rights of those with land in once-determined irrigation command areas. More direct engagement with field level issues would have undoubtedly brought these issues to the fore. Now the Act stays within a top-down, irrigation system focussed, water supply paradigm, a second 'machine' feature.

Because we have not investigated the policy formulation process in detail, we cannot say with certainty whether the above features have been decided consciously, or are the product of an internalized conventional policy implementation model. It may be the case that the reform managers have chosen to avoid too many complications at the start. However, it is unlikely in our view that a bureaucratic government machinery is able to address the new complex issues that are emerging effectively, no matter what the commitment may be at the higher levels. One would first have to reform the bureaucracies themselves completely, to generate the necessary flexibility and creativity, and overcome engrained resistance to reform. Bureaucratic reform is part of the Andhra Pradesh reform initiative, but it is a long route to addressing the local issues in the irrigation systems. It seems to us that the reform process here runs into a fundamental dilemma, which is not specific to Andhra Pradesh: to what extent are bureaucracies able to self-transform when there is no articulated 'demand' or 'pressure' for that transformation from the target population, in this case water users and Irrigation Department staff. What will be the mechanisms to overcome internal resistance within the Irrigation Department (Warwick's fifth point, implementer discretion) and the divisions within the local communities (Warwick's sixth point, the influence of clients)?

Again we can only speculate. The first point, Irrigation Department resistance, is in the hands of the government, though politically difficult: the constitution of Project Committees that would imply a further reduction of Irrigation Department powers, and the institutional reform of the Irrigation Department. The second, the local processes, is much less in the hands of the government, particularly because any political party depends on the support of the local leadership. The cynical view here would be that the problem of the reform being captured by the local elite will be left as it is, and that the government will focus on sufficient and efficient local revenue collection as its primary focus.[33]

To conclude, the government approach is not a full-blown 'machine model' of implementation, but neither does it fully incorporate the implications of a 'transactional model'. Though the Act that is the basis of the irrigation reform exercise on paper creates institutions of users that can demand services from the Irrigation Department, such a shift in the balance of power between the two, implying a relationship of negotiation on clear and equal terms, has not emerged so far. Further advance of the reform process in this direction depends on the strategies to overcome resistance within the Irrigation Department, and local divisions in the users communities, that is the capture of the policy by the local elite. Put more generally, and in terms familiar to irrigation professionals, the reform is still supply driven. Demand, and negotiating capacity on the 'clients' side has not yet emerged.[34] We suggest that this lack of articulated demand for reform (or 'ownership' of the reform) is not a phenomenon exclusive to Andhra Pradesh, nor to irrigation, but occurs more generally.

Notes

1 This paper was also published in the *International Journal for Water* Vol. 1, No. 3/4. pp. 360-379 (2001) with the title 'Participatory irrigation management in Andhra Pradesh, India: policy implementation and transformation in the Tungabhadra Right Bank Low Level Canal'. The present version has some minor updates added. We thank the editors of the journal for permission to reprint the paper.

2 The fieldwork was conducted between June and October 1999, with a short additional field visit in April 2000. The geographical location of the research is the Yemmiganur sub-Division of Tungabhadra Right Bank Low Level Canal, located in Kurnool district, Andhra Pradesh. R. Doraiswamy spent 34 fieldwork days in the location; Kim Engbersen's M.Sc. thesis fieldwork period was 10 weeks; Peter P. Mollinga made three short field trips and supervised the research.

3 The term is Kerkvliet's (Kerkvliet, 1990).

4 The term is Grindle's (Grindle, 1977).

5 See for example Ohlsson (1995).

6 The same can be said about policy formulation, though the range of actors may be different (and the 'beneficiaries' usually excluded).

7 One 19th century system in the Rayalaseema area is the Kurnool-Cuddapah (KC) Canal, and some systems were built in the 1920-1930s in the Telengana region (Dindi project, Nizamsagar project). The Tungabhadra system that is discussed in this paper was conceived in the 1860s, designed in the 1930-1940s, and largely built immediately after Independence (see Mollinga, 2003). Andhra Pradesh also has tens of thousands of so called tank irrigation systems. Tanks are farmer/user-managed small reservoirs formed by putting a bund across a small valley. They irrigate from several tens to several hundreds of hectares, or act as percolation tanks for well irrigation. This paper focuses on the government constructed and managed canal irrigation systems. Individual canal systems irrigate from thousands (medium irrigation) to hundreds of thousands (major irrigation) of hectares.

8 The Andhra Pradesh reform exercise seems to have acquired symbolic importance for some of the international funding agencies. When early 2001 the Andhra Pradesh government presented its budget with a substantial deficit, the World Bank and DFID quickly stepped in to fill the gap, suggesting that they are more than keen to make the reform experiment succeed. The image created of the reform process is that of a combination of market oriented economic reform and good governance, which suits the funding agencies' priorities.

9 The policy is implemented in steps, so not all elements of the new policy are implemented in one go. One example is the linking of payment of water rates and provision of maintenance funds, which will be implemented from 2001-2002 (see below).

10   The conception of the Act is an interesting story in itself, but will not be discussed here. It can be noted that after extensive debate and political canvassing the Act was supported by all political parties.
11   In the State 10292 WUAs were formed and 174 DCs. Around 2000 of the WUAs are located in larger-scale canal systems (major irrigation 1699, medium irrigation 413 as on 31 May 1999), the others in minor irrigation, particularly tanks.
12   In case elections are held each member has two votes, one for the TC member and one for the president.
13   In the first year, Rs.100 per acre was allocated: 50 for the WUA, 20 for the DC, 20 for the Project Committee (though non-existent) and 10 for the Village Council. In the second year this was changed to Rs.100 for the WUA and DC each, of which farmers had to contribute 15% themselves. In the third year the allocation for major irrigation projects was Rs.25 per acre for WUAs, and Rs.10 per acre for DC and PC, and for medium irrigation projects this was Rs.35, Rs.15 and Rs.15 respectively (1 hectare = 2.5 acres).
14   With the Kurnool Branch Canal added to this the total length becomes 374 km.
15   Other elements, like the levy of water rates are not discussed here. When this was first written, early April 2000, the Government of Andhra Pradesh had just announced to irrigation officers that the availability of maintenance/rehabilitation funds was going to be made dependent on the collection of water rates by WUAs. This was laid down in a Government Order in February 2001, and will be introduced in the 2001-2002 agricultural year. The rate will be Rs.200 per acre (Rs.500 per hectare), to be collected from farmers by the WUAs. Half of this will flow back to WUAs, DCs and a small part of it (10%) to the village council. The other half will go to the Irrigation Department. Our field information suggests that WUA presidents will be extremely reluctant to do the collection, which may imply that the Revenue Department may have to do it. The importance of the new system however lies in the creation of a connection between water rates paid and maintenance grants received, and who collects is not essential for that connection. However, who collects is important for internal collection procedures within the WUA.
16   Estimation is not easy as there are likely to be discrepancies between the official register of landowners, and the actual situation, due to backlogs in registration of sales and property divisions. Also there is a degree of (unofficial) tenancy. In 2000 the APFMIS Act was amended to the effect that tenants can also become members of WUAs.
17   The problem of low attendance of general body meetings of the WUAs seems to be a general phenomenon. The *Hindu* newspaper reports that in West Godavari district 'the attendance did not exceed five per cent questioning the legitimacy of the bodies' (25 March 2000).
18   A disadvantage of elections along party political lines is that a lot of money will be spent on campaigning, which somehow has to be recovered. This is what may happen in the next elections for WUA Managing Committees, upsetting the functional conception of the WUAs in the Act (see also section 6).
19   The works are conducted in the period that the canal is closed, that is May-June in the case of Tungabhadra Right Bank Low Level Canal.
20   Either as officially registered contractors, or as unofficial contractors working under the name of an officially registered contractor. The issue of how the works are exactly contracted out needs further research.
21   The money for the WUAs does not come at once, but in installments. A next installment is only released when a certain part of the works is finished. In practice

this means that presidents/WUAs have to advance money from their own pocket to finish the works, as the first installment is not more than 40%. Only well to do presidents can afford this, and it creates a logic for economizing on construction costs and quality of works. In the new system of water rates collection refunds to WUAs and DCs are anticipated on a three-monthly basis, in relation to the amount collected in that period.

22 We have collected oral evidence that suggests that in some cases the WUAs/presidents have to pay the same level of bribes as other contractors used to pay before. On corruption practices in this region, see Wade (1982).

23 The illegal head versus legal tail problem was further exacerbated by the circumstance that in the head reach of the distributary canal wells could be constructed in red soils, but not in the tail end, where black vertisols are found.

24 Details on the collection and sharing arrangements can be supplied on request.

25 We have statements of farmers that WUA presidents and TC members now sometimes take payments from illegal irrigators, but we have no first hand evidence on this.

26 A feature of some other major canal irrigation projects in Andhra Pradesh is that the actually irrigable command area is much lower than the planned irrigable command area. This means that the construction of the canal infrastructure is not completed. Because the dam/reservoir is completed, overall, water is not scarce in such systems. This in its turn may imply that area expansion is very likely when canal infrastructure is improved. The Tungabhadra Right Bank Low Level Canal may be one of the few systems in the State with actual overall water scarcity – and not only because of its peculiar design features. This hypothesis needs to be tested through further fieldwork.

27 What is striking is that despite a high degree of standardization there is a capacity to learn and adapt within the executing bureaucracy, and such learning and adaptation is actively pursued.

28 A more cynical interpretation would analyze the irrigation reform programme mainly as a political strategy by the present government to mobilize the rural vote (and successfully so, given the re-election of the TDP government in October 1999), but as an exercise with little substance otherwise. Though this consideration is undoubtedly there, it would be unfair to reduce the reform to that. These broader political support base issues may however well hamper further advance of the reform.

29 It can be noted that training activities for water users have first trained office bearers of WUAs. The idea may have been that such office bearers pass on information to their membership, but that seems to be a somewhat too optimistic assumption. The training may in fact have helped to increase the information gap between office bearers and membership. It can well be argued that training should have started with the normal members. There is no guarantee that the office bearers will be re-elected. Training of members is preparing the ground for democratically functioning WUAs. However, the logistics of training of WUA members are somewhat overwhelming.

30 For example, the Act has no provision for reserved seats on committees for women or Scheduled Castes/Scheduled Tribes, as in many other representative bodies.

31 The reference here is to experiments with volumetric supply in Maharashtra, organized by NGOs. It should be noted that in cases where this volumetric supply has been successful, it has been accompanied by intense efforts of local organization across the boundaries of existing social divisions (see Lele and Patil, 1994).

32 The latter also illustrates Warwick's second point, that formal organization structures are not deterministic in their impact. The functioning and performance of (standardized) WUAs seems to be diverse, and generally different from that prescribed in the Act.

33 The fact that the sitting TDP government did not do well in the local elections for district, *mandal* and village councils adds another element to the local dynamics. In the years leading up to the next general elections (scheduled for 2004) there is likely to be intensified struggle on local political control, in which the WUAs may also become important. The first five-year term of the WUA Management Committees ended in 2002. New elections are scheduled for April 2003.

34 Some individual WUAs and DCs are very strong agents, but the reference here is to the reform process in general, and articulated demand at the collective level.

References

Atchi Reddy, M. (1990), 'Travails of an irrigation canal company in South India, 1857-1882', *Economic and Political Weekly*, Vol. 25, no. 12, pp. 619-28.

Bolding, Alex, Peter P. Mollinga and Kees van Straaten (1995), 'Modules for modernisation: colonial irrigation in India and the technological dimension of agrarian change', *Journal of Development Studies*, Vol.31, no.6, pp.805-44.

GOAP (Government of Andhra Pradesh) (1982), *Report of the commission for irrigation utilisation. Volume I. Volume II*, Hyderabad.

GOI/PC/PEO (Government of India/Planning Commission/Programme Evaluation Organisation) (1965), *Evaluation of major irrigation projects – some case studies*.

Grindle, Merilee S. (1977), *Bureaucrats, politicians and peasants in Mexico. A case study in public policy*, University of California Press, Berkeley.

Jairath, Jasveen (1999) *Participatory irrigation management (PIM) in Andhra Pradesh: contradictions of a supply side approach*, Paper presented at the Researchers Conference 'The long road to commitment; a socio-political perspective on the process of irrigation reform', held in Hyderabad, December 1999.

Kerkvliet, Benedict J. Tria (1990), *Everyday politics in the Philippines. Class and status relations in a Central Luzon village*, University of California Press, Berkeley.

Lele, S.N. and R.K. Patil (1994), *Farmer participation in irrigation management: a case study of Maharashtra*, Horizon India Books, New Delhi.

Mackintosh, Maureen (1992), 'Introduction', in Marc Wuyts, Maureen Mackintosh and Tom Hewitt (eds), *Development Policy and Public Action*, Oxford University Press, in association with the Open University, Oxford, pp. 1-9.

Mollinga, Peter P. (2001) 'Water and politics: levels, rational choice and South Indian canal irrigation', *Futures*, Vol. 33, pp. 733-52.

Mollinga. Peter P. (2003), *On the waterfront. Water distribution, technology and agrarian change in a South Indian canal irrigation system*, Wageningen University Water Resources Series 5, Orient Longman, Hyderabad, India.

Narasimha Reddy, D. (1999), *Designer participation: politics of irrigation management reforms in Andhra Pradesh – India*, Paper presented at the Researchers Conference 'The long road to commitment: a socio-political perspective on irrigation reform' held in Hyderabad, December.

Oblitas, Keith and J. Raymond Peter (1999), *Commencing irrigation sector reforms through management transfer to farmers: the case of Andhra Pradesh, India*, World Bank Technical Paper.

Ohlsson, L. (1995), *Hydropolitics: conflicts over water as a development constraint*, Zed Books, London.

Raju, K.V. (1999) *Participatory irrigation management in Andhra Pradesh: a way forward*. Paper presented at the Researchers Conference 'The long road to commitment: a socio-political perspective on irrigation reform' held in Hyderabad, December.

Ramamurthy, Priti (1995), *The political economy of canal irrigation in South India*, Ph.D. thesis, The Graduate School of Syracuse University, Syracuse.

Rao, G.N. (1985), 'Transition from subsistence to commercialised agriculture. A study of Krishna district of Andhra, c.1850-1900', *Economic and Political Weekly*, Vol. 20, no. 25 and 26, pp. A60-69.

Thomas, John W. and Merilee S. Grindle (1990) 'After the decision: implementing policy reforms in developing countries', *World Development*, Vol.18, no.1, pp.1163-81.

Upadhya, Carol Boyack (1988), 'The farmer-capitalists of coastal Andhra Pradesh', *Economic and Political Weekly*, Vol. 23, no. 27, pp. 1376-82 and Vol. 23, no. 28, pp. 1433-42.

Wade, Robert (1982) 'The system of political and administrative corruption: canal irrigation in South India', *Journal of Development Studies*, Vol.18, no.3, pp.287-328.

Wade, Robert (1988), *Village republics: economic conditions for collective action in South India*, Cambridge South Asian Studies no.40, Cambridge University Press, Cambridge.

Wade, Robert and Robert Chambers (1980), 'Managing the main system: canal irrigation's blind spot', *Economic and Political Weekly*, Vol. 15, no. 39, pp. A107-112.

Warwick, Donald P. (1982), *Bitter pills. Population policies and their implementation in eight developing countries*, Cambridge University Press, Cambridge.

Chapter 9

The Politics of Irrigation Policy Implementation: Networks of Votes, Bribes and Coca-Cola in the Philippines

Joost Oorthuizen

Introduction

This chapter points out that to understand the outcomes of irrigation policy reforms it is necessary to trace and analyse the politics of its implementation. Policies entering local arenas of irrigation systems are not simply implemented or taken for granted by the different actors involved. They are often seriously contested, if not the subject of overt socio-political struggle. These policies thus become the subject of what Kerkvliet defines as 'everyday politics' (as contrasted to formal, party and state politics), which is about 'the debates, conflicts, decisions and co-operation among individuals, groups and organizations regarding the control, allocation and use of resources, and the values and ideas underlying these activities' (Kerkvliet, 1990: 11).

The most obvious resource at stake during irrigation policy implementation is water. Given its scarcity during critical periods of plant growth in many areas of the world and the open access character of the public canal systems in which irrigation is often provided to farmers, water is a seriously contested resource (Chambers, 1988; Mollinga, 2003). Irrigation policies often aim to alter existing water distribution and allocation practices in order to attain efficiency and/or equity goals. These policies tend to become contested, as they potentially provide ways and means for local actors to improve their control of water at the expense of others.

Both case studies of policy implementation discussed in this chapter stress this point. Discussed first is an experiment conducted in the Philippines during the 1970s. It tried to improve water supply to the tail end areas of canal systems by changing main system management practices. It became an internationally famous experiment that provided the early evidence for the emergence of a new set of beliefs about the causes and solutions for poor performance during the second half of the 1970s. As we will see, a powerful local actor, whose control over water was at stake, contested the experiment. He tried to stop or at least neutralize the policy during the implementation process. However, this is not reported in the scientific and policy literature.

The second case study is about the world-famous turnover programme of the National Irrigation Administration (henceforth the NIA) of the Philippines. This programme consists of a set of policies to increase the involvement of farmers in the management of the system. As the programme provided intended and unintended opportunities to change the status quo in water control, it also brought new dimensions to the everyday politics of water allocation and distribution. The second policy reform also shows that water is not the only reason for policies to be contested. In the case of the turnover programme, irrigation service fees were a major issue. Both the local NIA managers and its personnel and farmer leaders of the newly established irrigation institutions tried to reshape the turnover policies in order to get control over the collection and allocation of this precious resource. Finally, public positions were at stake. The turnover programme calls for the establishment of locally elected farmer leaders to represent water users of irrigated areas. Though this was largely disregarded in one of the areas of the case study, it became a political issue in another area.

This 'policy as a political process' perspective on policy implementation contradicts the conventional understanding of policy interventions dominant during the 1960s and 1970s, and still prevalent today. This conventional view conceptualizes the policy process as essentially linear in nature. It implies some kind of step-to-step progression from policy formulation to implementation to outcomes, after which one could make an *ex post* evaluation to establish to what extent the original objectives had been achieved. This is however a gross oversimplification of a much more complicated set of processes which involves the reinterpretation or transformation of policy during the implementation process, such that there is in fact no straight line from policy to outcomes. Implementation is not simply the execution of a particular policy, but rather implementation should be viewed as a transactional process involving negotiation over goals and means between parties with conflicting or diverging interests (Long and van der Ploeg, 1989, see also Mackintosh, 1992; Mooij, 1996; Clay and Schaffer, 1984).

The case studies show the shortcomings of customary monitoring and evaluation studies. It is demonstrated that policy processes usually remain black boxes in such exercises. Based on the conventional view on policy models, these evaluation studies are limited to before-after measurements and thus only provide information on the outcomes of an intervention. Such measurements tell us nothing about the process that connected the output to the inputs and therefore provides no insights why a particular outcome was achieved (Korten, 1989). Also, the scope of orthodox monitoring and evaluation studies is a very restricted one. The studies only measure the achievement of the original policy objectives. They usually do not take into account existing irrigation practices and experiences of the local actors, nor do they try to understand the ways in which the evaluated policies interact with other governmental or locally induced policies (Long and van der Ploeg, 1989: 230).

Clearly, there is a need for a more careful and inclusive monitoring of policy implementation processes. This is not only required for analytical purposes, but also for improved policy implementation processes. It allows the implementers to learn from on-going developments in the field and to take appropriate measures to improve policy outcomes. In the conclusion of this chapter we discuss an example of irrigation policy interventions in the Philippines, in which so called Process Documentation Research methods were used as an essential element of such a 'Learning-by-Doing' policy environment.

This chapter also emphasizes the need for a thorough understanding of politics in the more traditional sense of the word, if one wants to understand policy outcomes in irrigation systems. Anybody who has been involved in large-scale irrigation can provide examples of politicians interfering in irrigation management practices. These experiences are however usually discussed over a drink and considered as exciting, separate incidents that are ignored in regular policy analysis and implementation programmes. The local arena of irrigation systems is considered to consist of two actors only: the irrigation agency personnel and the farmer water users. The case studies in this chapter show that this might be an important flaw in policy analysis in irrigation. Whether one likes it or not, politicians at different levels of the political system often play a large role in irrigation management (Wade, 1982; Repetto, 1986; Mollinga, 2003).

In more general terms, the political system is often very much embedded in the local arena of water management and thus needs to be an integral part of policy analysis and policy action. This chapter tries to show how the particularities of the Philippine political system shape irrigation policy interventions. Filipino politicians have great latitude to interfere in the public domain. Their power is moreover based on a combination of political patronage and their capability to act violently. As we will see, the political system is very much a family-based system, which allows families to operate in both the political domain as well as the administrative domain of irrigation management (McCoy, 1994).

The Location of the Case Studies

Both the policy reforms discussed in this chapter were implemented in the Penaranda River Irrigation System (PRIS), a large-scale canal irrigation system in Central Luzon, the Philippines. Originally constructed in the 1930s, the system was expanded and modernized during the 1970s when it became one of the four districts of the Upper Pampanga River Integrated Irrigation System (UPRIIS). The potential service area of the district is 24,000 hectares. The water is diverted from a river and supplemented by the UPRIIS reservoir (Figure 1). Farmers almost exclusively grow rice in the district during both seasons. It is a flat and low-lying area, which drains into a neighbouring swamp. During the wet season, large parts of the tail end areas are water congested. Given a design water duty of 1.5 litres per second, it should be possible to grow rice in the whole district during both seasons. In reality however, supply per hectare does not meet demand in most of the dry seasons. Water is scarce and unreliable during especially the latter half of the cultivation period of rice, which runs from October until April.[1]

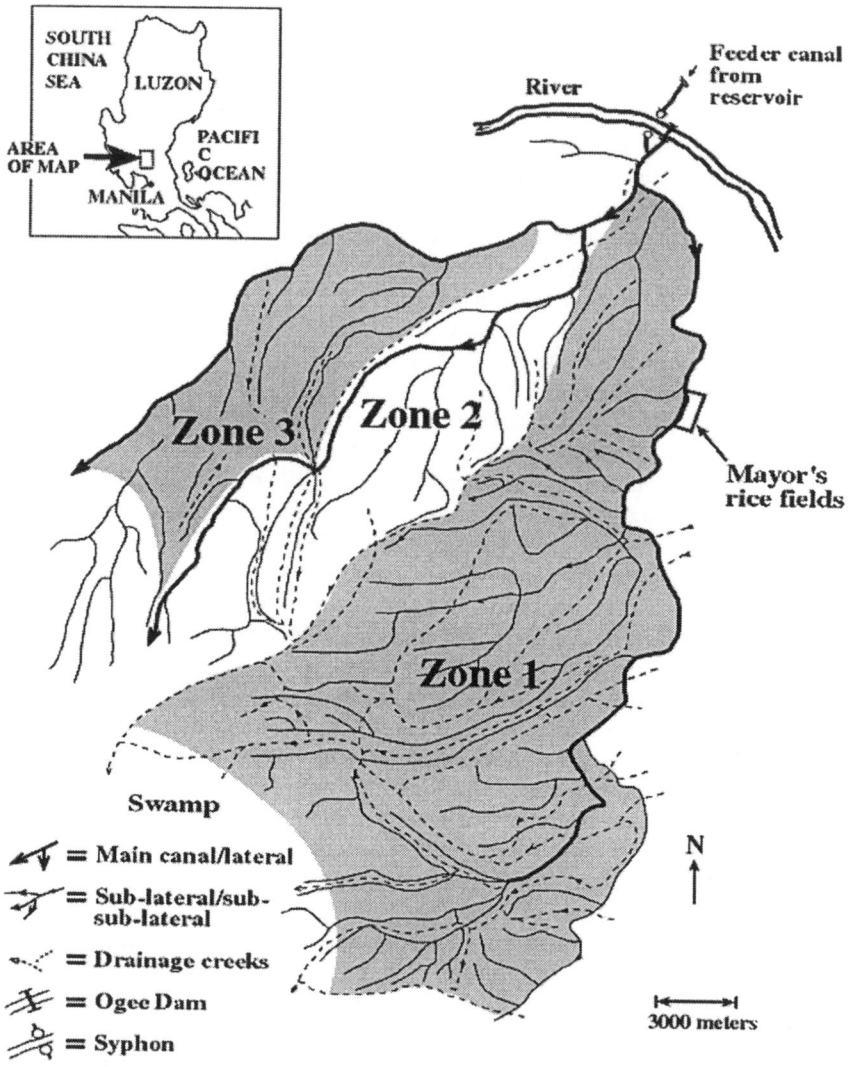

Figure 9.1 The Penaranda River irrigation scheme

The agency in charge of the management of the district, the National Irrigation Administration (NIA), tries to solve the water scarcity problem by limiting the planted area during the dry season. Before the start of each season the NIA determines the so-called 'programme area', which sets the size and the boundaries of the areas to be planted. The NIA usually excludes the downstream parts of the district from the programme area. However, farmers tend to ignore these guidelines and plant and try to irrigate their paddies despite the fact that their fields are located outside the programme area. Consequently, water becomes extremely scarce during most dry seasons, and crop losses due to water stress are a recurrent phenomenon in especially the tail end areas of the district.

The chapter starts with a discussion of the first case study. It is about an experiment to improve water supply to the tail-end areas of one of the three zones of the district through changing main system management practices. It is firstly shown that the implementation process remained a black box in the evaluation studies of the experiment. We will then unpack this black box, and discuss its politics. Next, the second case study discusses the politics of turnover policy implementation in two of the three zones of the district, and provides an extensive analysis of the complex ways in which the administrative system is embedded in the wider socio-political structures of the area.

This study is based on one-and-a-half years of intensive fieldwork during the years of 1996 and 1997. The presented data were gathered in hundreds of semi-structured, in-depth interviews with farmers, irrigation agency staff and politicians involved in water distribution. Also secondary data were collected at the district office of the National Irrigation Administration. The study is part of a wider Ph.D. research titled, published as Oorthuizen (2003).

Unpacking Black Box No.1: The Implementation of a Main System Management Experiment

The first case of policy implementation discussed in this chapter is about a famous experiment to improve main system management practices conducted in the beginning of the 1970s. It challenged common beliefs on the causes of and solutions for poor performing large-scale canal systems. The Water Management group of the International Rice Research Institute (IRRI) carried out the experiment, in collaboration with the NIA. It was conducted on the main lateral of Zone 1 of the system during the 1974 and 1975 dry seasons.[2] The purpose of the project was to develop, implement and evaluate a package of management practices to improve the distribution of water along the main lateral, and thus to improve the irrigated area and average yields (Valera and Wickham, 1976: 3).

The experiment was based on two interventions. First, water scheduling was implemented for the main lateral in times of scarcity. For this purpose, the main lateral of Zone 1 was divided into four sections, each covering a part of the service area. For instance, the first section – including its sub-laterals – would be allowed to take all available water from the main lateral in the first few days of the week. The next sections would follow during the rest of the week. Second, it was tried to provide the downstream sections with water for land preparation and transplanting before supplying the upstream sections. The upstream sections were not allowed to take water from the main lateral for the first three weeks of the season. A technician, who was also responsible for communicating about these measures with the farmers, supervised each section (Wickham and Valera, 1978: 72).

The outcome of this experiment was impressive. From the first (head) to the fourth (tail) sections of the lateral, yields declined from 2.5 (head) to 0.4 (tail) t/ha in 1973 before intervention. But, in 1974 and 1975 after intervention, only from 2.6 (head) tot 2.1 (tail) t/ha and from 3.0 (head) to 2.3 (tail) t/ha respectively. The percentage increases in production (area x yield) for the four sections from 1973 to 1975 were 23, 69, 154 and 1994 respectively. Though water was more abundant in the later years, these positive outcomes were also attributed to the improved canal management (Valera and Wickham, 1976, cited in Chambers, 1988: 109).

This experiment provided the early evidence that altered common beliefs, in both national and international irrigation policy-making bodies. Until that time, it was widely believed that negligence and the lack of co-operation of farmers caused poor water management. Farmers did not follow modern practices of rotating water below the outlet and applying water moderately in line with crop water requirements. Instead, they persisted with their inefficient old practices of continuous field-to-field irrigation of their paddies, and checking water in the lateral canals whenever they felt it to be necessary. Policy interventions therefore focussed on on-farm development, organising farmers below the outlet, and training farmers in rotational field methods. The impressive results of the experiment gave room for a new paradigm that puts its focus on the main system. It supported the idea that poor water management is first and foremost to be solved at the level of the main system rather than at the field level (Wade and Chambers, 1980). It suggested that farmers are not at fault. As long as farmers are confronted with erratic and insufficient water flows in the main system, their 'uncooperative and negligent behaviour' is logical and hard to avoid (Wickham and Valera, 1976: 72-75).

This experiment is a good example of what Korten (1989) classifies as 'conventional summative evaluation practice'. Based on a before-after measurement of results it provides information on the outcomes of the experiment. The disadvantage is that this type of study tells us nothing about the process that connects the outputs tot the inputs and therefore provides no insights why this particular outcome was achieved. The study gives very limited information on the implementation process itself. It only tells us that they had many discussions with farmers about the experiment. Some of them initially resisted the project, but in the end 'farmers in the upper reaches of the lateral gradually came to support the new scheme once they were assured of an adequate share of water even in times of water shortage' (Valera and Wickham, 1976: 4). There is no reference at all to existing or previous main system management practices, let alone to the socio-political conditions which allowed this rather drastic experiment to succeed. In line with orthodox intervention models, this experiment is seen as a spatially and temporally discrete project. It, as it were, takes out history, thus implying that memory and learning from the past are in fact superfluous (Long and van der Ploeg, 1989: 230). Thereby it can be taken to be replicable elsewhere, by a donor and policy community looking for 'best practices' and 'models'.

Obviously, the people of Zone 1 are not without a history, as Brian Fegan's (1979) anthropological study of peasant strategies of a village located in the tail-end of this lateral aptly shows. He points out that water is the major risk factor for the rice cultivation in the downstream part of the lateral. During the wet season, waterlogging and flooding is the major problem, while water scarcity threatens cultivation during the dry season. In both cases, farmers had developed sophisticated strategies to cope with the risk, and complex social relations had emerged around the construction and maintenance of waterways in the wet season and the distribution of irrigation water in the dry season. During the dry season, irrigation water was a major source of struggle. Farmers had developed a clever communication system (including the use of information from bus drivers passing upstream irrigation canals) to get up to date information about water availability in the main system, and used multiple strategies to get access to this water.

The struggle over irrigation water shaped social relationships.

> Long series of violent clashes between families or kindreds, parts of villages and whole villages if plotted on a map of the irrigation system, indicate that though the incidents occurred at drinking parties, festivals, basket-ball games, courting, transplanting, etc., the enemies belong to adjacent areas upstream and downstream from each other on the irrigation system (Fegan, 1979:132).

Farmers made every effort to evade what controls and inspection exist, to get access to the water, and put pressure on the NIA's water masters ranging from persuasion based on ties of kinship, friendship, locality and political alliances, through offers of bribery, to occasional threats of violence. Fegan also described bribery and the use of powerful politicians as strategies applied by local political leaders to influence the main system management practices of the NIA's district engineer. Fegan moreover shows that the experiment to schedule water in the main system is not new to the system at all. In the beginning of the 1970s it was an established practice by the NIA to rotate water between the three zones of the system. The rotation was based on a system of priority. Priority was rotated over the three zones, so that each zone had first priority every 3rd year. A priority zone would get water first for land preparation and transplanting, while the water supply to the other two zones was delayed, to stagger demand peaks. In case of scarcity, the demand of the zone with top priority supposedly was to be satisfied first (*ibid.*, 128-129).

Given these tense socio-political relations around water distribution, it is hard to believe that this experiment did not encounter any serious resistance during its implementation. And indeed such resistance did exist. My own inquiries into the implementation process revealed that the IRRI-NIA people responsible for the experiment ran into a serious conflict with the mayor of a municipality covering the upstream area Zone 1. At the time, this mayor was not yet the key-player in the management of the main system of later years (see further below). He was not amused by the effort to streamline water supplies to his constituents in favour of a downstream municipality. But there was no reason to intervene, as the relatively favourable water supply conditions during the two dry seasons of the experiment enabled the IRRI-NIA staff to improve water supply in the downstream areas, without endangering crop production in the upstream municipality. The conflict was however caused by the fact that the IRRI experiment hampered the water supply to the rice estate of the mayor's family. At the time they owned around eighty hectares of rice fields located directly along Zone 1. This elevated area was located on the left bank, and as such was outside the service area of the system. Nevertheless, these fields were irrigated (illegally) by diverting water from the main lateral of Zone 1, which was then pumped into the mayor's fields. This required one of the major checks in the main lateral to be fully closed for several days in a week to allow the water level to build up for the diversion and the subsequent pumping of sufficient water for this large rice area. Though officially the check is under the control of the NIA, it was in practice considered the territory of the mayor's family, not to be touched without their consent. The IRRI experiment challenged his control over this major check.

One of the former section supervisors appointed by NIA to implement the checking schedule along the main lateral of the zone told me, the mayor could not simply tell the NIA's engineers to stay away from 'his' check, as these engineers could not easily disregard IRRI's experiment. He thus had to look for another way. At one night during the dry season of 1975, the mayor's brother – who was considered to be a trigger-happy 'tough' guy – and his gang, entered the quarters of the four section supervisors. Being drunk, and unhappy with the scheduling practices of the supervisors, he engaged them in a fistfight. The mayor saw his opportunity and acted upon it. The next day, he sent policemen and put the supervisors in jail for molesting his brother and his gang. The NIA engineers and IRRI researchers had to negotiate with the mayor for their release. They were transferred to another job. The IRRI experiment could continue, but the scheduling had to be adapted so it would no longer threaten the mayor's interests.

After over a quarter of a century, too little data is available to fully unpack this black-box experiment, and to give a detailed account of the social interactions emerging out of the implementation process. The above however shows that the experiment was implemented in an area where water distribution is a seriously contested matter, while the experiment itself was successfully resisted by a powerful actor. Another indication of this resistance is that the experiment did not turn into a routine practice in the management of water in Zone 1 after the IRRI experiment stopped in 1975, despite the claims of its success. As we will see below, the NIA formally still uses a similar kind of scheduling, but in practice they are not able to implement it during times of scarcity, as local power holders prevent them from doing so. The effort of IRRI to prioritize water deliveries during the beginning of the season to the tail end has failed to survive the pilot-experiment. Some of the farmers and NIA officials still remember the experiment as 'operation dulo' ('dulo' means 'below' or 'far-end'), but consider it something of the past which is not feasible given the socio-political power configuration in the area.

Why did Valera and Wickham not mention this political intervention in their evaluation of the experiment? It may have escaped their attention. It is more likely however, that although it seriously affected the outcomes of the experiment, they considered the political intrusion as irrelevant for their findings on improved main system management scheduling. Such reasoning is typical for conventional monitoring and evaluation studies of policy interventions. The experiment is seen as a closed system, in which the relationship with the wider social and institutional environment is typically under-recognized. These relations are treated (if at all) informally, and are viewed as a source of problems and misunderstandings, rather than as an essential part of the experiment requiring explicit attention. It is in line with a 'blueprint approach' towards project planning, that intentionally limits its scope to areas that can be predicted, planned and controlled by project management (after Mosse, 1998: pp. 5 and 11).

By limiting the evaluation of the experiment to de-contextualized (technical and managerial) inputs and outputs, it became a model for future interventions all around the world.[3] This approach is obviously not without danger. First, the empirical basis for the successful claim of the model is questionable, as the experiment was already altered during the implementation process, and was not even adopted in the case study area in subsequent years. Second, the wider conditions that made this experiment possible (e.g. a powerful implementing party, favourable weather conditions) are lost for future users of the model. They thus do not know whether the policy model can be replicated in their own area of operation, and/or what kind of managerial measures should be taken to maximize its changes of success. It makes one wonder: what would the recent history of irrigation management reforms have looked like if case Wickham and Valera had discussed the political angle of their main system management experiment?

Unpacking Black Box No. 2: The Implementation of Turnover

Following the success of its participatory approach for communal, farmer-managed irrigation systems, the NIA developed the so-called turnover policy for agency-managed systems during the eighties. The objectives of this policy were to engage farmers in irrigation management in a substantive manner in order to solve both managerial and financial problems. Farmer participation in irrigation management would improve water distribution and maintenance and therefore increase the output of the irrigation systems in terms of irrigated area and yields. Also, the involvement of farmers in irrigation management was to help the NIA to become financially viable for its operational budget. It was expected to increase the income of the agency, as the turnover policy would make farmers more willing to pay irrigation fees. And it would allow the NIA to reduce its operational expenses by cutting the number of field personnel, as their duties would be taken over by farmers.

In UPRIIS, the turnover programme started with the implementation of the first part of the World Bank funded Irrigation Operation Support Program (IOSP-1), which ran from 1989 to 1991. The design of the policy implementation was as follows. Newly hired Institutional Development Organizers were to familiarize the NIA's field water masters with the idea of turnover. In turn, they were to organize farmers into associations. Locally respected farmer-leaders (so-called FIO's, Farmers' Irrigators Organizers) were to be hired by the agency for a year to assist the water masters in organising the irrigation associations. The irrigation associations should cover an area between 200-500 hectares. It should consist of elected Boards of Directors, each representing an irrigated area of 20-50 hectare.

The members of the association should also elect one of the board directors as president of the association. The associations were furthermore federated into federations at sub-zone level (covering the area of a sub-lateral), at zone level and at district level. At each level they were to co-operate with and support respectively the NIA's water master, zone engineer and operational engineer in the field of water management and system improvement works (World Bank, 1988).

The associations were to become legally recognized entities, which would enter into a turnover contract with the NIA. Associations could enter into a type 1 or type 2 contract. Under a type 1 contract, the association is responsible for routine maintenance of a certain length of a canal, for which it is to be remunerated by the NIA. Under the type 2 contract, the association undertakes operations and supports the NIA staff in the collection of irrigation service fees (ISF) from its members. It is remunerated for this activity through a percentage of the collected fees. It is supposedly stimulated to achieve high collection efficiencies, as the percentage given to the Irrigation Association increases with higher collection efficiencies (Wijayaratna and Vermillion, 1994).

All board members of the associations were to be trained extensively by the NIA. They could follow multiple day courses on system management, financial management and leadership development. Every year there should be re-elections of the board members, and new members are again entitled to follow these trainings.

By and large, this policy implementation design was followed by the NIA. The district employed six Institutional Development Officers, who became the primary agents for the implementation process. The FIO's were selected and employed and after two years 78 Irrigation Associations were established and legally recognized, as well as federations and confederations at the different levels of the system.

Black-Boxing the Process: The Evaluation of the Turnover Policy

Several conventional evaluation studies have been conducted on this turnover policy. These are all limited to information on the outcomes of the policy, and do not provide insights in why this particular outcome was achieved. At the level of the district, the district administration provided yearly reports on the implementation of the policies. This is however limited to statistics only. It tells us the number of Irrigation Associations and federations (re-)organized in a given year, and the number of meetings held between these institutions and the management of the districts. It also provides data on fee collection and irrigated areas. It does not provide any insight in the role of these institutions in the reality of day-to-day management of the district. It seems that the data are used to satisfy demands of the national office and the international donors who financed the programme. The data put forward were manipulated to provide a good picture of the turnover programme. Like one of the IDO's admitted: 'whenever I renew a contract of an Irrigation Association or federation I have to report it as if it was reorganized. Every year I report that all the associations are re-organized. In reality this only happens in a few cases.'

At the national level, the turnover policy was evaluated by two separate large-scale evaluation studies. The first study, which had a strong focus on the UPRIIS system, was conducted by a group of researchers from two international research institutes (IFPRI and IIMI) (Svendsen, Adriano and Martin, 1990; Svendsen, 1993). This study compares data of the same set of systems before and after the implementation of the turnover policy. The second study – which also included UPRIIS as part of its database – was conducted by NIACONSULT, a subsidiary corporation of the NIA (NIACONSULT, 1993). Both studies claim that the turnover programme resulted in increased fee collection, reduced personnel costs, a higher irrigated area and higher dry season yields. The second study, which compared systems in which turnover was implemented to a group of systems without such policy programme, uses the data to claim that farmer participation is instrumental in cost-effective management of irrigation systems. Both studies claim that the need for NIA to collect fees from water users to finance its own administration is an important mechanism behind these positive results. Unfortunately, neither of the two studies provides any insight in what really happened during the implementation process of the turnover policies in the different irrigation systems, and how these policies were given shape in real life. The implementation process remained a black box. Their claims about what happened during implementation are therefore not based on any direct evidence, but only considered to be probable, since the outcomes of the intervention are in line with their hypotheses.

We will now try to unpack the black box by taking a closer look at the implementation process of turnover in Zone 2 and Zone 1 of the fourth district of UPRIIS. We will specifically look at the ways in which local actors used the turnover policy to re-shape the socio-political arena of water control. We first take a look at Zone 1, and then turn to Zone 2. We will show that the policy had very different outcomes in the two zones, reflecting a different socio-political set-up in the two zones. This discussion again shows the shortcomings of the conventional way of monitoring and evaluation. The suggested positive relationships between fee payments, turnover and irrigation performance suggested in these studies are simplistic, and fail to come to terms with reality.

Zone 1: Networks of Power, Violence and Bribes

In Zone 1 the turnover policy was implemented as discussed above. Irrigation Associations, and federations at sub-zone and zone level were established, while all Irrigation Associations entered into a fee collection contract, and a few entered into a canal maintenance contract with the NIA. Nevertheless, the policy had little impact on water control. Though the water management institutions do exist and function one way or the other, they were largely disregarded and/or neutralized by the powerful players in the distribution of scarce water in the zone. Politicians have been major players in the distribution of the water. They used their political power and threat of violence to control the water, in order to gain politically (votes, allies) and/or economically (bribes). First we will take a look at the role of these politicians in water management, their objectives and their bases of power. We will then discuss the implementation of turnover in this politicized context.

Politicians became central actors in the water management of the zone in the dry season of 1983-84 – several years before turnover was introduced in the area – and remained so ever since.[4] Water control became the subject of struggle between two rival networks. They were fighting over the political control of the municipality of San Jose,[5] and its villages, that is largely located within the boundaries of Zone 1 (see Figure 9.2).

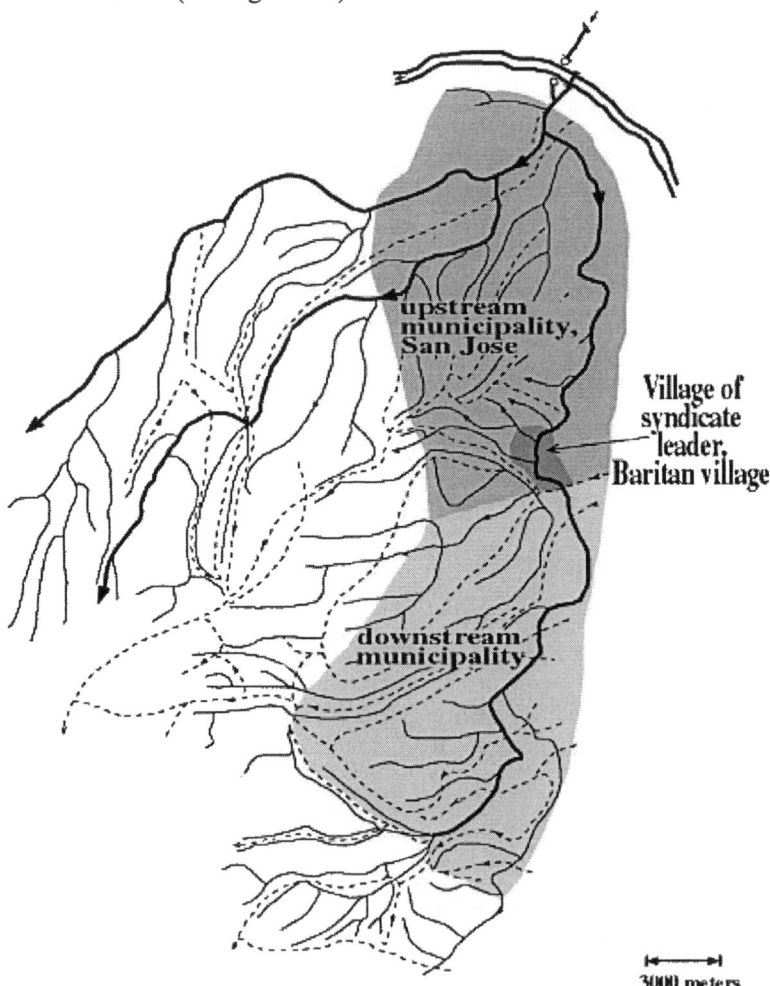

Figure 9.2 Layout of the PRIS system, indicating municipal boundaries of Zone 1, as well as the boundaries of Baritan village

From 1983-87, the group of secretary Juatco was in control of the water of Zone 1. His power in irrigation was based on the fact that he was a close ally to a mayoral[6] candidate. This candidate appointed him as the municipal secretary, after his faction was able to win the mayoral elections in 1983. Juatco moreover owned a large landholding around one of the important head gates of the zone, and had political control in his home-village, which is one of the four tail-end villages of San Jose. Furthermore, a close ally, the 'barrio captain'[7] capitan Dizon, controlled one of the other tail-end villages of San Jose, i.e. the tail-end village of Baritan (see Figure 2). This village is crucial in the water control of the tail end of the zone, as it is the 'bottleneck' of the main lateral of the zone. Three head gates of this lateral, physically controlling the water for the tail-end area of around 8,000 ha, are located within the boundary of this village. The group of Juatco was that strong that – at the peak of its power – nobody dared to come near the main gates of Baritan without his consent, including NIA's field-staff and engineers.

The situation changed when the archenemy of the Juatco-group, the Meneses-clan, re-took the position of mayor of San Jose in 1988. Juatco could no longer control the gates of the main lateral. Capitan Dizon, still remaining in power in his village, continued to manage the gates under his jurisdiction, although his rule over the irrigation water in the tail end of San Jose was weakened by occasional interventions of the now-hostile mayor. In 1995 however, Dizon rule in irrigation came to an end when the mayor put one of his sons in charge of the management of water in Zone 1. From then onwards, the Meneses clan is in control of water management in the zone.

These politicians cannot simply control water management in the zone by sitting behind their desks and making a few phone calls to the NIA engineers. On the contrary, it is a labour intensive matter. It took the son of mayor Meneses for instance around 25 days of hard, around-the-clock work during the dry season of 1996-97 to distribute the water to the tail-end villages. A year earlier, he spent almost three months to do the job. Moreover, most of the work of operating the canal structures is done at night. During the night it is relatively easy to bring water downstream, since the farmers in upstream villages are not present in the field. Also, he had to be present at night to stop rival groups of tail-end farmers from changing gate-settings and 'stealing water'. But even during the day he could hardly get any sleep, as he had to oversee gate-settings and to entertain large numbers of tail-end farmers who visited his house, seeking help and information. Both day and night he contacted people to guard gates, and to discipline farmers in water use, to seek advice, to solve conflicts and to counter-act moves of rival groups trying to control the gates. Hence, his operation requires a vast intelligence network, to know the events taking place in the field and the moves of their enemies. And it required a fair amount of logistics. During the night, Meneses used at least two vehicles, a few motorbikes and a minimum of 15 farmers to control the water in Zone 1.

The reasons why politicians put this much of their time and money in water control are clearly political in nature. Firstly, it is an important means for political patronage. This is especially true in the case of San Jose, where the majority of the lands are potentially served by the irrigation system, and where the livelihood of the vast majority of the voting population depends on rice farming. Or in the words of secretary Juatco:

> The first time I became involved in irrigation was simply because I wanted to save my own fields from drought as well as the fields of my good 'cumpare'[8] capitan Dizon. At the

time I knew nothing about irrigation and the names of canals and structures. In later years however, more and more people started to come to my house and asked me to help them to get water to their field or village. That is when I discovered the political virtue of irrigation. Once you enable farmers to harvest, they will have strong feelings of 'utang-na-loob' [debt-of-gratitude, JO], and their family will vote for you in the next election.

Also his archenemy, the Meneses family, is clear about the political virtues of in irrigation.

My father asked me to make sure that the whole of San Jose is able to get a good harvest during the dry season. My father depends on the 'poor man's vote'. His platform is a very simple one: to provide electricity, security and irrigation water. He asked me to take care of the latter promise.

Secondly, these politicians use their control over water to build and/or maintain their network for electoral purposes. The Meneses-clan supported their candidates for the village elections, by making them instrumental in getting water to their respective villages. Also the social gatherings needed for water distribution, like the late night meetings with farmer leaders, the nightly dinners in the villages, and the endless waiting at head gates, were used for political planning and strategizing. This is reinforced by the fact that elections usually take place in the month of May, just one or two months after the period of water scarcity.

Thirdly, controlling water allows these politicians to build or maintain a wide socio-political network of powerful friends and allies. This is clear when we look at the network of capitan Dizon. His network is much bigger than what one would expect for a leader of a remote, small and resource-poor village. Irrigation is the key to this peculiar network. As said above, the three head gates located on his territory are like a bottleneck for a huge area usually suffering from water stress. The larger part of this area consists of villages located in a neighbouring, downstream municipality. Due to his control over these gates, he was able to establish relationships with many local power holders from these areas, who were seeking his patronage for water. He became such an important figure in this neighbouring downstream municipality, that he even started to play a role in its electoral politics. Candidates for the mayoral elections of this municipality were seeking his endorsement, since his control over water allows him to influence the electoral behaviour of farmer leaders of these tail-end areas. He even cracks jokes about it: *'This municipality consists of 50 villages: the 49 located within its boundaries, and mine...'*

Fourthly, controlling the water can provide for a good source of income, as farmers are willing to pay bribes in exchange for (the promise of) water. For capitan Dizon, who controlled the physical bottleneck of the tail-end areas of zone one, bribery became big business. He is locally known by the farmers as the leader of the (water) 'syndicate'. At the height of his power, this politician was able to make deals with tail-end farmer leaders, representing an area of approximately 300-1000 hectares. These deals were made in the beginning of the season. Water would be delivered to these areas, in return for the right of this captain to thresh the harvested paddy at the end of the season.[9] These leaders were sometimes required to also promise a payment of two sacks of rice per hectare after harvest. In some cases, groups of tail-end farmers in a canal section were given water in return for the right to farm a few hectares of land free of rent in the area.[10] Also, 'gifts' were provided to the captain at his birthday party, and to his field-men, which were operating the water during the time of water crisis. In later years far fewer deals were being made with this politician, since he lost his control over this bottleneck to the earlier mentioned mayoral political candidate.

Legally, the politicians have no say over water management, as it is the sole responsibility of the NIA's district office. In reality, however, public resource management in the Filipino countryside is about 'little administration and much politics' (Fegan, 1994: 93). The Philippine state is often characterized as a 'weak state' (McCoy, 1994: 10), meaning that the capabilities of the state bureaucracies to manage public resources are seriously undermined by 'a strong society', consisting of elite political families. Ironically, much of the power of these families is derived from their control over the political institutions of the state (Sidel, 1995).[11] Being modelled after the American polity, these political positions provide for large discretionary powers, which in the Philippine context has led to the emergence of local 'bosses', '(...) who exercise considerable regulatory powers over the legal and illegal economies in their jurisdiction' (ibid: 230). The strength of mayors is based on their formal say over taxation, licensing and regulation, as well as law-enforcement (Sidel, 1995: 222-392). To the NIA district engineers, whose office and most of their houses are located within the municipality of San Jose, it is therefore unwise and practically very difficult to enter into a conflict with these power holders, and counteract their moves.

But even at the level of the village, the irrigation engineers might find it difficult to stop local politicians (the barrio captains), as everybody expects the ruling political family in the villages '(...) to take a leading role whenever (...) official or non-government organizations are making decisions that directly affect the village and its environs; and to prevent, openly or covertly, implementation of decisions about which they were not consulted or that they do not like' (Fegan, 1994: 101).

It might also be a very risky thing to do. Politicians are not only feared for their regulatory powers, but also for their readiness to use violence. To many politicians, (the threat of) violence – along with patronage – is the key instrument for seeking and maintaining power, and to many it is a way of life – and death (see for instance Cullinane, 1994, 1995: xx). In popular terms, the power of politicians is based on the three 'G's': Goons, Guns and Gold. Being tough, or even better being a known killer is a crucial asset in political careers (Fegan, 1994). The fear for violence is an important mechanism in the distribution of water in the main system of the district. Although there are only two clear cases in which people were murdered in the recent history of water management in the district, coercive acts and show of force take place in every dry season, and it is thus not without reason that people fear for their lives and act carefully. Most field personnel of NIA were faced with physical threats when doing their duty. This is just one of the many experiences which were shared with me by the field personnel.

> Several years ago, I was stopped on the road by a tough guy, who was the son of a notorious village leader. He was angry about the way I distributed the water. He drew his

gun and shot a hole in my helmet, which was at the carrier of my motorbike. He told that me that he would put a hole in my head the next time I would enter his village again (Excerpt of interview with former zone Engineer.)

The most dangerous job is the job of a gatekeeper of a main distribution gate. Although they are not paid to do so, they are still expected by their superiors and farmers to guard the gates at night during times of water crisis. One particularly brave gatekeeper, in charge of a trifurcation gate in the tail-end of the system, had been shot at three times during recent dry seasons.

For the politicians interfering in the management of the zone, the fear of violence is an asset. Their reputation as killers, their possession of powerful arms, and their use of tough men during the nightly raids, allow them to control the gates. Only the brave and powerful dare to challenge the authority of these politicians. Mere farmers and even NIA field personnel do not touch the main gates without prior approval of these politicians involved, especially at night, even if the gates are left unattended. This fear allows politicians to control the gatekeepers of NIA as well. Although they loose their neutral position and will be branded as a 'bata'[12] of a politician by farmers and their colleagues, many of these gatekeepers seek the protection of a dominant politician involved, to avoid being hurt. The above-mentioned gatekeeper did seek the protection of the son of mayor Meneses. The latter provided the gatekeeper with a high power rifle to defend himself at his gate. After it became known that he had become a 'bata' of the son of the mayor, nobody challenged his authority over his gate at night anymore. The threat of force also reinforces the reasons for the district engineers responsible for water management in the zone, to stay out of it. They hardly ever go out into the field at night time, and can therefore not control the water management of the zone. This is an advantage to politicians, because their control over the gates is less contested, when the NIA engineers are not present in the field.

Also the behaviour of farmers is structured by the threat of violence. Farmer leaders of the tail-end villages employ several tactics of risk avoidance. When they go out at night to look for water upstream, they bring along a group of armed co-villagers. They also make sure, however, that none of them is 'hot headed' in temperament. Furthermore, they only enter upstream villages at night to remove checks in the sub-laterals and close turnouts, after they have made a courtesy call to the local village captain or a farmer leader of that particular canal section. They usually first find out during daytime whether the water needs of the upstream area are more or less satisfied already (Svendsen, 1983).

The threat of violence strengthens the need for the different stakeholders involved in the water management of the zone to invest time and money in building trust and social networks by befriending people. Bribes and gifts not only serve to 'buy' water, but also to avoid being shot at during the long hours of the nightly raids.

The Implementation of Turnover in Zone 1

Turnover did not drastically change the rules of the game in Zone 1, as the policy was largely disregarded by the different actors. In the beginning of the implementation process, some of the institutional development organizers and the

NIA's field men were keen to establish a strong zone federation of Irrigation Association leaders which would be in charge of water management and would therefore stop political interference in water management. Though they indeed found some local leaders who supported the idea and joined the federation, this idea failed to materialize. In most cases, the newly established Irrigation Association leaders and federation representatives did not challenge the control of local political leaders over the issue of water. To this latter group, acting as a broker of water for their village is an important source of power, which they would not easily give up. They continued to seek water for their areas through their political network and/or illicit payments of bribery.

The turnover policy was however not disregarded altogether, as it did provide some opportunities for the different actors in the zone. The farmer leaders of the tail-end villages are opportunistic in their strategies. They explored every available way of getting access to water. One potential way is to acquire the sympathy of the NIA. Clearly, the NIA is not the major player in water management. It can however still make a difference, as it is in charge of determining the programme area, and as it can at least partly control the water distribution during daytime. Some well-organized tail-end villages therefore used the turnover programme to develop friendly relationships with the NIA's district managers. They stepped into the Irrigation Associations and the federation, invited the NIA's leadership to their Irrigation Association meetings during which they were entertained with food and drinks, and above all they ensured a high fee collection in their area.

To these villages, this is just one of the multiple strategies employed. Leaders of such a village might employ a combination of formal, informal and illegal strategies. One leader might become the Irrigation Association president and be in charge of entertaining the NIA by attending formal meetings, organizing drinking sessions with the NIA engineers and collecting irrigation fees. Another would be in charge of collecting illicit payments for the illegal distribution of water, while a third leader would be in charge of a gang of village men that patrolled irrigation canals during nightly raids. In some other villages, water was one of the subjects of interfactional political fights. In such a case, one faction would try to secure water for their followers by mobilizing their political network, while another faction would seek water by developing their relationships with the NIA.

The NIA's district management tried to use the turnover policy to be perceived as concerned and responsible engineers, while at the same time trying to stay out of trouble. They tried to stay out of the water management of Zone 1 as much as possible, to avoid conflicts with the involved powerful and potentially violent politicians. But they could not leave it altogether. Firstly, they badly needed the collection of irrigation fees to satisfy the demand of their staff for salaries and benefits. Hence, the NIA's claim for irrigation fees needed to be legitimized among farmers. Secondly, once in a while they had to respond to farmer demands for water, in order not to lose their confidence altogether, and to avoid being petitioned for transfer by farmer (political) leaders with connections in the NIA's central office.

They tried to employ the turnover programme to meet these two interests. The organising and training process and the occasional meetings of the Irrigation

Associations and federations provided them with the opportunity to legitimate the NIA's (poor) performance and its needs. They used the gatherings to argue that the low fee collections jeopardized the NIA's capability to properly operate and maintain the system, which in turn further reduces the willingness of farmers to pay their fees. This 'vicious circle' could only be transformed into a 'virtuous circle' when the farmers would help the NIA in managing the system and above all would duly pay their irrigation fees. They stimulated the farmers to join the Irrigation Associations and the federations and thus to help the NIA in managing the water, instead of seeking help from the politicians. In their day-to-day practice then, the NIA leadership tried to maintain their public image without running into conflicts, by following an opportunistic strategy. Instead of co-ordinating the implementation of a rigid water distribution schedule, they would simply entertain every request of these Irrigation Association leaders (as well as any other farmer leader or politician visiting their office). These leaders were given a paper slip that ordered a NIA gatekeeper to meet the request of a particular farmer leader. During one day, a gatekeeper was often confronted with several incompatible paper slips from different farmer leaders. Though this might be inefficient in terms of water use, the irrigation engineers to avoid conflicts and to satisfy as much farmer leaders as possible considered it the optimal strategy.

Apart from these general strategies of the NIA's district management, the engineers developed their personal strategies that shaped the turnover policies as well. The operational engineer for instance, became a close friend with capitan Dizon, the one who was deeply involved in the water management of the zone, in return for bribes and political contacts. It was widely believed that this engineer was involved in these clandestine practices. But whether he did so or not, he did use this friendship to build up his own socio-political network, among others to become the future district chief. His dealings with this local power holder obviously did not help the effort of others to develop the federation in order to organize water management without the involvement of politicians. The district chief himself also pursued his personal projects. He became a close friend with a federation president from one of the very tail-end villages of Zone 1, located in the home province of the chief. They set up a fishing business in converted rice ponds. It was said that they used the NIA's equipment to develop fishing ponds, and that he diverted irrigation water to this very tail-end area to satisfy the need for fresh water in the fishing ponds. Whether these allegations are true or not, it strengthened farmers' beliefs that getting access to scarce irrigation water is best served through a strategy of wheeling and dealing, rather than through paying irrigation fees.

The Implementation of Turnover in Zone 2

The implementation of the turnover policy has produced very different outcomes in the case of Zone 2. Any conventional evaluation study would claim the turnover policy in this zone to be very successful. Roads and canals are in relatively good shape, if compared to Zone 1. Fee collections in most of the Irrigation Associations of the area reach around 70-95% of the collectible (compared to a much lower collection in most Irrigation Associations of Zone 1). While thousands of hectares in the tail end of Zone 1 are usually excluded from the programme area, a much lower percentage of the irrigable areas are excluded in the second zone. Crop damages due to water stress is much less of a problem in this zone. The Irrigation Associations and the federations have become well-established institutions who manage water in the main system and collect fees in close collaboration with the NIA field staff. The Irrigation Association presidents and the NIA field staff became a close group, who have reinforced their intensive working relationships through numerous drinking sessions, and who in many cases became cumpadres and cummadres by means of their children's baptism or marriage.

How to explain this remarkable outcome? Any conventional evaluation study would fail to do so. Surely, turnover has triggered these relationships to develop. But conventional analysis would fail to see the important historical factors and the peculiar socio-political set-up that enabled the local actors to accommodate the turnover policy in such a productive and positive manner.

A key factor in this success story is the close relationship between the NIA district leadership and a large tail-end village of the zone, the village of San Anton (see figure 9.3). The operational engineer, the zone engineer, the maintenance engineer and a former zone engineer, all belong to families that originated from this village. The reason is that a former top official of the NIA's central office came from this village as well. When UPRIIS was looking for engineers in the mid-1970s, it was through his patronage that these newly graduated engineers were hired by the NIA and given a position in the district. Since that time, and long before turnover was introduced in the area, the farmers from this tail-end area developed a close working relationship with the district office, and were favoured by the office in terms of the size of the programme area, water and maintenance activities to roads and canals. The turnover policy provided the opportunity for different actors to further develop this relationship.

The key actor in shaping the turnover policy in this zone was the zone engineer. He was a very active zone engineer, who did his very best to bring water to the zone and to satisfy demands of especially the tail-end area of the zone. He spent a lot of time, energy and money to create a tight network of NIA's field men and Irrigation Association presidents taking the lead in water management and fee collection. During times of water scarcity, he would deploy his own truck and motorcycles to allow this group of men to patrol the gates and to search for water in the main system. He spent a lot of money on food and drinks to feed these people during these gatherings, and would even entertain this group in beer houses, after a day or night of managing water or collecting fees.

This zone engineer did so for both political and economic reasons. I first discuss the political angle. He is the eldest son of a long time vice-mayor of the municipality of Baliwag, which is located in the district. San Anton village – as one of the larger villages of this municipality – is the bailiwick of this vice-mayor. His son has been a great help to his (re-)election campaigns. By bringing water to this tail-end village, the vast majority of the villagers support the candidacy of his father. Moreover, the Irrigation Association leaders – which were selected with the consent of the zone engineer – all belong to the political clan of this politician, and

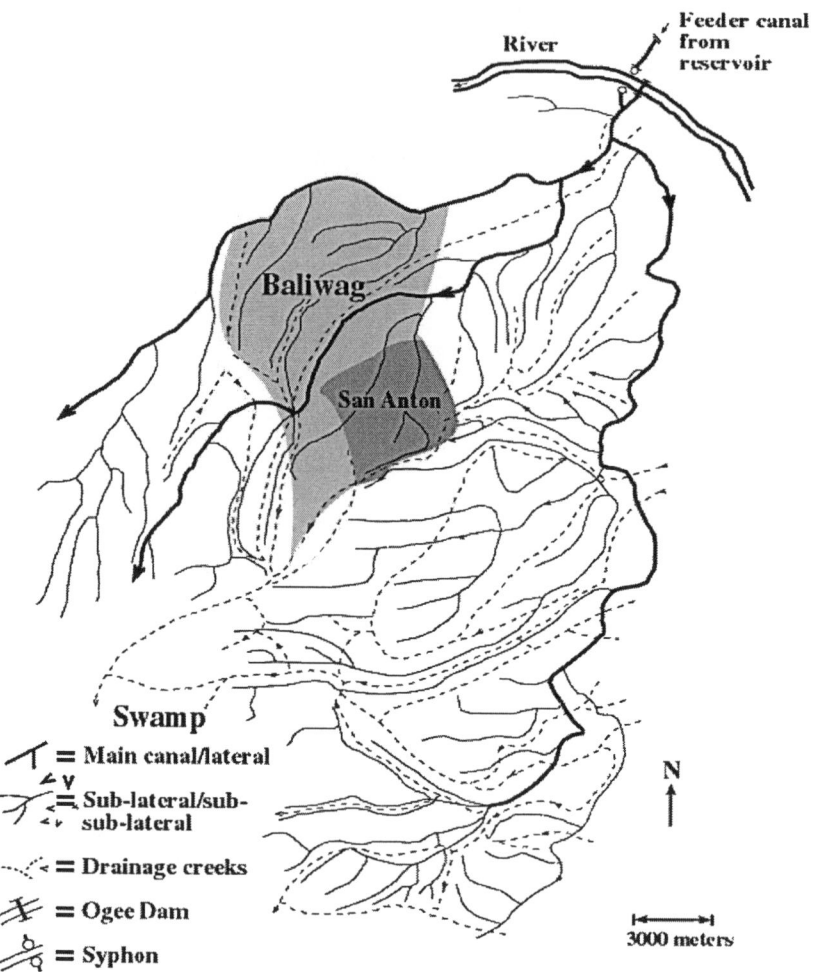

Figure 9.3 Layout of PRIS, indicating San Anton village as part of the municipality of Baliwag

were or became active campaigners during election time. Turnover thus enabled this zone engineer to reinforce the political machinery of his family. It is obviously no coincidence that this son of the vice-mayor became the zone engineer of this zone. Apart from the good relations with the engineers in the district office who originated from the village, the vice-mayor became a close friend and cumpa o the district chief. He is above all well connected to a very powerful political family in the area, who in recent years provided the congressional candidate of the district. He thus used these relationships to allow his son to become the zone engineer of Zone 2.

The zone engineer also used the turnover policy to further develop his economic projects. He runs one of the largest grocery shops in the municipality, which by far is the most favourite shop for his co-villagers. Moreover, he was able to acquire the exclusive distributing rights for both Coca-Cola and San Miguel beer to all of the selling points in the tail-end villages of Zone 2. He thus had a clear stake in supplying the rice fields of this tail-end village with sufficient irrigation water. He would like to win the sympathy of these potential buyers, but also to ensure the income of these consumers.

> My sales during the dry season very much depend on the yields of the farmers. If they have a good harvest, they will spend much on food and drinks. During April and May [this is the harvest season of the dry season, JO] most villages celebrate their barrio fiestas. When yields are low, my sales go down dramatically (Excerpt of interview with the zone engineer, October 1997.)

He moreover developed an economic stake in fee collection. Many farmers choose to pay their irrigation duties in kind, i.e. by a number of sacks of unhusked rice. The NIA provides the collectors of these sacks an allowance for carrying and hauling. This can be a lucrative activity, provided one can collect large quantities in relatively easy accessible areas in a relatively short period of time. This zone engineer was able to meet all these conditions. As he was well known to the farmers and as they were in majority pleased with his performance, it did not take much time and effort to persuade these farmers to pay their dues, and they often paid in full. He could make a substantial income from carrying and hauling large quantities of fee payments in kind, since he already possessed a truck for his grocery and distributing businesses. Moreover, he was able to use his connections in the district office as well as among big-time politicians to keep the canal roads in good condition.

To the leaders of the irrigation associations of especially the tail-end area of the zone, the turnover programme enabled them to develop their own interests. As part of the irrigation group headed by this zone engineer, they could build up their socio-political position in the village. It connected them to the important political family of the zone engineer, and allowed them to become public figures as small-time brokers of irrigation water and maintenance activities. Moreover, it provided them with an income from irrigation fees. Some of them were able to acquire an income from the carrying and hauling allowances, as the zone engineer provided them with the right to do so.

More importantly, some of them were able to acquire a substantial income out of the fee collection incentives given to the Irrigation Associations. As said above, the turnover policy enabled Irrigation Associations to enter into a fee collection contract. The contract stipulated that the Irrigation Association leaders should conduct campaigns among the farmer members of the association in which they should in order propagate the payment of irrigation fees to the NIA. They should also assist the NIA's field personnel in actually collecting the fees. In return, the Irrigation Association was entitled to a percentage of the collected fees in its area. Moreover, the higher the collected percentage, the higher the percentage given to the Irrigation Associations. This could lead to a substantial amount of money.[13] The Irrigation Association presidents did not see this money as belonging to the association, but considered it at least partially as a personal remuneration for their doings. This practice was criticized in several Irrigation Associations, but to no avail as the NIA tolerated it, and did not act upon it.

The zone engineer focussed on the creation of a tight network of NIA field personnel together with the Irrigation Association presidents. He did not see the need to involve the board members of the Irrigation Associations. The NIA district also exclusively focussed on the Irrigation Association presidents. It only invited them for training seminars and meetings at the district office. Moreover, the NIA field personnel also benefited from this collection money, and thus did not want to change the practice. In several cases they received a share of the collection incentive intended for the Irrigation Associations. They claimed the share as they did most of the work. The incentive to the Irrigation Association was based on the assumption that the Irrigation Association board members would spend a great deal of time for campaigning and collection of irrigation fees. In reality the Irrigation Association Board Members and even the Irrigation Association president did not do a great deal, as most of the campaigning and collection work continued to be done by the NIA field personnel. The collection incentive thus became a closed deal between the NIA field personnel and the Irrigation Association president (in which a share of 20-50% was given to the NIA field personnel).

To conclude, the case study of turnover policy interventions in Zone 1 and Zone 2 has shown that the de-contextualized way of conventional monitoring and evaluation studies fail to come to grip with reality. Such studies cannot explain the very different outcomes of the turnover policy interventions in two zones, which are caused by differences in the socio-political set-up. It has been shown that politics played a large role, as powerful politicians were deeply engaged in the water management of the two zones. In zone 1 the turnover policies were largely disregarded. In zone 2 politicians actively supported and gave shape to the turnover policies, in line with their own political, commercial and bureaucratic interests.

Conclusion

This chapter has tried to show the virtues of a 'policy as political process' perspective. Conceptualising policy implementation as a transactional process rather than simply the execution of a plan, allows one to look at the ways in which local actors shape and respond to certain policies. Such a perspective is helpful to understand why particular policy outcomes are achieved. It provides insight in the local arena of irrigation systems, and thus is useful for outsiders to understand what is really going on in these systems, and what is actually relevant and irrelevant for these local actors. This helps both policy makers and policy implementers to search for more appropriate policies and more sensible ways and means to implement these policies.

The focus on politics is useful in two ways. Firstly, it bends one's mind to the fact that policies are the subject of contest and strategic action. Irrigation policies do not just consist of ideas subject to consensual open discussion, but potentially create new dimensions to the local struggles over the control of such precious resource like irrigation water, fees or public positions. This helps one to understand why policies often fail to succeed. Changing main system practices or setting-up federations in Zone 1 were not bad ideas, and probably in the abstract accepted by all actors involved. The point is that these policies were effectively resisted or disregarded by powerful actors who did not want to give up their control over irrigation related resources. It also helps one to see that policy implementers need to develop a strategic programme to get their policies implemented the way they want to, and thus to be able to counteract moves of local parties resisting the agenda of the intervening parties.

Secondly, it centres on the role of politicians in irrigation management. In most policy programmes, politicians are ignored as these programmes only take water users and irrigation personnel into consideration. Given the public character of irrigation water and canal systems and the great importance of irrigation water for the livelihood of a high percentage of the voting population in the irrigated areas of the Third World, the more likely starting point of any analysis would be to assume the presence rather than the absence of politicians as key actors in irrigation management.

At the more substantive level, this chapter tried to determine the structures of socio-political relations of water control in the Philippines. We have seen that political patronage is an important mechanism in water distribution rather than formal administrative procedures. This is in line with general theories of resource control in the Philippines, which argue that the administrative state apparatuses are weak in relation to the powerful elite political families. Moreover, we pinpointed the importance of (the threat of) violence as a mechanism in water control. As the capability to act violently – and to get away with it – is mostly found among the political families, it further explains their dominant position in water control. Bribery was found to be an important mechanism as well, but we saw that the capability to bribe rested with a local politician rather than with the irrigation administration. This phenomenon calls for further comparative study of the interrelationships between the political and the administrative system, as it seems to contradict the findings in other countries. Wade's famous study of water control in South India did for instance show that irrigation officials rather than politicians were primarily responsible for bribing farmers in exchange for water (Wade, 1984).

Lastly, we have seen examples of the family as an important socio-political phenomenon in water control in the Philippines. The zone engineer in Zone 2 could not simply be regarded as an individual working in government, but he belonged to an elite political family, with interests in the administrative, the political as well as the business domain all at the same time. A sharp distinction between the administrative and the political domain in the analysis of water control clearly misses the point. The intimate intra and interfamilial relationships between farmers, irrigation officials and politicians (either based on kinship or on cumpadre/cummadre ties) are to be understood when analysing the politics of irrigation policy intervention in the Philippines (see for a general discussion of the concept of family and its role in the Philippine political economy, McCoy, 1994).

At the more practical level, this chapter shows the need for a careful monitoring of the dynamics of irrigation policy interventions. Both case studies furnished evidence of the shortcomings of conventional monitoring and evaluation methods. These methods – which are based on before-after measurements making use of a pre-determined set of limited indicators – did not allow the evaluators to understand why particular outcomes were achieved. Their findings were presented in a de-contextualized manner. The findings of these conventional studies are therefore useful for policy development purposes, nor for the implementation of the policy under different circumstances.

'Process documentation' or 'process monitoring' are new approaches to information management in policy implementation programmes, which try to overcome these limitations. The most intensive method is usually referred to as Process Documentation Research (PDR). PDR involves village level participant observation and record keeping by trained long-term resident researchers. Field level activities, meetings, negotiation, decision and implementation problems are meticulously recorded. Less intensive adaptations of PDR rely more on structured interviewing, reconstruction of events and the use of existing documentary materials. Field staff may undertake monitoring rather than specialized researchers, while review workshops and verbal reporting may replace detailed documentation (Mosse, 1998:9-10).

The general characteristics of these approaches are as follows. First, they involve continuous information gathering to be able to understand the dynamics of intervention processes. Second, they are oriented to the present that is 'the intimate relationship with what is happening right now'. Third, it is action oriented. The outputs of the process monitoring provide feedback to policy makers, project managers or researchers that helps them to respond immediately to events. The information is used to analyze failure, adapt approaches and in other ways facilitate more rapid (managerial) responses to events, lead to 'course corrections' and stimulate modification of project objectives and strategies in the light of implementation experience. Fourth, the approaches are inductive and open-ended. The focus of process monitoring and documentation is not narrowed down to expected outputs or impacts, but includes an account of events and relationships and diverse impacts including those which fall beyond the project as officially understood. It thus draws attention to the areas falling beyond management control, and thus allows critical reflection of the projects own definition of problems and solutions (ibid.: 10-11).

This type of information management is thus helpful for both policy makers and project managers to better understand what is going on, and to act upon it. It is used in learning process approaches, which make use of policy, or project designs that are flexible and change as a result of learning from implementation. It is also used as a means for engagement of different parties involved in the implementation process and consensus building. The approaches recognize that different stakeholders have different interests and that their ownership and commitment is important to achieve successful outcomes. They give new emphasis to the understanding and monitoring of institutional interests and relationships, and to inter-agency communication and consensus building (i.e. making learning a *joint* process) (ibid.: 7).

Ironically, one of the first systematic efforts to apply Process Documentation Research in a 'Learning-by-Doing' policy environment was carried out in the irrigation sector in the Philippines. PDR was an essential element in the pilot phase of what is now known as the world famous NIA's participatory approach towards communal irrigation systems. NIA previously provided assistance to communal systems at no cost to farmers. However, in line with a new government policy, since 1975 NIA had required farmers to pay for the construction costs in building or improving communal systems. After, the construction, the farmers are expected to operate and maintain the system. Implementation of this new policy required the organization of farmers into an association capable of complying with the policy conditions. NIA therefore started a number of pilot projects that were seen as 'learning laboratories' in which NIA wanted to learn how to involve farmers in irrigation development.

PDR was essential to the process. It took the NIA four years to develop this new participatory approach. PDR was applied in five out of the 18 pilot sites, which amounted to 73 months of research. Full time researchers worked in each pilot site covering 200-300 hectares. The field researchers were trained and supervized by a group of social scientists from Ateneo de Manila University, who represented the PDR field team in a central policy making body in charge of developing the new participatory approach, the so called 'Working Group'.

The conduct of PDR involved a systematic documentation of the activities and concerns of the farmers and the agency personnel. The documentation was done through participatory observation of project specific activities and interviews to clarify matters. Detailed monthly reports were sent to the Working Group. This group consisted of top-level policy makers of NIA, management experts, social scientists and a representative of the Ford Foundation that funded the pilot programme. For them, the monthly reports were 'a clear window into the rich detail of uncensored field experience' (Korten, 1989: 15). Moreover, these process documenters became important key informants to the working group. In many instances, it was not the process documentation itself that became the basis of action, but rather the discussions in which the process documenter or supervisor participated as a key informant (ibid.:17).

The working group used this information to continuously evaluate and refine the field interventions, which eventually resulted in a comprehensive set of tools and methodologies. It became the basis for the new participatory approach, adopted by the NIA throughout the Philippines. It also allowed the NIA officials in the working group to take a high number of appropriate measures within their own bureaucracy to overcome the observed problems in the field. Finally, the hands-on information was used by the working group to organize discussions and seminars on the developments in the field, and thus to engage the NIA management at various levels of the hierarchy into the process. PDR thus facilitated a systematic search for improved operational procedures and the gradual building of new capacities within the irrigation agency.

Obviously, this successful case of the use of process documentation and monitoring in shaping a policy programme should not be de-contextualized as well. There were specific conditions that allowed this experiment to succeed. Moreover, its success has been less pervasive than it is sometimes claimed to be. A number of the key players of the working group have portrayed their experiences as a successful way to 'transform an irrigation bureaucracy'. Unfortunately, it is not that easy. This new approach towards policy interventions was hardly adopted by the core of the agency: the departments in charge of the construction and management of large-scale, NIA managed systems. Irrigation agencies thus show different faces. On the one hand the NIA has thus been responsible for a very progressive policy programme using experimental and open-ended monitoring and evaluation techniques. On the other hand, it was also in charge of the turnover policy programme discussed earlier in this article, which made of use of conventional monitoring and evaluation methods.

Notes

1. The main cause of this problem is the limited supplemental supply from the reservoir. The larger part of the allocated water to the district is usually diverted to an upstream district during the dry season. This is possible, because the feeder canal of the downstream district is at the same time a main lateral of the upstream district. The encroachment of several hundreds of hectares of rice land located outside the boundaries of the service area in the upstream parts of the district is another reason for the scarcity of water. Effective rainfall in the dry season is negligible.
2. At the time, the system was called the Penaranda River Irrigation System. A few years later, the system was upgraded and the irrigable area was enlarged, while a feeder canal connected the system to the Penaranda reservoir. It thus became the fourth district of the Upper Pampanga River Integrated Irrigation System (UPRIIS).
3. In the present vogue of the international donor community, the experiment would probably have been classified and propagated as a 'best practice', to be learned from by irrigation policy makers and project managers.
4. The season of 1983/84 was the first season in which farmers experienced serious problems of water scarcity, since the Penaranda river irrigation system was rehabilitated and integrated into the larger UPRIIS system by the end of the 1970s. Water scarcity, causing water stress and yield reduction or crop failures up to several thousand of hectares have remained a problem ever since.
5. The names and places mentioned are fictional, to protect the privacy of the actors discussed.
6. The mayor is the elected leader of a municipality, covering a number of villages (called 'barangays' or 'barrio's').
7. A 'barangay (or 'barrio') captain' is the elected leader of a village.
8. The 'cumpare' custom is an important way of building networks in the Philippines. 'Cumpare' relationships are established during weddings and the birth of children, and signify close friendship, mutual support or a patron-client type of relationship.
9. The captain charged the regular fee of 6 sacks for every 100 sacks of threshed paddy. The disadvantage to farmers is that they loose their access to credit or farm inputs, which is usually provided by local thresher owners in return for the right to thresh their paddy after

10 harvest.
All the three kinds of bribe do not require any payments from farmers at the start of the season. The reason is, of course, that these tail end farmers do not possess the capacity to pay at the beginning of the season. They are often very poor, because of – among others – occasional crop failure due to water stress. This situation often meant that local leaders could not keep their promises made to this captain in terms of area to be threshed or fee payments to be made, because the farmers could not or would not stick to their promise. It also happened that these local leaders pocketed part of the fees collected from the farmers. Hence, this 'water market' was full of politics as well.

11 There are some major academic controversies on the characteristics of the Philippine state, as well as on the mechanisms and sources of political power, which however are beyond the scope of this paper (see for a discussion, Sidel, 1995, chapter 1).

12 'Bata' literally means child, but is used in the pejorative sense, indicating a 'messenger-boy'.

13 For instance, the total collectible amount of fees in an IA with an average area of 350 hectares during dry season is around 370,000 Pesos. If over 90% of that amount is indeed collected, the IA is entitled to 15% of the collected amount. This amounts to 50,000 pesos for one season (at the time, the equivalent of 1700 dollars).

References

Chambers, Robert (1988), *Managing canal irrigation. Practical analysis from South Asia*, Oxford and IBH, New Delhi and Calcutta.

Clay, Edward and Bernard Schaffer (1984), 'Introduction Room for Manoeuvre: The Premise of Public Policy', in Edward J. Clay and Bernard B. Schaffer (eds), *Room for Manoeuvre. An Exploration of Public Policy in Agriculture and Rural Development*, Heinemann Educational Books, London, pp. 1-13.

Fegan, Brian (1994), 'Entrepreneurs in Votes and Violence: Three Generations of a Peasant Political Family', in Alfred W. McCoy (ed.), *An Anarchy of Families: State and Family in the Philippines*, Ateneo de Manila University Press, Quezon City, pp. 33-109.

Fegan, Brian (1979), *Folk-Capitalism: Economic Strategies of Peasants in a Philippine Wet-Rice Village*, Ph.D. Thesis, Yale University.

Kerkvliet, Benedict J. Tria (1990), *Everyday Politics in the Philippines. Class and Status Relations in a Central Luzon Village*, University of California Press, Berkeley.

Korten, David C. (1989), 'Social Science in the Service of Social Transformation', in Cynthia C. Veneracion (ed.), *A Decade of Process Documentation Research: Reflections and Synthesis*, Institute of Philippine Culture, Ateneo de Manila University.

Long, Norman and Jan Douwe van der Ploeg (1989), 'Demythologizing Planned Intervention: An Actor Perspective', *Sociologia Ruralis*, Vol. XXIX, no. 3/4, pp. 226-49.

Mackintosh, Maureen (1992), 'Introduction', in Marc Wuyts, Maureen Mackintosh and Tom Hewitt (eds) *Development Policy and Public Action*, Oxford University Press, Oxford, pp. 1-13.

McCoy, Alfred W. (1994), 'An Anarchy of Families: The Historiography of State and Family in the Philippines', in Alfred W. McCoy (ed.) *An Anarchy of Families: State and Family in the Philippines*, Ateneo de Manila University Press, Quezon City, pp. 1-33.

Mollinga, Peter P. (2003), *On the Waterfront: Water Distribution, Technology and Agrarian Change in a South Indian Canal Irrigation System*, Wageningen University Water Resources Series 5, Orient Longman, Hyderabad.

Mooij, Jos E. (1996), *Food Policy and Politics. The Public Distribution System in Karnataka and Kerala, South India*, Ph.D. thesis, University of Amsterdam.

Mosse, David (1998), 'Process-Oriented Approaches to Development Practice and Social Research', in David Mosse, J. Farrington and A. Rew (eds) *Development as Process. Concepts and methods for working with complexity*, Routledge, London and New York, pp. 3-30.

NIACONSULT (1993) *An Evaluation on the Impact of Farmers' Participation on NIS Performance*. NIACONSULT, INC, Quezon City.

Oorthuizen, Joost (2003), *Water, Works and Wages. The Everyday Politics of Irrigation Management Reform in the Philippines*, Wageningen University Water Resources Series 3, Orient Longman, Hyderabad.

Repetto, 1986 *Skimming the Water: Rent-Seeking and the Performance of Public Irrigation Systems*. Research Report No.4. World Resources Institute.

Rocamora, Joel (1995), 'Introduction. Classes, Bosses, Goons and Clans: Re-imagining Philippine Political Culture', in Jose F. Lacaba (ed.) *Boss: 5 case studies of Local Politics in the Philippines*, Philippine Center for Investigative Journalism, Metro Manila, pp. 7-31.

Sidel, John (1995), *Coercion, Capital and the Post-Colonial State: Bossism in the Postwar Philippines*, Ph.D. thesis, Cornell University.

Svendsen, Mark (1993), 'The Impact of Financial Autonomy on Irrigation System Performance in the Philippines', *World Development*, Vol. 21, no. 6, pp. 989-1005.

Svendsen, Mark (1983), *Water Management Strategies and Practices at the Tertiary Level: three Philippine Irrigation Systems*, Ph.D. thesis, Cornell University.

Svendsen, Mark, Marietta Adriano and Edward Martin (1990) *Financing Irrigation Services: A Philippine Case Study of Policy and Response*. Second Draft. International Food Policy Research Institute, Washington D.C. and the International Irrigation Management Institute, Colombo.

Valera, A. and Tom Wickham (1976), 'Management of Traditional and Improved Irrigation Systems: Some Findings from the Philippines', *Choice in Irrigation Management*, Paper for ODI Workshop, Canterbury, 27 September – 1 October, ODI, London.

Wade, Robert (1982), 'The System of Administrative and Political Corruption: canal irrigation in South India', *Journal of Development Studies*, Vol. 18, no. 3, pp. 287-328.

Wade, Robert and Robert Chambers (1980), 'Managing the Main System: canal's irrigation blind spot', *Economic and Political Weekly*, Vol. 15, no. 39, pp. A107-112.

Wickham, Tom. H. and A. Valera (1978), 'Practices and Accountability for Better Water Management', *Irrigation Policy and Management in Southeast Asia*, IRRI, Philippines, pp. 61-75.

Wijayaratna, C. M. and Douglas L. Vermillion (1994) 'Irrigation Management Turnover in the Philippines: Strategy of the National Irrigation Administration'. *Short Report Series on Locally Managed Irrigation.* No. 4, Colombo, IIMI.

World Bank (1988) *Staff Appraisal Report. The Philippines Irrigation Operations Support Project.* Washington D.C., Report No. 7102-Ph.

Chapter 10

The Politics of Irrigation Reform: Research for Strategic Action

Peter P. Mollinga and Alex Bolding

Unsurprisingly, the objective of this concluding chapter on the politics of irrigation reform is explicitly political.[1] One main objective of the workshop at which the first versions of the papers were presented was to make 'politics' a legitimate topic on the agenda of global irrigation management and reform discourse. Through the publication of this book we hope to broaden that effect from the participants/authors to the irrigation/water resources community.[2] However, one book will hardly suffice, and therefore this chapter frames a proposal for further research. Framing a research agenda is not just an academic exercise: we want to argue that research that explicitly addresses the political dimensions of irrigation management and reform can and should inform strategic action by the different interest groups concerned. Irrigation is a field with historically close connections between research and practice, which means that in principle there should be opportunities for such linking of research and strategic action. An additional starting point is that, as this collection abundantly shows, strategic action is and has to be context-specific. This calls for more grounded and detailed research. We are advocates of irrigation reform, though much debate is possible and required on the nature, objectives and methodologies of reform.

We address the following topics in this chapter. We start with a look at the notion of 'models' for reform. The book shows that the history of irrigation reform efforts is characterized by the succession of a series of 'model' approaches: from the Philippines at the start, to Mexico, to Andhra Pradesh, India at present. An empirical critique of these models is that their success in most cases needs to be heavily qualified. But there is a broader question to be addressed: the problems related to abstraction from context and history.

The second section argues the importance of a contextualized look at irrigation reform processes. As an example it discusses how (populist) electoral politics often undercut on-going reforms in the irrigation sector. Together the first two sections emphasize the need to take seriously the complexity of irrigation reform situations, the dangers of simplification and the importance of context.

The third section discusses the notion of 'alignment strategies' as the proposed central focus of further research on irrigation reform processes. When reform processes are conceptualized as social processes in which different interest groups negotiate institutional transformation through a variety of strategies, using an array of resources, resulting in intended and unintended consequences, and with other issues than strictly institutional ones implicated, it becomes crucial to understand, particularly for advocates of reform, how all these differences come together (or fail to come together), at the different levels and stages of the reform process. How does this mediation of interests take place, and what does 'mediation' mean in concrete terms? Part of the discussion is how social power 'works' in irrigation reform processes: in the processes of negotiation and other forms of interaction, as discursive power, and through the role of politicians in irrigation reform.

After this we discuss three concrete research themes for further work. These are the resilience of irrigation bureaucracies, the role of international development funding agencies and the capture of irrigation reform implementation.

The fourth section discusses the resilience of irrigation/water bureaucracies. This refers to the phenomenon that in reform processes the irrigation/water bureaucracies that are the subject of institutional transformation (in addition to institutional changes at users/field/system level) show remarkable ingenuity to maintain their primary orientation towards physical works (construction, rehabilitation, maintenance). This raises a number of questions regarding professional re-orientation in the irrigation/water-engineering domain, and when and how new insights may get internalized in the profession.

The fifth section treats the global dimension of irrigation/water sector reform, notably the role of international development funding agencies (like the World Bank, Asian Development Bank, and bilateral donors). There are questions about the power and influence they actually exert, and on the effectiveness and implications of that influence.

In the sixth session we discuss how irrigation reform policies processes tend to be captured and re-shaped in the implementation process. Both irrigation agencies and local actors attempt to remould reform policy to fit their own perspective and interest. If irrigation reform is to achieve resource redistribution special efforts may be required. Analysis of the intricacies of capture and transformation may provide new entry points for irrigation reform initiatives.

In the seventh section we ask the question what kind of substantive, explanatory theory could be aimed for regarding irrigation reform processes. We answer the question by arguing for context-specific, structural plus historical causal explanation.

The last part of the chapter addresses the question: what to do next? We start with a reflection on the politics of research in the eighth section. The organization of the workshop and the editing of the book provide evidence that 'politics' is a sensitive topic indeed. Why is it that so little research is done on the politics of irrigation management and reform? Why is it important to do it? How and by whom can it be done, given the sensitivity of the subject?

In the last, ninth section we discuss our drivers for further research and how research can contribute to strategic action in relation to the way social power 'works' in irrigation reform.

Models for Irrigation Reform: A Critique

After reading the two chapters on the Philippines (Chapters 4 and 9) it should be clear that songs of praise on the Philippine model need considerable qualification. In this case achievements of the reform seem to be much less than is usually suggested in international discussions.

The Philippine model has travelled far. It was the inspiration, for example, of a number of irrigation reform initiatives in different Indian States. The propagation of the 'model' in India in the 1980s and 1990s was facilitated by the Ford Foundation, by supporting several NGOs to undertake pilot experiments with water user organizations, notably in the States of Maharashtra, Gujarat and Andhra Pradesh. One positive effect of these initiatives and this facilitation has certainly been that the concepts of farmer participation and water user organization have been introduced and become part of the policy discourse. A second positive effect is that the experiments have shown the possibility of higher levels of farmer control over canal management. However, with the benefit of hindsight it could be argued that some issues could have been recognized and addressed earlier when a closer look at the conditions of possibility and the practice of the Philippine 'model' had been taken. One such condition was strong political support from the top, which was absent in the Indian cases. One practicality similar to the Philippines situation was that implementation in agency managed canal irrigation systems was very difficult. All the Indian experiments ran into the 'scaling up' problem, and most of them had a tendency to black-box 'farmers' into the singular categories of head-enders and tail-enders, without too much attention to the finer aspects of the inequities in local social relations. Individuals involved in these experiments may well have been aware of some of these issues at the beginning, but this did not translate into changes in approach (Joshi and Hooja, 2000; Mollinga, 2001a).

At this point it may be asked what exactly acts as a 'model' in such situations. The word is not an ascription by us, but is actively used in the policy discourse. For example, irrigation professionals and others were talking about an 'Andhra model' three years after the irrigation reform process had started in the state of Andhra Pradesh, India in 1997.[3] Models in this incarnation are never very precisely defined, and only partly exist in a formal written form (they do exist however, as bullet points in power point presentations that are used in training and 'sensitization' sessions). The components of a 'model' in this sense are a belief that things can/must be done in a particular way (the participatory approach in case of the Philippines, a government-written Act covering a whole State with a lot of 'political will' supporting it in the Andhra Pradesh case) plus a limited set of concrete policy instruments (like the role of the facilitator/community organizer in the case of the Philippines, and a particular concept of a Water Users Association in the Andhra Pradesh case): the 'essence' of a successful reform process. This is the kind of simplification that circulates as a 'model' and enthuses and enrols people, but, as we argue, also suggests too easy roads to 'success'. At a more detailed and practical level of policy and project reports much more content can be found, with a much lower level of simplification. But in such documents 'power and politics' and 'contextualization' (see below) are conspicuous by their absence.

The gist of the analysis reported in the two papers on the Philippines could also have been reported in 1980, and certainly in 1990 – and maybe it was somewhere.[4] About the reasons why such documentation is rarely done, or not acknowledged, we refer to our speculative discussion in Chapter 1. We identified three possible reasons: 1) models fit the engineer's preference for blueprints and prototypes, 2) models fit the bureaucrat's disposition towards general solutions fitting a variety of cases, and 3) models are part of the 'promise of success' syndrome in the donor world, where they serve to convince decision-makers that success is possible in short time frames. We add one element to the list of attraction factors for the 'model' perspective. Models promise or suggest that there are procedural solutions to substantive issues. 'Iniquitous development can be solved by using PRA methods', would be a rather crude example. Without wishing to argue that method is irrelevant, quite the contrary, the high level of enthusiasm for method, procedure and 'tool-boxes' should, we suggest, be regarded with some scepticism. People and projects advocating such methods do sometimes get carried away, as is also increasingly recognized in the literature on the subject.[5] There is a lot of affinity with the engineers' desire for 'scientific' approaches, to keep nasty, murky issues like politics out of the equation (see Chapter 1). Whatever implications can be derived from the material, we hope to have provided convincing evidence that it would be wise to be much more modest in proclaiming and implementing 'models'.[6]

But modesty regarding 'models' is not only required because impact of reform is usually less or different than desired. The second example in Chapter 9 on the Philippines, where the engineer appointed through political relations achieves high collection rates of fees and reasonably effective water distribution while running several businesses, shows that also success is highly contextual and needs to be explained. And this context-dependence is the more fundamental point. When it is not understood why a particular reform initiative does what it does in a particular situation, replicability may be suggested where it doesn't exist. A more realistic concept would be to see other experiences as intellectual and practical resources for addressing context-specific issues in new situations, rather than as 'models'. This makes the life of policy makers and implementers a lot more complicated, but all the chapters in this book show that that complexity cannot be wished away, and that uncertainty and strategic action are necessary parts of the game.

Context is a synchronic concept, but it has a diachronic dimension as well: history. The Mexican case study (Chapter 3) provides an example of the importance of history in understanding irrigation reform. The drastic irrigation reform in Mexico in the early 1990s seemed very sudden and unexpected to many observers (and it has been termed a 'big bang' approach to reform). These observers analyzed that the overall neo-liberal perspective of the Mexican government and financial crisis triggered it. Closer study shows that a long process of debate and manoeuvring that consolidated when a certain configuration of circumstances and relationships occurred preceded the apparently sudden change. This suggests several general points regarding the notion of reform 'models'.

Firstly, it adds to the modesty argument. Models not only are contextual but also have a history. They grow and emerge in historical processes, sometimes slower, sometimes faster. The Mexican case shows that that process can partly be shaped, but that there are also many contingent factors. It could easily have gone otherwise. The Philippines and Pakistan cases (Chapters 4 and 7) add to this that historical decisions may have both intended and unintended consequences for irrigation reform at later dates. 'Today's problems were yesterday's solutions.' In Pakistan the creation of the WAPDA (Water And Power Development Authority) and OFWM (On-Farm Water Management) agencies as separate from the Irrigation Department had strong implications for the later reform process. For example, WAPDA proved useful for speedy, centralized construction of big river works in the 1970s and 1980s, but later became part of the 'institutional problem' and is perceived by many as an obstacle for institutional reform. In the Philippines some of the structures, procedures and experience of earlier reform efforts hamper progress in later ones. For example, it can be argued that the financial autonomy of NIA (the National Irrigation Administration) led to cuts on operation and maintenance subsidies, which triggered a process where NIA officials spent more time on collecting irrigation service fees at the expense of attending to maintenance, ultimately resulting in World Bank/Asian Development Bank loans for rehabilitation, thus reinforcing the construction bias of loans and activities of the agency.

The Zimbabwe case (Chapter 6) shows that not only institutions are inherited from earlier periods and processes, but the irrigation systems themselves may be unsuitable because not economically viable under new conditions (pumping systems that become too expensive to operate when government withdraws financial support). The Zimbabwean case also shows that internalized management styles from the colonial period, characterized by centralized control, may provide a hurdle for reform in the sector.

Secondly, it will be very difficult to implant, transfer or enforce approaches developed in one place in or to another. This only works when the new ideas and approaches resonate with domestic agendas and developments. Pakistan is a clear case where that resonance is not very apparent, and Indonesia in the 1987-1999 period was a similar case (Chapter 5). In the Philippines and Mexico, in contrast, there was a situation of receptiveness for new ideas in which external agents could insert new perspectives (accompanied or not by financial arm-twisting).

A policy question that emerges from this discussion is how to find a middle-path between the extreme 'model' approach of standardization and instrumentality on one hand, and the extreme form of context-specificity ('every village or irrigation outlet requires its own, individual and unique approach') on the other. The paper on Andhra Pradesh, India (Chapter 9) suggests that the reform managers in that case seem to be trying to find a balance between these two extremes. The best answer we have to this is probably something that did emerge from the Philippine experience: the learning process approach. The trick seems to be to organize the policy process in such a way that it becomes responsive to context and changing issues and needs. This is not a spectacular conclusion, but a rarely practiced one, and difficult to implement. However, the South Africa case (Chapter

2) gives some reason to be optimistic about the possibility of participatory policy formulation processes, though nowhere are history and context more poignantly present.

Irrigation Reform as Part of Wider Economic and Political Change

A logical implication of the critique of the simplification of reform situations by thinking in terms of models is that an alternative approach would have to contextualize in both analysis and intervention and squarely address historical and structural complexity. Irrigation reforms do not come out of the blue, nor solely emerge in response to specific problems that beset the irrigation sector. Irrigation reforms are often tied in with or part of wider processes of economic and political change. As most chapters in this book highlight, the avalanche of irrigation reforms that started during the 1980s and early 1990s can be linked to the emergence of the neo-liberal paradigm of economic development. This stresses the need for more market, less state and more civil society in almost all sectors of the economy. Furthermore, changes in the political regime of a country, like shifting from authoritarian forms of governance to democratic or popularly elected government, often entail changes in the configuration of state agencies and their relationship with the populace at large. Therefore, there is a need to look at irrigation reforms in a wider perspective than the aims and transformations related to the irrigation sector.

One important, because recurring, element of political context is that the electoral concerns of populist politics may undermine the financial viability of reformed irrigation agencies. An example is the Estrada led government in the Philippines that came to power in 1998 on the back of promises to do away with irrigation service fees. The resulting cut in irrigation service fee rates by 50% actively undermined the viability of newly configured water user associations and the service-oriented NIA. Equally, in Zimbabwe the promises of free government services made by ZANU(PF) during the independence struggle, resulted in under-financed and badly maintained irrigation systems after independence, because of the stubborn refusal by many irrigators to pay water rates and the failure of a cash-ridden government to provide the necessary resources on its own account.

Alternatively, concerns over electoral control of the countryside by ruling political parties sometimes leads to contradictory policy directives in the irrigation sector, depending on the latter's importance on the voters' roll. Thus the redirection of maintenance funds from the irrigation department to new water user associations in Andhra Pradesh (Chapter 8) can be conceived as a deliberate attempt by the ruling political party to construct a foothold in the vast irrigated *hinterland* of this Indian state. Even in areas where irrigation does not count for many votes, like Zimbabwe, fear of giving rise to opposition parties can lead to the active undermining of disciplinary actions by irrigation agencies as demonstrated in Chapter 6.

These findings call for a qualified and diversified arsenal of reform strategies that take into account the importance of wider economic and political concerns, rather than the recipe orientation of many reform advocates that originate from the irrigation sector and focus exclusively on the problems that hamper efficient operations in the sector (*cf.* Vermillion and Sagardoy, 1999). The room for countering or addressing such wider political concerns very much depends on the nature of the policy processes involved and the inclusion or exclusion of certain interest groups from this policy process, an issue that we address in the next section where we discuss the alignment of interests in the policy process.

Analyzing the Process: Alignment Strategies

After having established the need to address the complexity or irrigation reform processes and the need to contextualise them, we now turn to the approach and themes for further research. Under approach we argue for, in this section, a focus on alignment strategies in irrigation reform processes. In the following sections we identify three important research themes that are suggested by the country case studies for further research: the resilience of irrigation bureaucracies, the role of international development agencies, and the capture or reform policies by local actors. In the last section we reflect on the development of explanatory theory on irrigation reform, and discuss the strategic value of further research once again.

Several chapters in the book give conceptual elaboration of the 'policy as process' and 'politics of policy' perspectives. These are Chapters 3, 6, 8 and 9. From the references in these chapters it is clear that the politics of policy and policy as process notions are not new. In fact, they are well-established frameworks, which mostly emerged in the 1980s, with different variants extant.[7] What is new is its application to irrigation reform. We will therefore not use the irrigation cases to 'prove' that these frameworks are useful for policy analysis. More case studies are needed to test the usefulness of the different variants of policy analysis frameworks for irrigation reform. Here we only highlight some of the more specific conceptual issues that can be found in the chapters, to introduce the discussion on 'alignment'.

In Chapter 3 the authors of the Mexico case argue that the dichotomy between policy formulation and implementation should be transcended, and replaced by a vocabulary that uses 'articulation' of policies as a central concept. This allows them to see hydraulic bureaucracies not only as policy implementers, achieving or not achieving policy objectives, but as historical agents that try to articulate their 'overriding concerns' throughout the policy process.

Once the point of the essential continuity of policy processes is absorbed, the discontinuities in that process also have to be recognized. The authors of Chapter 6 argue that in Zimbabwe it is important to distinguish policy discourse and policy practice, and show that the two are quite disjointed. The chapter makes the important point that 'reform' may also take place by default as they put it, in the absence of any consolidated policy. In Zimbabwe the financial crisis of the state triggered local processes of institutional transformation in irrigation.

These observations raise the question of what is meant by policy and public policy, and how reform is linked to them. In the Zimbabwe case it may be questioned whether there is 'reform' at all, when that is understood as purposive action (through policy) to change the structure and dynamics of functioning of irrigation agencies and/or the relationship of the agencies with water users, or just 'transformation'. We have deliberately used a very broad concept of 'reform' in this book, that is, not tried to very precisely define and demarcate it. Finally, the interesting point is through which mechanisms changes in irrigation agencies and their relationships with irrigators occur, rather than the definitional question of what constitutes reform and when it becomes just change or transformation.

Chapters 8 and 9 on policy implementation in Andhra Pradesh, India and the Philippines make a similar (double) point as the Zimbabwe chapter. Policy formulation and implementation are linked elements of an overall process of institutional transformation (and reproduction without significant change in some cases), but at the same time there is a discontinuity between formulation and implementation. In both cases there was considerable political support for the policies, but the realities of implementation created serious problems. One dimension of this is that the actors in the arena of policy formulation and the arena of policy implementation only partly overlap. Irrigators and field staff were not part of the formulation process, while the politicians that sanctioned the policies were not part of the implementation arena (or had double agendas from the beginning).

The first conceptual element of analyzing policy processes is thus to model the structure of the policy process at hand. Though very different vocabularies are used, the different approaches have the following main elements. Firstly, they all model the structure of the policy process by identifying arenas or domains in which policy practices take place.[8] Secondly, they posit that the embeddedness of the policy process is important, and that policy processes have both intended and unintended transformative effects (physical, economic, institutional, political, and so forth).

The second conceptual element addresses the 'politics of policy' focus of our discussion.[9] Such a perspective obviously has to unpack the notion of 'politics'. When politics is the mediation of social relations of power, the central issue in understanding politics is to understand how social power 'works' in policy processes. We tentatively suggest the following three ways in which irrigation reform processes are 'powered'.[10]

The *first* understanding of social power is that of politics as negotiation, struggle and contest. These are processes in which different (policy) actors mobilize different resources, interact in different ways and (re-)shape policies. There are questions of inclusion/exclusion, open/hidden agendas, front/back room interaction, and the like. Nevertheless the processes are relatively explicit, and at least partly public in most cases. There is a wide variety of forms of interaction. One classification that resonates with the editors' own experience of irrigation reform processes is to distinguish the politics of blame (or conflict), the politics of negotiation, the politics of collaboration and the politics of partnership.[11] The general point is that there are qualitatively different modes of engagement. There may or may not be an evolutionary sequence. On the side of practice, methodologies for institutional development assume that it is possible to induce and facilitate such evolution. The present attention for multi-stakeholder processes, institutions or platforms and polycentric governance in the water sector posits the desirability of 'collaboration' and 'partnership' approaches. This book suggests that the actual practices may be located more in the domains of blame (or conflict) and negotiation. In this book South Africa is the only example where there seems to have been a politics of collaboration in the formulation phase, but maybe this is a too optimistic interpretation.

The *second* way in which irrigation reform processes may be 'powered' is in the more indirect way of how policy processes discipline meaning, perception and behaviour. This refers to a number of things. One is how through policy discourses power is exerted by the way problems are defined, how interest groups are labelled, and other simplifications are constituted.[12] It might be argued for example that the power that the irrigation sector has enjoyed within the government administration of some countries is partly due to its ability to make the technical discourse on irrigation the dominant discourse. This involves (at least in the Indian context) defining irrigation development as the creation of infrastructure, a water supply enhancement orientation in addressing water resources problems, and emphasizing the connection irrigation-agricultural development-economic growth-nation building.[13]

Another example is the irrigation management turnover concept (IMT). It posits that there is a government domain of management that is too large and which partly or fully has to be handed over to farmers-irrigators. One can easily see the resonance with the 'rolling back the state' dimension of the neo-liberal development paradigm. An empirical criticism of this notion, supported by several chapters in this book, is that the government and farmer domain may be separated on paper, but in reality they are highly inter-penetrated. Another criticism would be that IMT assumes a two-actor model: managers and irrigators (state and farmers). In reality more actors are involved in irrigation management, and these are unlikely to be sidelined easily. A third aspect of the IMT concept is its functional orientation. It looks at irrigation management as a circumscribed functional and technical activity. In reality water managers, water users and others are concerned about much more than irrigation and system performance. These empirical criticisms suggest a need to re-think the whole concept of IMT, looking at a much broader set of possibilities for irrigation management organization than the implicit 'joint management' system of irrigation agencies and water users associations.[14] The point in this context is that the concept defines the problem in a particular way, thus directing its users to particular solutions.

Ferguson (1990) has analyzed the depoliticizing effect of development assistance discourse and practice, something that could also be applied to the irrigation sector. The technocratic and depoliticized 'scientific' discourse that has been dominant in the irrigation sector is under challenge, particularly through critiques from an environmental perspective, and to some extent through poverty and governance oriented perspectives (see also Chapter 1 on sources of attention for irrigation reform).

The *third* way in which irrigation reform processes are 'powered' is by the role that politicians play in them. This might be considered as a subset of the first and to some extent the second way of being 'powered' described above. We single it out because of its apparent significance in irrigation reform implementation contexts. The type of power concerned here could be called the accumulation of political capital (and sometimes rents) in policy implementation by brokerage activities of bearers of state political authority. In India these brokers would be primarily MLAs (Members of the Legislative Assemblies), in the Philippines the mayors are central actors in this respect. Their activities are set in the context of overall populist politics, which creates a set of specific mechanisms for political office bearers to exert power as brokers in policy processes.

These three different ways in which social power 'works' in irrigation reform processes are not alternative views (though they may be associated with different social science schools of thought). All three may be relevant in every case, but to a different degree and in different ways.

The different forms of power are not only conceptually different but also have different forms of strategic (public) action associated with them. To the first are associated efforts to establish 'level playing fields' in policy formulation and implementation, as in attempts to establish multi-stakeholder processes or platforms. These focus on increasing the inclusiveness of policy processes, to get fairer and more constructive bargaining.

The strategic action associated with the second form of power is policy critiques. These are (de)constructions of policy discourses like for example emerging discussions on water and privatization from the 'anti-globalization' viewpoint[15] and a 'mainstream' effort like the Global Dialogue on Water, Food and Environment.[16] A large part of the global politics of water is about the (re-) definition of meanings and problem statements, the development of innovative concepts and approaches, in short, setting the terms of debate and policy priorities.

The strategic action associated with the third way in which power 'works' in irrigation would logically focus on transparency and accountability. Though it does to our knowledge not exist as yet, one could imagine an organization like Transparency International[17] for the water/irrigation sector. More generally, public access to information, public awareness raising and better auditing procedures could be part of such an approach. A basic issue in this context is how the desirable relation between politicians and water managers – government managers or farmer managers – is conceived. The standard model of a democratic setup is that of a separation between policy decision-making by people's representatives and implementation by managers. In many cases the boundary between legislative and executive power is highly permeable – hence the notion of 'political interference'. In many irrigation reform situations it seems more realistic to try to reshape the role of the bearers of political authority in implementation, rather than to eliminate them from the process, as this would touch the heart of really existing populism.

Finally, how does one investigate the 'working of power' in the different phases and domains of the policy process? We propose that the study of the 'alignment of interests' is a useful starting point. With this phrase we try to summarize the substance of policy processes as processes. Interests need to be aligned to bring policy into existence and to make it 'work', but also the activity of interest alignment (or non-alignment) is what keeps the policy process going (or stops it). We prefer the word alignment to articulation because it expresses better the purposiveness of the process (though there may be different and competing purposes).[18] As the authors of the Zimbabwe chapter remark, purpose and objectives need to be part of the concept of (public) policy, no matter how many qualifications are necessary with regard to consistency and other factors. It is only when there is purpose that alignment becomes relevant, otherwise there would only be interaction. 'Alignment of interests' seems to us a useful entry point for further analysis: where does it (not) take place (what is the network of arenas in which it is located), who does (not) participate (what are their interests, overriding concerns, identities), how does it (not) happen (what are the strategies and resources (not) used), and what are the outcomes or effects?

'Alignment' applies to all three ways in which power 'works'. This is obvious for the first meaning, where negotiation and related concepts are central. In the second way there is cognitive and ideological alignment,[19] or alignment through cognitive and ideological means. Such processes may be only partly conscious (*cf.* Bourdieu's notion of people's dispositions). In the third way it is again more obvious, though alignment perhaps often amounts to juggling.

The chapters in this book show that there are substantial differences in the way policy processes unroll. Contrast for example the South Africa case (Chapter 2) with the Mexico case (Chapter 3). In South Africa there was a very open consultation process organized as part of policy formulation. A central characteristic of the Mexican political system is that such processes happen(ed) in a very small circle of senior politicians and bureaucrats. South Africa is also the only case in which irrigators were involved in the policy process, though there are questions regarding the capacity to engage of different groups (notably well-organized white commercial farmers *vs.* poorly organized smallholder black farmers). The paper makes the interesting observation that it matters a lot whether representation means that each *perspective* is represented in the consultations, or that the *groups* holding a perspective are also physically present in the process. Chapter 9 on policy implementation in the Philippines shows that (the threat of) violence may be an important resource in achieving 'alignment' at field/system level, a factor not often emphasized, but probably quite common. Regarding outcomes, Chapter 4 on the Philippines provides a nice example of how the changes in the Type II and Stage II turnover contracts express different alignments of interests in different phases of the reform process. Another interesting observation is made in the chapter on Mexico, where it is argued that the infamous 'political will' is the outcome of a 'successful' policy process, and not a pre-condition, that is, it expresses alignment achieved.

The chapters also underline the importance of contextual and historical factors in the alignment of interests. It is remarkable that in both South Africa and Mexico, two cases with rather drastic reforms, the percentage contribution of agriculture to the GDP is low: 5% to 6% according to the papers. It is not difficult to imagine that the relative importance of agriculture can make a lot of difference for the feasibility of reforms with strong implications for the agricultural sector, though contribution to GDP is obviously not the only factor contributing to the power of the agricultural interest. The proliferation of vested interest groups, in the Pakistan case within the government administration through creation of new institutions in the past, has already been referred to above. In Zimbabwe the irrigation question is strongly connected with the land question, which has its own dynamics. The feudal agrarian structure in Pakistan is an important factor in the lack of progress in reform, while in the Andhra Pradesh, India case study the existing agrarian structure explains why the rural elite so easily captures the reform. Some contextual and historical factors are temporary. The windows of opportunity created by the end of the *apartheid* regime in South Africa and the end of the Suharto regime in Indonesia will only exist for a limited period.

Alignment is more than pragmatic or opportunistic compromise, though that there should be gains for everyone included in the process, is certainly a bottom-line characteristic. Alignment has an element of internalization of the compromise by the actors involved. The South Africa chapter points out that the consultation process through which alignment took place re-shaped the views, self-perception and relations of the different actors involved. In the Mexico case, the process also touched the parties involved – the views of the hydraulic bureaucrats changed in the process. In many of the other cases there was a degree of 'plain bargaining' to reach strategic compromises. Other elements of alignment are finding the proper language to express compromise, personal rapport of negotiators, and understanding and appreciation of others' interests. Whether alignment that is enforced can be called alignment is an open question, as well as whether alignment can be enforced.

Though the papers in the book give considerable information on how alignment of interests has taken place or failed to do so, more detailed analysis of negotiation processes in policy formulation, resistance strategies within and by hydraulic bureaucracies, and capture and transformation processes in policy implementation would be highly valuable. This would allow a more precise discussion of the meaning of 'alignment'. In the next three sections we discuss three themes for further research suggested by the preceding chapters that could throw more light on the question of alignment strategies, and on other matters.

The Resilience of Irrigation Bureaucracies

A theme that is present very strongly in a number of chapters is the resilience of irrigation bureaucracies. Irrigation bureaucracies in the Philippines, Indonesia and Mexico have been very good in maintaining their construction orientation, and the flow of funds for that, in conditions of increasing emphasis on institutional reform and operation and maintenance rather than construction of new systems (Chapters 3, 4 and 5). The Indonesian irrigation agency converted a system rehabilitation-before-turnover approach very efficiently into a physical business-as-usual rehabilitation programme. The Philippine National Irrigation Administration kept its physical works orientation throughout the era of institutional reform and after the end of political protection. One of the devices was to subsidize its financial sustainability, which policy said should be based on collection of water fees from irrigators, from the capital account, that is, the budget for infrastructure projects. Another strategy was to selectively support and obstruct government policies affecting the organization. The irony, and paradox, of the Mexican case is that the irrigation agency regained its bureaucratic and financial autonomy and focus on construction/physical works by striking a bargain about irrigation management turnover with the other actors in the policy arena: the government and the international funding agencies (concentration with decentralization).

These were all proud, big and powerful agencies, mostly controlled by a single discipline of professionals, civil engineers, who sought to retain their status, influence and room to manoeuvre, in environments increasingly critical of 'conventional' engineering approaches. These histories generate a number of observations and questions.

It is clear that irrigation bureaucracies stand to lose power in irrigation reform processes: they may be down-sized, their budgets for physical works may be cut, and they have to re-orient professionally towards new issues like management and environment. The latter is not necessarily a loss, but is often perceived as such.[20] The reactions of different agencies to this seem to vary. One pattern is the defensive strategy: how can we mould things in such a way that we can continue what we are already doing? The Indonesian case study (Chapter 5) is an example of this, and the Pakistan case (Chapter 7) could be added, as well as many other irrigation agencies not discussed in this book. The Philippines and Zimbabwe cases (Chapters 4 and 6) show that decentralization and devolution of O&M tasks to local government agencies may be actively resisted by the central irrigation agency by means of its monopoly on the required technical expertise and the greater job security offered.

The second strategy seems to be to (partly) incorporate new concerns and to (partly) re-invent the agency. The Mexico and Philippines cases (Chapters 3 and 4) illustrate this pattern. In the Mexico case the re-invention is more far-reaching. In comparison with the Philippines the turnover of irrigation systems to users is much more substantial. The trajectories of institutional development of irrigation agencies in the world are poorly documented, while these agencies are commonly considered a major part of the problem of irrigation reform (and, perhaps paradoxically, also of the solution). It would therefore be sensible to do more research on such trajectories of transformation and non-transformation.

One dimension of such research would be a thorough investigation of professionalism, professional culture, and professional identities in irrigation agencies. The Pakistan case makes reference to the 'caste-like' relations between different disciplines in an organization (civil and mechanical engineers). Another aspect of this is the extreme gender bias in most irrigation agencies: they are male bastions *par excellence*, perhaps only outdone by the army (from which many public works and irrigation departments in fact originate). The association of technology and masculinity has been problematized in the feminist literature on science and technology, and may hold some relevant clues for the irrigation engineering profession also. The practical background to this concern with professional identities is the thought that unless these change it is difficult to see how irrigation agencies will be able to internalize new demands on the sector.

The Role of International Development Funding Agencies

Another paradox that emerges from the case studies of the irrigation bureaucracies is that international development funding agencies have over the past decades contributed substantially to the power and status of the irrigation agencies through providing large loans for infrastructure creation. These same funding agencies are now calling for, and are trying to enforce, bureaucratic reform of these hydrocracies – a term used in Chapter 4 – aiming at a very different, more modest, distant and service provision-oriented role in water resources management. There is also a contradiction in the 'tagging on' of institutional reform to smaller but still substantial loans for physical works, nowadays more focused on rehabilitation of systems. This induces on-paper acceptance of institutional conditionalities with an eye to obtaining the loan for the rehabilitation package. Within the international funding agencies there is also no full unanimity on the institutional reform objective. At a personal level quite a few of the funding agency staff have their roots in these same construction-oriented irrigation agencies, and in related construction companies. The construction bias is also present in the funding agencies, even when a real shift in emphasis has taken place from big new infrastructure creation to rehabilitation and operation and maintenance.

This is one of the issues that emerge from the different chapters regarding the role of international development funding agencies in the process of irrigation reform. Some others are the following.

Above it was already indicated that one role these agencies play is to be the carrier of new ideas. The World Bank literally calls itself a knowledge bank. Part of this is information and reporting of experience (often in 'lessons learnt' and 'best practices' formats). Another part is more paradigmatic in nature. Apart from information and knowledge, messages about development strategies are also communicated. In the 1990s these have often been packaged as conditionalities to loans that are part of structural adjustment policies. However, the force of these conditionalities can easily be exaggerated. The Pakistan case (Chapter 7) shows that the World Bank and Asian Development Bank do not shy away from financial arm-twisting to get institutional reforms formally accepted. At the same time the case shows that the Pakistan government can get away with non-implementation of quite a bit of it. The funding agencies have geopolitical considerations and their pressure to disburse to deal with, which seems to make them sometimes accept watered-down reform proposals. Chapter 7 on Pakistan states that the project cycle is the 'Achilles heel' of the funding agencies. That cycle embodies the need to disburse and spend. Not doing so creates many practical problems for the funding agency staff, the consultant-implementers and the government officials involved. One mechanism is that all these institutions have a 'target orientation' in their work planning. Projects' operational procedures usually do not fit a learning process approach to policy implementation. Despite this bureaucratic logic, one really wonders in the Indonesian case (Chapter 5) why the funding agencies continued their support in the face of blatant non-adherence to terms agreed. The problem of conditionalities and their effectiveness is further compounded by the relative difficulty of assessing institutional changes compared with the ease of monitoring execution of physical works.

A process we do not know very much about is the negotiation process between governments and the larger funding agencies. The papers on Mexico, the Philippines and Indonesia (Chapters 3, 4 and 5) give some clues in this respect, but much remains to be documented. In the case of the World Bank, the so-called task manager of a loan seems to be an interesting point of entry for further study. This person is the day-to-day interlocutor between governments and World Bank, and – sometimes – has considerable influence on the shaping of the loans (see George and Sabelli, 1994). Some are highly skilled strategic actors, and we can only hope that some of that experience gets written down.

The Zimbabwe paper (Chapter 6) provides some interesting insights in the behaviour of smaller, bilateral donors. On the positive side, a donor agency like Denmark's DANIDA has contributed to an emerging policy discourse on irrigation management turnover in Zimbabwe. On the less positive side it has contributed to the reproduction of the 'dependency syndrome', that is the expectation and calculation of farmers that the government, in team with donor agencies, will come to their rescue when their system threatens to collapse. This 'syndrome' stems from the populist compulsions that the government feels it should answer to, and donors' desire 'to do something' through a long series of pilot projects. One unhappy consequence is that there is now a group of senior government staff that is employed as consultants by the bilateral donors that has a vested interest in the continuation of a situation without a clear irrigation reform policy. Existence of a clear policy framework and consistency of government and donor approaches would substantially reduce the need for advisory work.

Another interesting point is raised in Chapter 7 on Pakistan, where it is observed that funding agencies have invested very little in 'advertising' and explaining planned reforms to those affected by it. This raises the question of whether there is any 'demand' for reform on the irrigators and field staff side. 'Even now, more than five years into the process, there is still no evidence of prominent political, agriculturist, or community leadership support in Pakistan for the irrigation reforms, much less an active group or lobby that advocates their effective implementation on merit.' At the same time interesting initiatives are happening at local level, by communities and by NGOs. Perhaps external agencies like funding agencies could play more of a facilitating role in bringing together the 'top' and 'bottom' perspectives in reform.

The Capture of Irrigation Reform Policy in Implementation

In none of the chapters in this book that describe policy implementation that process is straightforward. Implementation does generally not move neatly from stated objectives to projected outcomes. This is because policy implementation is a series of actions and interactions in which different individuals, groups, networks, lobbies and organizations attempt to mould the implementation to fit their own perspective or interests. In the Mexico and Philippines cases (Chapters 3 and 4) as well as the Indonesian case (Chapter 5) the irrigation bureaucracies are the main remoulding actors, as already discussed above, for purposes of organizational survival and continuation of their 'core business'. In terms of outcomes this meant that in the Mexican case the Irrigation Management Transfer objectives were implemented in all irrigation systems, in the Philippines it meant a successful implementation of the participatory management model in farmer managed systems but a very problematic process in the national large-scale irrigation systems. In these users organizations are in place but effectively no control has been delegated to them. In Indonesia agency remoulding led to a total neutralizing of the institutional component of the irrigation reform package and an exclusive focus on physical rehabilitation.

In cases where the irrigation reform policy actually reaches the field so to speak local actors may play an important role in its reshaping. The Andhra Pradesh, India and Philippines cases (Chapters 8 and 9) provide examples of how local elites and politicians captured reform policy, redefined its objectives and changed its outcomes. Andhra Pradesh is a case of straightforward capture. Elites occupy the board positions in the new water users organizations and control their functioning. Elections are an insufficiently strong tool to avoid this. The office bearers then become the contractors for the physical works that are part of the reform package, and build new relationships with the Irrigation Department for managing this. The issues of water distribution and water fee collection are left untouched, despite the empowerment of water users organization through an Act to take control of them. The Andhra Pradesh case suggests that unless special efforts are made regarding the difficult issue of resource redistribution, policy implementation is likely to reproduce existing inequalities or even accentuate them by providing new resources to local elites.

In the Philippines the process is more complex and has varied outcomes in different parts of the irrigation system. An intense process of struggle and other forms of interaction emerges around water, physical works and water fees (which are linked to the payment of agency staff). The irrigation associations that are an essential element of the reform package, and around which a fundamental shift of control relations was conceived, do not play a very important role in practice. Or more precisely, their role and effectiveness depends on a range of other factors and processes. The Philippines case also shows the crucial role of the street-level bureaucrat of irrigation systems: the gatekeeper or ditchtender. S/he is the person at the interface of users and the bureaucracy, at the division points where the interaction of these two and other actors are concentrated. The gatekeepers and ditchtenders usually have low social status and prestige, but are often the unsung heroes of canal irrigation management. Through their actions an important part of the policy reshaping takes place.

Understanding the outcomes of irrigation reform policy implementation requires an understanding of the intricacies of the capture and reshaping processes at different levels. Such analysis may provide new entry points for irrigation reform initiatives.

Substantive Theory on Irrigation Reform

The discussion above has focused on the general approach to the study of policy processes in irrigation (the need to address complexity and to contextualise, the analysis of alignment strategies as entry point) and identified three research themes that may be fruitful areas for further work (the resilience of irrigation bureaucracies, the role of international development funding agencies, and the capture of reform in implementation). Whether does this leave substantive, explanatory theory on irrigation reform?

There are several reasons why we have not endeavoured to propose a specific theoretical framework for explaining the dynamics of irrigation reform processes. In the first place we believe that the domain of irrigation policy reform studies would benefit from a diversity of approaches and frameworks. This book may thus be read as an invitation to start developing explanatory theory on irrigation reform processes and the working of hydrocracies. Secondly, we feel it may be a bit early to venture into more general theorizing, and that production of additional detailed case studies should get priority given the scarcity of material available.

But these are relatively weak reasons. A more pertinent reason stems from the nature of the irrigation policy discourse in the 1990s. This has been dominated by a number of explanatory frameworks that seek to generalize on the dynamics of policy processes and hydrocracies. These are the rent-seeking perspective on the functioning of irrigation bureaucracies (Repetto, 1986), the economic theory on the role of financial incentives in improving irrigation and agency performance (Small and Carruthers, 1991; Winpenny, 1994), and, to a lesser extent, the theory of design principles for self-governing irrigation systems (Ostrom, 1990, 1992).

All are flawed or have strong limitations. A conceptual critique common to all is their limited concept of human agency (of individual utility maximization or optimization) and the absence of a concept of social power in the frameworks (Mollinga, 2001b). The economic theory of the impact of financial incentives on institutional behaviour has been shown to be wrong or at least too simplified (Oorthuizen, 2003). The application of the theory may actually lead to reduced levels of performance and service provision.

Repetto's perspective has created or consolidated a commonly shared negative image of irrigation bureaucracies as rapacious rent-seekers, and thus underwrites neo-liberal structural adjustment policies that seek to slim down agencies and reduce their role in public service provision. This has put the agencies even more on the defensive than they already were as a result of environmental critiques, and explains, in our view, part of the very limited achievements of bureaucratic reform in this sector. Close study of irrigation management practices tends to show that there is a multiplicity of motivations and strategies of irrigation bureaucrats, and that there is a considerable level of commitment to 'good performance' at certain locations in the institutions, further expression of which is, however, heavily constrained (Mollinga, 2003, Oorthuizen, 2003). Tapping this potential would give a different entry point for improvement of public service provision.[21]

One probably unintended effect of Ostrom's work on self-governing irrigation institutions has been the strengthening or confirmation of the belief that irrigation institutions can be engineered in the same way as irrigation hardware can. The idea of 'design principles' for institutions is the basis of this; the notion resonates strongly with the engineer's as well as the bureaucrat's professional outlook (see Chapter 1 and above). The hopefully more lasting effect of this work would be the emphasis on governance, and the differentiation of governance and management, and related to that the question of allocation and rights.

Given this record of general theoretical frameworks in the irrigation policy discourse, we are extremely hesitant to propose another one. Though there is no necessary relation between nomothetic theory for explanation and the use of 'models' in designing policy intervention, it is not difficult to imagine a certain resonance between the two.

Even more pertinently, and more fundamentally, we do not believe that explanatory theory of this kind and at this level of generalization is very helpful. These are structural approaches that privilege certain elements or aspects of reality and ignore other aspects or elements. More importantly, they lack history. The irrigation systems that the theories seek to understand and the societies they are part of are complex, open systems whose features can only be understood with some degree of comprehensiveness when their historical evolution is analyzed, and whose structure changes over time. Understanding of processes like irrigation reform thus always needs a combination of structural and historical explanation.[22]

The implication of this is that explanatory accounts that incorporate both structural and historical elements would have to be situated at the level of the societies to which they refer, that is, be context-specific, rather than at the level of general social theory. At the latter level of abstraction there can be comparison and reflection on concepts employed for analysis, but not generalized causal explanation of the structural plus historical kind.[23] Even when universal categories for the analysis of human behaviour could be identified (as for example frameworks like game-theory and cultural theory, both used in water resources studies, posit), the historical and context-specific element would still have to be added for the understanding of concrete processes of transformation, and for making decisions on strategic intervention in these processes.

To conclude, we opt for context-specific, structural plus historical, explanatory accounts of concrete irrigation reform processes, for which the unit of analysis should be defined in relation to the research question. At a higher level of abstraction we would encourage comparative discussion of irrigation reform processes and reflection on the conceptual and methodological tools used in the different contexts. General theories of human behaviour, as individuals or as institutions, can be tools in such an undertaking, but there is the danger that 'applying the nomothetic approach to human affairs' will yield 'no more than trivial generalities' (McAllister, 2002: 27).

The Politics of Research

In the last two sections of this concluding chapter we look at how to move ahead with research on the politics of irrigation or water sector reform. In this section we first look at the sensitivities involved in doing research on the topic. In the last section we outline our drivers for further research and how that research could contribute to strategic action in irrigation reform processes.

To introduce the point of the politics of research, we look at the genesis of this book. When we started this workshop and book project we were not sure that we would be able to get the kind of papers that we wanted: detailed accounts of irrigation reform processes 'from the inside'. The objective was not to uncover gossip and expose organizations and individuals, but to provide grounded analysis of concrete policy formulation and implementation processes. We anticipated that this might be a sensitive undertaking, because people would perceive it as such. 'Insiders' might censor themselves as their livelihoods depend on continued employment in the sector, and organizations might not be happy to be made the subject of analysis.

And indeed this happened. The author of one paper was not allowed to attend the workshop by his employer and present the paper that he had written. The employer argued that there were substantive and methodological problems in the paper, and that the paper would damage the position of the organization in the country where it was working. As organizers and editors we only very partially agreed with the former, and the second reason was outside our purview. We could not but accept the cancellation of the presentation.

In other instances the process was subtler. Several authors have clearly indicated up to where they were willing to go in writing, even when in oral and email discussions before, during and after the workshop more information and more audacious interpretation came to the fore. This means that reading between the lines is necessary in some papers to capture the full argument. For example such concerns are one of the reasons why the Indonesia paper focuses on a structural analysis of institutions and interests, while not going into much detail about individual personalities or particular incidents in bureaucratic struggles.

Yet another instance was the discussion at the workshop of papers commissioned on the reform process in Andhra Pradesh, India. As this workshop on the politics of irrigation reform immediately preceded the fifth international INPIM (International Network for Participatory Irrigation Management) workshop, hosted and co-organized by the Government of Andhra Pradesh, there was great sensitivity towards criticisms of the Andhra Pradesh approach to irrigation reform. There was heated discussion at the conference between paper authors and government representatives (which was highly beneficial to the conference as it stimulated discussion also in later sessions).

These examples are not meant to point fingers at or criticize individuals and organizations. It is quite remarkable what has been written, and people active in the sector have provided a lot of new information and insights. As organizers and editors we are therefore more than satisfied. However, the examples do illustrate that a discussion on the politics of irrigation reform when located within, and with professionals from the sector participating, is not a neutral affair. Interests, personal and institutional, do shape the discussions to some extent. This was not unexpected, as indicated, and we don't want to be naïve and moral about it, but it does raise issues for further research on the topic.

We first want to argue that part of the anxiety to discuss and write about the everyday politics of irrigation reform policy formulation and implementation is misinformed. One could very well argue that reform processes that claim to be about participatory irrigation management should preferably be participatory not only in implementation but also in formulation and evaluation. The now much cited *mantra* of multi-stakeholder participation points in this direction. Research on the politics of irrigation reform is then nothing but a systematic reflection on that participatory process.

Another reason to be less anxious is the fact that 'non-official' and 'off the record' debate on the politics of irrigation reform is rampant. In the corridors of meetings, over dinner or drinks, a lot of analysis of the ins and outs of these processes takes place. Furthermore, the papers provide several examples of conscious strategizing by policy actors during formulation and implementation. One interesting example of this can be found in Chapter 7 on Pakistan, where the Provincial Irrigation Department makes an effort (but fails) to gain control over the On Farm Water Management section in the Agriculture Department, using the vocabulary of the reform policy it is opposing.

Our contention is that the sector and the cause of irrigation reform will gain by 'professionalizing' this debate. That is, by addressing the political issues in public debate, through systematic analysis based on grounded research, the quality of strategic analysis and decision making on intervention may be improved. Given the limited success rate of institutional reform processes this may not be an unwarranted investment. Not that socio-political analysis will immediately provide more 'success', but it may provide the basis for more informed intervention practices. It opens up for debate the political economy of the 'how' question, in addition to the 'what' question that is now the focus of most public debate.

How then to do more research on the politics of irrigation/water sector reform? In the implementation of such research the limitations that arise for people working within the sector need to be recognized. We want to suggest that more opportunities need to be created for practitioners to write about the practices they are engaged in, in a reflective manner. These opportunities can be facilitated or moderated by institutions that do not have an immediate interest in the matter, notably universities and independent research institutes (though, of course, individuals within these institutions may be equally 'bound up' when they individually benefit from consultancy assignments in the sector for example). For institutions that do have a direct interest in the matter of irrigation reforms, like the international funding agencies, professional debate could be strengthened by studies looking beyond 'best practices' and 'lessons learnt' in the so-called success cases. Our feeling is that meticulous contextual analysis, monitoring and reporting on/of cases of 'policy failure' would lead to more profound debate on the obstacles and opportunities for irrigation reform than the 'selling of success' strategy presently in vogue.

This is all argued from the perspective of those active in and committed to the sector. What does the topic hold for social scientists and academics more generally? The case of irrigation/water sector reform provides excellent material for social theory development in a number of fields. The general umbrella theme is the changing role of the state in the context of globalization, liberalization and/or decentralization policies, and the implications of that for the livelihoods of resource users, in this case water. But more specific themes can be derived from the discussion above and the preceding chapters.

To conclude this section, we argue that despite its sensitivity as a topic, research on the politics of irrigation/water sector reform needs to be expanded and strengthened. This would benefit water sector practitioners and policy makers, as well as the academic community. Intellectual interfaces need to be created where reflection on concrete irrigation/water sector reform processes can take place, combining contributions of different types of 'insiders' with those of independent scholars. Such intellectual interfaces need not be considered new, as the prolific practice of writing memoirs and holding public lectures by retired political actors demonstrates.[24] What we advocate are organized forms of self-reflection by practitioners whilst they are still *in office*.

Conclusion: Research for Strategic Action

In a recent essay in the *Economic and Political Weekly*, a leading journal in India, Ramachandra Guha comments on the social science scholarship on South Asia as done in the region as against that undertaken by the quickly growing community of South Asian scholars based in North American academia.[25] He summarizes that research in the humanities is generally based on one or more of three motivations. Firstly, it can be driven by the criterion of 'relevance' (influencing policy or 'correcting the injustices of history'); secondly, solving 'intellectual puzzles' to understand complex social systems may be the driver; and thirdly, a person may simply be following 'intellectual fashion' (Guha, 2003:1122). The first two define research in relation to the 'wider world', the third in relation to the 'printed word'. He suggests that the regional South Asian research is primarily moved by the first motivation, that Europeans tend towards the second and that the research done from North America is 'just a little more likely to be driven by fashion'.[26]

The drive and perspective of the editors of this book is mainly based on the first motivation. This orientation towards practice is a structural characteristic of research on irrigation. As observed earlier, that research has been historically linked very closely with the practice of irrigation policy.[27] This close link creates limitations, as discussed in the previous section and in Chapter 1, but in principle also allows constructive engagement of (reflective) research and policy practice. The composition of this book is an effort to strengthen such constructive reflective engagement. Most of its authors are active as water professionals and have been based in both academic and professional locations in their working lives.

The overall objective of future research would in our view – therefore – have to be to get a better grip on the notion of strategic action in order to improve capacity for such action. We have the optimistic assumption that more explicit reflection on modes and processes of strategic action may contribute to more productive and democratic irrigation reform processes. It is difficult to support this optimism by empirical evidence, but the same is true for the opposite position. Our main (political) interest lies in improvement of the negotiation capacity of those now generally excluded from the policy process: the non-elite farmer-irrigators and other water users, and others whose livelihoods depend on irrigation (reform), that is, the constitution of inclusive processes of irrigation governance and management.

From this particular standpoint, the generic contribution that research on irrigation reform can make to the capacity for strategic action derives from the three ways that social power 'works' in irrigation reform (see above).[28]

The first contribution of research to strategic action would be to provide the conceptual tools for gaining insight in the negotiation and other interaction processes in irrigation reform processes. We have already suggested that the study of alignment strategies in the different phases, stages or arenas of the irrigation policy reform process would be a useful starting point. Social scientists working on public policy (reform), negotiation processes and conflict resolution would find a lot of interest in the irrigation domain.

The second contribution would be the provision of tools to critically examine the terms of debate in the irrigation policy discourse. A contextualized understanding of the different actors' discursive categories in relation to their interests and strategies is crucial for this.

For example, in Indonesia a lot of effort was invested in developing a classification of irrigation systems. This was not simply a scientific exercise in taxonomy, but was laden with political meaning, because it determined where policy intervention, i.e. rehabilitation activities, would be permissible.

Another example is debates on the merits of financial autonomy of irrigation agencies. This is generally discussed in terms of the incentive structures it creates for better performance, a conclusion derived from general economic theory on this topic. This may miss the point altogether, as the Mexican case shows. There the interest of the hydraulic bureaucracy in cost recovery and fee collection stems not so much from a concern with performance, but from the possibility it creates to gain control over resource flows from users.[29]

The analysis of irrigation policy narratives and discourses would seem to us a fascinating area for further enquiry.

The third contribution of research from a strategic action perspective would be analysis of the role of politicians in irrigation reform processes. This is likely to be the contribution that would be least welcomed by the dominant actors in the present irrigation policy discourse.

In the early 1980s Robert Wade published his now seminal articles on 'the market for public office' and the 'system of administrative and political corruption' (Wade, 1982, 1985), but very little additional documentation on these issues has emerged since then, even when Repetto's programmatic paper on the rent-seeking behaviour of irrigation bureaucracies has become very influential (Repetto, 1986). Even on the more positive aspects of the role of politicians in irrigation (reform), like mediation of conflicts and mobilization of resources, very little has been documented.[30] There would seem to be a large challenge for the disciplines of political science and public administration at this point.

Herewith we have sketched what we hope to work on in the coming years, in collaborative efforts of irrigation professionals and professional academics. When this book contributes to more serious discussion on the political dimensions of irrigation reform, and how to translate that into practical guidelines for action, our objective will have been more than achieved.

Notes

1   This chapter has benefited from detailed comments by Bryan Bruns and Flip Wester.
2   We do not intend to suggest that no one has written about power and politics in irrigation, or no one has produced 'contextualized' irrigation analyses. Examples of such work are the rich anthropological literature on irrigation, rent-seeking perspectives like that of Wade and Repetto, Chambers' critique of 'professionalism' in the sector, a political economy literature that discusses irrigation in terms of its role in agrarian change, and several other contributions. One needs to carefully distinguish between what is written and discussed in the academic literature on irrigation, and what is debated in the irrigation policy discourse. We want to suggest that the latter is constrained in scope as compared to the former (a comment that is meant to reflect as much on academics as on practitioners). This book could not have been written however if there would not be people active in both discourses simultaneously. We *do* argue that there is very little work available, both in academic and in policy literature, looking at the dynamics of irrigation (reform) policy processes – a subset of irrigation studies – and that 'politics' is a topic and term that is generally avoided in irrigation policy discourse.
3   This speed in declaration of model status probably was a convergence of the World Bank's desperate need for positive development examples in India to underpin its development paradigm, and the eagerness of the Andhra Pradesh government to portray a particular image of good governance, but that is another story.
4   One irony is that the early stages of the Philippine reform experience are extremely well documented, including the process dynamics, notably in the work of Korten (1982) and Korten and Siy (1989), and with much additional detail in the 'greyer' literature. The notion of process documentation research in irrigation originates here (see Veneracion, 1989). However, this literature covers the process in the communal irrigation systems, that is, the small, effectively farmer-managed systems. The subsequent and much less successful experience in the national systems (the large-scale canal systems) is hardly documented. The Philippine 'model' was extracted from this literature by others, and it is this selective reading of the Philippine experience that traveled around the world, and inspired others. This act of simplification and de-contextualization is all the more worrying, and surprising, exactly because of the existence of rich documentation of the first phase of the reform process. It illustrates not only the dangers of such simplification and de-contextualization, but also the apparently enormous desire or attraction to commit these acts.
5   See for example Cooke and Kothari (2000) and Mosse, Farrington and Rew (2001). The former has the telling title *Participation: The New Tyranny?*, and is a book that according to the publisher 'challenges participatory practitioners and theorists to reassess their own role in promoting a set of practices which are at best naïve about questions of power, and at worst serve to systemically reinforce, rather than overthrow, existing inequalities'. This illustrates the emerging process of reflection on the status given to method.

314   *The Politics of Irrigation Reform*

6   Our presentation to the INPIM conference was titled 'cautionary tales from the research front'. Thomas Panella and the authors of this chapter did the presentation.
7   Some of that variety can be found in this book, but there is quite a bit more. One example can be found in a recent PhD thesis on what could be called the politics of project implementation, in two donor-funded institutional reforms of irrigation projects in Mali and Madagascar (Arne Musch *The small gods of participation*, Twente University, the Netherlands, 2001). The author uses a vocabulary from general management science, but arrives at a very similar approach to policy analysis.
8   In an institutional analysis of the management of salinisation in public irrigation systems Scheumann distinguish five 'action arenas': 1) planning and design, 2) investment decisions, 3) executing investments, 4) operation and maintenance, and 5) groundwater and salinisation control at the farm level (Scheumann, 1997:214-220). Such series or networks of arenas can also help to situate policy processes in the wider context of agricultural practices, general politics, (world) markets et cetera. There is methodological affinity with 'commodity chain' analysis (see for example Ramamurthy, 2000). Musch, in the Ph.D. thesis referred to above, operationalizes his framework by analyzing the flows of resources (like for example knowledge) through the system.
9   Though we haven't distinguished the two very sharply so far, it may be emphasized at this point that 'policy as process' and 'politics of policy' are not synonyms. 'Policy as process' is the broader term; 'politics of policy' the narrower one. We do not claim that a 'politics of policy' as a *specific* perspective will uncover all elements or dimensions of policy processes.
10 We thank Jos Mooij for her ideas and discussion on this issue.
11 This typology was presented by Ramachandra Guha in a discussion during the *Rethinking Environmentalism* conference held in Delhi, 6 December 2001. All four types are responses to and departures from prescriptive policy approaches of state agencies.
12 *Cf.* Foucault's distinction between coercive and discursive power, and see Scott's analysis of the way states simplify and discipline (Scott, 1998).
13 Historically irrigation has been associated with colonization, expansion of the agrarian frontier, and the 'settling of unruly tribes' as Stone (1984) was quoted in chapter 1. Other examples of this include the USA, Sri Lanka, India, Indonesia and many others. This suggests a strong relation between irrigation and nation-building, and the state's efforts to 'civilize', that is, create disciplined citizens. At lower levels of aggregation this disciplining has been rather less successful. State bureaucracies' control over and capacity to regulate irrigation systems is often rather limited. The debate on irrigation reform partly emerged because of this incapacity.
14 This broader scope of 'institutional options' is starting to enter the global irrigation management discourse (see Johnson, Svendsen and Gonzalez, 2002). Also Vermillion's emphasis on more clearly distinguishing governance and management function in irrigation systems is helpful in this regard (see www.iwmi.org on IWMI's research initiative on effective governance).
15 See www.canadians.org/blueplanet and www.brettonwoodsproject.org for example.
16 See www.cgiar.org/iwmi/dialogue.
17 See www.transparency.org.
18 We also feel 'alignment' is more adequate than 'balancing'. Alignment has fewer normative connotations. Balancing tends to imply some form of justice or equity. These are not necessary components of alignment.

19 In the literature on hydropolitics interesting analyses on water policy discourses are emerging, for example employing the notion of 'sanctioned discourses' (Jagerskög, 2002). This type of analysis has hardly found application in irrigation studies.

20 An example where an irrigation agency has internalized new concerns is the incorporation of Environmental Impact Assessment as a new type of professionalism in the USBR (United States Bureau of Reclamation). For a fascinating analysis of this process see Espeland (1998).

21 A similar, but much faster development seems to take place in the domain of urban water supply. The early 1990s saw a strong push for privatization and rationalization of public water supply agencies, in developing countries as elsewhere, but in the early years of the new millennium some of the multinational water companies have started to withdraw their investments in developing countries (for example in Manila, the Philippines, in 2003, which was a show case of private sector involvement some years earlier). The companies now concentrate on European and North American markets. At the same time this seems to have put new vigour into the public utilities for water supply and some have started to reform themselves as a defense against privatization. Irrigation has not seen the introduction of the private sector in irrigation management in the same way, but also in this case there seems to be some re-thinking of the 'market' and 'rolling back the state' perspective of the 1990s.

22 See McAllister (2002) on the divergence of historical and structural approaches in the natural and human sciences, and how and why the desired combination rarely happens. For general treatment of the ontology and epistemology suggested here, see Sayer (1984). We thank Hans Mooij for pointing us to McAllister's paper.

23 More precisely put, the 'level' at which explanatory theory applies depends on how the research object is defined (and the possibility and appropriateness of certain types of explanation depends on the nature of the research object – here the reference is the complex systems in and through which irrigation reform takes place). Spatial metaphors like 'levels' have limitations, because sometimes the object is, for example, a 'column' – another spatial category with limitations: if one would want to investigate how a policy or project of a multilateral development agency translates at field level the research object crosses different 'levels'. Each 'column' (for each different case) would exhibit context-specificity, which would put limits on generalization. Irrigation reform processes generally derive a lot of their specificity from the nation state and the national political economy in which they are located, which is the basis for the possibility to do country case studies.

24 Examples from the water resources sector include the journals of David E. Lilienthal, first head of the Tennessee Valley Authority and senior advisor on the water sector for the World Bank (Lilienthal, 1964), as well as some of the process-oriented chapters on the NIA reforms by Korten and Siy (1989).

25 See Ramachandra Guha (2003),'The Ones Who Stayed Behind', in *Economic and Political Weekly*, Vol.38, nos. 12 & 13, pp. 1121-1124.

26 And particularly the North American emphasis on and variants of cultural studies and post- structuralism, which Guha finds 'trends of dubious intellectual worth' producing mainly essays that 'for the most part (...) are merely extended literature reviews, parasitic assessments of other people's works according to the winds of theoretical fashion and the canons of political correctness' (Guha, 2003:1123). He expresses the hope that the 'absorption with the self' of this category of scholarship might be turned towards a 'serious study of the history and politics of the west' (Ibid.:1124). He basically argues for empirically, politically and geographically grounded research – the first being a must – on South Asia (and other places for that matter).

27 We have also noted (see chapter 1) that there is an academic literature on irrigation (but hardly on irrigation reform) that is not driven by practical policy concerns, but more likely by motivation 2, academic curiosity. This literature has however had very little impact in the sector.

28 A second, situation-specific, contribution would be that of concrete analysis of the dynamics of certain cases. At this level there would have to be a move towards substantive, explanatory theory, to explain and interpret the processes at hand. As suggested above, we believe the conceptual structure of such explanations and interpretations might well be case-specific. In any case, openness towards and generation of explanations and interpretations from different frameworks would enhance the process of critical reflection that we are after. A third contribution would potentially be the design of methodologies for institutional transformation as processes of 'social learning' or 'change management' (the reference is methodologies for institutional learning and transformation as developed in management science; *cf.* Checkland and Scholes, 1990; Caluwé and Vermaak, 2002). However, this field is largely hypothetical as it assumes (at least) a (limited) demand for such transformation processes on the side of the irrigation agencies. This book clearly suggests that such demand is generally wanting. Efforts at this mode of bureaucratic reform have been hardly tried in the case of larger irrigation bureaucracies and have been unsuccessful when tried (see Merrey, 1998 on Egypt). It therefore seems warranted at this stage to conceptualize irrigation reform processes as processes of negotiation, contestation, struggle, pressure and the like, rather than as processes of institutional learning by design.

29 The book provides several more examples. There was considerable debate in Indonesia on whether rehabilitation should be done before or after turnover. This issue has also been discussed in detail in the international irrigation reform community. When such debate is abstracted from local contexts, and is treated as an 'objective' question to which there should be one clear, if necessary conditional, answer, the political meaning it has in concrete situations may easily be lost. Yet another example is that the creative handling of different accounts within its budget by the National Irrigation Administration in the Philippines should not be understood from a chartered accountant's perspective, but from the perspective of institutional survival strategies.

30 For an exception, see Oorthuizen (2003), which analyses the role of mayors and other politicians in Philippine irrigation management in detail. For elaboration and comments on Wade's findings, also see Ramamurthy (1995) and Mollinga (2003).

References

Caluwé, Leon de and Hans Vermaak (2002), *Learning to Change. A Guide for Organisation Change Agents*, Sage, Thousand Oakes.

Checkland, P.B. and J. Scholes (1990), *Soft Systems Methodology in Action*, John Wiley, Chichester.

Cooke, Bill and Uma Kothari (eds) (2000), *Participation: The New Tyranny?*, Zed Books, London.

Espeland, Wendy Nelson (1998) *The Struggle for Water. Politics, Rationality, and Identity in the American Southwest*, University of Chicago Press, Chicago.

Ferguson, James (1990) *The Anti-Politics Machine: 'Development', Depoliticization, and Bureaucratic Power in Lesotho*, Cambridge University Press, Cambridge.

George, Susan and Fabrizio Sabelli (1994), *Faith and Credit. The World Bank's Secular Empire*, Penguin Books, London.
Jagerskög, Anders (2002), *The Sanctioned Discourse – A Crucial Factor for Understanding Water Policy in the Jordan River Basin*, Occasional Paper no. 41, Department for Water and Environmental Studies, Linköping University.
Johnson, Sam H., Mark Svendsen and Fernando Gonzalez (2002), *Options for institutional reform in the irrigation sector*, Discussion Paper prepared for the International Seminar on Participatory Irrigation Management, 21-27 April 2002, Beijing.
Joshi, L.K. and Rakesh Hooja (eds) (2000), *Participatory Irrigation Management. Paradigm for the 21st Century*, (2 volumes), Rawat, Jaipur.
Korten, Frances F. (1982), *Building National Capacity to Develop Water Users Associations: Experiences from the Philippines*, Staff Working Paper no. 428, World Bank, Washington D.C.
Korten, Frances F. and Robert Y. Siy (1989), *Transforming a Bureaucracy: The Experience of the Philippines National Irrigation Administration*, Kumarian Press, Hartford.
Lilienthal, David E. (1964), *The Journals of David E. Lilienthal: Volume 1: The TVA years, 1939-1945*, Harper and Row Publishers, New York.
McAllister, James W. (2002), 'Historical and Structural Approaches in the Natural and Human Sciences', in Peter Tindemans, Alexander Verrijn-Stuart and Rob Visser (eds), *The Future of the Sciences and Humanities. Four Analytical Essays and a Critical Debate on the Future of Scholastic Endeavour*, Amsterdam University Press, Amsterdam, pp. 19-54.
Merrey, Douglas J. (1998), 'Governance and Institutional Arrangements for Managing Water resources in Egypt', in Peter P. Mollinga (ed.), Douglas J. Merrey, Martin Hvidt and Lutfi S. Radwan, *Water Control in Egypts's Canal Irrigation. A discussion of Institutional Issues at Different Levels*, Liquid Gold Paper 3, ILRI and Wageningen Agricultural University, Wageningen, pp. 1-22.
Mollinga, Peter P. (2001a), *Power In Motion. A Critical Assessment of Canal Irrigation Reform with a Focus on* India, Working Paper 1, IndiaNPIM, New Delhi.
Mollinga, Peter P. (2001b), 'Water and Politics. Levels, Rational Choice and South Indian Canal Irrigation', *Futures*, 33: 733-52.
Mollinga, Peter P. (2003), *On the Waterfront. Water Distribution, Technology and Agrarian Change in a South Indian Canal Irrigation System*, Wageningen University Water Resources Series 5, Orient Longman, Hyderabad.
Mosse, David, John Farrington and Alan Rew (2001) *Development as Process. Concepts and Methods for Working with Complexity*, Indian Research Press/ODI Development Policy Studies, New Delhi.
Oorthuizen, Joost (2003), *Water, Works and Wages. The Everyday Politics of Irrigation Management Reform in the Philippines*, Wageningen University Water Resources Series 3, Orient Longman, Hyderabad.
Ostrom, Elinor (1990), *Governing the Commons. The Evolution of Institutions for Collective Action*, Cambridge University Press, New York.
Ostrom, Elinor (1992), *Crafting Institutions for Self-Governing Irrigation Systems*, Institute for Contemporary Studies Press, San Francisco.
Ramamurthy, Priti (1995), *The Political Economy of Canal Irrigation in South India*, Ph.D. thesis, Syracuse University, Syracuse.
Ramamurthy, Priti (2000) 'The cotton commodity chain, women, work and agency in India and Japan: the case for feminist agro-food systems research', *World Development*, 28(3):551-78.

Repetto, Robert (1986), *Skimming the Water: Rent-Seeking and the Performance of Public Irrigation Systems*, Research Report no.4, World Resources Institute, Washington D.C.

Sayer, Andrew (1984), *Method in Social Science. A Realist Approach*, Hutchinson, London.

Scheumann, Waltina (1997), *Managing Salinization. Institutional Analysis of Public Irrigation* Systems, Springer, Berlin.

Scott, James C. (1998), *Seeing like a State: How Certain Schemes to Improve the Human Condition Failed*, Yale University Press, New Haven.

Small, L.E. and I. Carruthers (1991), *Farmer-Financed Irrigation, The Economics of Reform*, Cambridge University Press (in association with IIMI), Cambridge.

Veneracion, Cynthia C. (ed.) (1989), *A Decade of Process Documentation Research: Reflections and Synthesis*, Institute of Philippine Culture, Ateneo de Manila University.

Vermillion, Douglas L., and Juan A. Sagardoy (1999), *Transfer of Irrigation Management Services. Guidelines.* FAO Irrigation and Drainage Paper no. 58. Food and Agriculture Organisation, Rome.

Wade, Robert (1982), 'The System of Administrative and Political Corruption: Canal Irrigation in South India', *Journal of Development Studies*, 18(3): 287-328.

Wade, Robert (1985), 'The Market for Public Office: Why the Indian State is not Better at Development', *World Development*, 13(4): 467-97.

Winpenny, J. (1994), *Managing Water as an Economic Resource*, Routledge, London.

Index

Notes: page numbers in italics refer to tables, boxes and maps; numbers in brackets preceded by *n* are note numbers.

Aboites, L. 60
access
  to decision-making process 42-3, 51-2
  to information 300
  to legal bodies *26*
  to water 19, 51, 179
  commercial interests 23, 24
accountability 3, 148, 161, 255
  lack of 175, 201(*n*8)
Acts of Parliament *see* legislation *under* individual countries
Agri-SA *see* South African Agricultural Union
agricultural sector 12, 14, 18, 68-9
  centralisation of 60, 66
agronomists 67, 68, 73, 82
Alday, Federico N. 114, *115*
Alemán, President 61
alignment strategies 291-2, 297-302
Anderson, James E. 167, 168
Andhra Pradesh 2, 4, 9, 240-62, *243*, 298, 309
  canal system
    described 247, *247*
    irrigated areas *252*, 260(*n*26)
    maintenance of 246, 251-2
    capture & transformation in 253-5, 306
  Irrigation Dept 249-50, 252-3, 255, 257
  legislation 253
    Farmer Management of Irrigation Systems Act (1997) 246, 254, 256-8
  model approach in 293, 295
  party politics in 253-4, 260(*n*28)
  PIM in 254
    impact of 247-53
  reform policy 244-6
    implementation of 241-3
  rural elite in 253-4, 255, 306
  WUAs in 245, 247, 254, 255, 293, 296
    accountability of 250-1
    composition of 248-9, *248*
apartheid 12, 14, 17, 45, 56(*n*11)
Aquino, Corazon 114
Asia 96, 242, 245, 311
Asian Development Bank 3, 98-9, *99-101*, 101, *102*, 111, 210
  compared with other lenders 103-4
  influence of 147-8, 158, 222
  loans stopped 136
  and reform 125, 127, 129, 139
  role of 292, 295, 304
Asmal, Kader 17, 45-6, 56(*n*10)
Australia 24, 43

Bagadion, Ben 110, 112
banking sector 19, 20, 61
Bautista, Honorio 110
Beetham, D. 59
boreholes 27, 208-10, 211
bribery 269, 273, 276, 278, 279
  see also corruption
British Empire, function of irrigation in 1
Budlender, Geoff 37
Buras, N. 69, 71

bureaucracies, irrigation 43, 57-9
  decentralisation of 73-4
  emergence of 1
  resilience to reform 8, 292, 297, 302-3
    bribery in 269, 273, 276, 278, 279
  themes of reforms in 2
Business South Africa 20

Calles, President 60, 61, 63
canals 3, 128, 147, 170, 182, 209, 218, 220, 221, 230, 240-62, 263
  and colonialism 1
  Distributary Committees for 245
  diversion of 151-2, 270
  localisation system of 244
    clearing 110, 120, 131, 141(n11), 187, 189
    lining 150, 155, 157, 190
  O&M 210, 218-19, 220, 221, 226, 227, 230, 251-2
  privatisation of 213
  characteristics of *247*
Cape Water Programme 21
Cárdenas, President 61
Catchment Management Agencies (CMAs) 18, 20, 21, *25*, 51
  and decentralisation 25-6, *26*
centralisation 97, 104
Chavunduka Commission (1982) 175
China 4, 9(n1)
Chitsiko, R.J. 175
civil engineers *see* engineers
clientilism 59
Coca-Cola 283
colonialism 1, 170, 314(n13)
compensation 24, 40
conservation 172
construction programs 60-1, 63, 65, 145, 146, 150
  bias towards 69-70, 104, 150-1, 302, 304
    shift away from 71-2
  local control over 160-1

context-specificity 294, 295-6, 308
corruption 24, 105, 145, 158, 194-5, 225-6, 250-1, 312
  bribery 269, 273, 276, 278, 279
cost recovery 166, 173-4, 197, 211, 214
Council of South African Banks (COSAB) 20
crop-input loans 192-3
crops
  cash 34, 170, 171, 182, 194
  costs of production 192-5
  losses 266
  selection *25*, 40, 129, 172, 188-9, 194-5, 244
  yields 2, 267-8, 271, 283
Cummings, R.G. 69-70

dams 53-4, 150, 179, 187-8
  building 12, 102, *106*, 178, 190
  social impact of 27, 39
  night storage 182
  storage capacities 60
  World Commission on 17
DANIDA (donor agency) 180, 191, 192, 193-4, 196, 305
De la Madrid, President 68-9, 71, 72, 75, 76, 79, 90(n30)
decentralisation 18, 24, 96, 113, 158-61, 178-9, 214, 303, 311
  CMAs and 25-6, *26*, 52, 53
  IMT and 74-5, 80, 82, 83, 85
  of provincial irrigation depts 207
Didiza, Thoko 36
Distributary Committees (DCs) 245, 246, *247*, 254
  composition 248-9
diversion systems *134*
donor agencies *see* agencies *under* loans
drainage projects *106*, 215, 222
drinking water 19, 61, 82, 84, 85
drip systems 170, 178
droughts 12, 18, 191-2, 193
dry land farming 190, 199
Dublin Conference 6

Echevarria, President 70, 76
education *see* training
electricity costs 192, 194, 195
engineers 5, 65, 68-70, 190, 277, 294, 308
　role in hydraulic bureaucracy 104, 108, 114, 209, 225, 253
　role in IMT 75, 80, 82, 281-4
environmental concerns 3, *15*, 16, 141(*n*13), 175
　protection of habitats 18, 21, 27, *106*
　sustainability 19, 33, 34
equipment
　budget for 123-4, 191
　lack of 156
　modern technology 170, 230
　ownership rights 178
　rental *106*, 108, 138
Estrada, Joseph 96, 134, 136, 296
European Union 190
evaporation losses 27

family-based system 265, 269, 285-6
famine relief 172, 201(*n*9)
FAO 178, 180
farm/field/village level development 2-3, 6, 9, 147
farmers 186-7, 265, 268, 271-2, 279
　commercial 22-7, 42, 221, 233(*n*22)
　contracts with horticultural companies 188-9, 190, 194-5, *195*
　cost-sharing 152, 153
　empowerment of 157, 160, 163, 199
　see also Irrigation Management Transfer
　ethnic groups *26*, 27-9, 31, 42-3
　interests of 145
　and Irrigation Depts 247, 252-3
　leaders (FIOs) 271, 272, 279
　loans for 192-4
　networks 199, 269
　participation of 3, 108-9, 121, 125-33, 154, 177, 183
　　lack of 250, 256-7
　　in NIA approach 264, 287
　　outlet committees 244-5
　　see also Farmers Organisations *under* Pakistan
　rain-fed 125, 136
　redistribution of land to 61
　rights 173, 176, 177, 197
　small-scale 22, 27-9, 40, 42, 172, 233(*n*20)
　　vulnerability of 64, 217
　subsidies for 137, 150-1
　training 126, 132
　violence of 273, 275, 277-8, 285, 301
FARMESA project 180
Fegan, Brian 268-9
Ferguson, James 299
field level funding 156
financial sustainability 2
flood 12
　control 61, *106*
food processing companies 194
Ford Foundation 109, 110, 157, 158, 287
forestry sector 21, 49
funding agencies *see* agencies *under* loans

game theory 235(*n*35), 308
garden cultivation 184-5, 199
gender issues *26*, 28, 39, 42-3, 141(*n*13), 260(*n*30)
Gilmartin, David 230
Global Dialogue on Water, Food & Environment 300
global dimension 6, 241, 292, 300, 311
Global Water Partnership 6
Golan, Amnon 102
González-Villareal, Dr 68, 69, 70, 73-4, 75, 77-8
　career 76-7
　and CNA 82

Gortari, Salinas de 58
governance 9(*n*)
gravity-fed systems 96, *107*, 138
Greenberg, M.H. 60, 62, 64
Grindle, Merilee S. 6, 58, 61, 64, 89(*n*21), 168-9
groundwater 21, 23, 209, 211, 232(*n*8), 238(*n*49)
   legal status of 18, 27
GTZ (development agency) 178
Guha, Ramachandra 311

horticultural companies 188-9, 190, 194-5, *195*, 199
Howlett, M. 167
Hunt, A.F. 174
hydroelectricity 12, 61, 74, *106*
hydropolitics 6, 241

IFAD (International Fund for Agricultural Development) 118-19, 196, 269
IMT *see* Irrigation Management Transfer
India 2, 5, 45, 255, 285, 293, 300
   National Water Management Programme (NWMP) 3
   see also Andhra Pradesh
Indonesia 2, 8, 145-65, 295, 303, 305-6, 309, 312
   construction programs 146, 160-1, 302
   decentralization in 158-61
   economic crisis 147
   funding agencies in 145-6, 147, 157, 304
      World Bank 1, 146, 147, 152-3, 154, 158, 162
         Water Resources Sector Adjustment Loan (WATSAL) 159, 161
   government 145, 153-4, 158
      Directorate General of Water Resources Development (DGWRD) 146, 154, 158

      Irrigation Operation & Maintenance Policy (IOMP) 147-8, 155, 157
         projects *149*, 157
      Ministry of Public Works 145, 146-7, 158
      National Planning Board (BAPPENAS) 147, 151, 158
      role in irrigation management 147
   irrigation schemes 150
      Advanced Operation Units (AOUs) 149
      size of 146, 147, *147*
   management transfer in 316(*n*29)
   preconditions of 150-3, 155-7
   O&M reforms 146-50, 151, 152, 154
      funding for 154-7
   Water Users Training Project (WUTP) 149
   WUAs in 148, 152, 153-4, 159-63
Indus Basin 207-11, 212, 216, 220, 230
industry 12, 17, 18, 19
information management 286-8, 300, 304
infrastructure 150, 190, 212, 280, 299, 303
   deterioration of 57, 64, 73, 145
   user rights 177
   see also operation & maintenance
Institutional Revolutionary Party (PRI) 59
Inter-American Development Bank 66
International Fund for Agricultural Development (IFAD) 118-19, 196, 269
International Network for Participatory Irrigation Management(INPIM) 309
International Rice Research Institute (IRRI) 109, 267
International Water Management Institute (IWMI) 221-2, 234(*n*28)

irrigation
  agencies 2, 3, 145, 146, 225-8
    autonomy of 8
    interests of 150-1
    staff 9, 265
  Boards 28, 51-2
  defined 1
  policy analysis of 265
  reform
    debate 4-5
    implementation, capture of 292, 297, 305-6
    models *see* models
    reasons for 2-3
    and social power 298-302
    substantive theory on 307-8
Irrigation Management Transfer (IMT) 71-2, 74, 76, 97, 127-30, 139, 166-206, 299
  case studies 180-97
  cost recovery in 166
  issue in political campaigns 77-81
  preconditions for 150-3, 217-18
  and World Bank 146, 158, 159
  and WUAs 150, 159-63
irrigation service fee (ISF) 145, 148, 153, 155, 156, 160
  see also under Philippines
Irrigator Associations 108, 111-12, 116, 120
Irrigator Associations *see under* Philippines

Java 146, 162
joint system management (JSM) *see* Irrigation Management Transfer
Juinio, Alfredo L. 98, 102, 112, *115*

Kerkvliet, Benedict J. 6, 263
Kerr, Donna H. 168
Kissan Board of Pakistan (KBP) 213
Korten, David C. 268

land reclamation *106*
land resettlement *see* resettlement
landowners 61, 213

and water rights 21, 23, 24, 28, 302
LBDC (Lower Bari Doab Canal) 212-13, 234($n25$)
legislation 2
  see also under individual countries
licences
  cultivation 200
  water use 20, 21, 24, 34, 40, 42
  witheld 173
lift irrigation 255
loans 85, 95, *99-101*, 105
  agencies 3, 9, 109, 111, 145, 179-80, 190-4, 196, 199, 210
  influence of 102-3, 118-19
  role of 292, 297, 304-5
  conditions of 3
  for farmers 191-2
    crop-input 192
  for irrigation construction 63
  negotiations for 64, 65
  and reform 139-40
  rehabilitation 125, 127
local elites 9, 263-4, 269-70, 277, 300, 306
Long, N. 168
López-Portillo, President 66
Lower Bari Doab Canal (LBDC) 212-13, 234($n25$)

Mackintosh, Maureen 242
Magat River Integrated Irrigation System (MRIIS) 110, 131, 132, 137
Mandela, Nelson 13, 36
Marcos, Ferdinand 96, 97, 98, 102, 105, 106, 108, 113, 136
  self-sufficiency objective 104
market forces *25*
mayors *see* local elite
media 14, 44, 232-3($n15$)
Mexico 8, 43, 57-94, 222, 297, 305-6
  agricultural sector in 68-9
  constitution of 60
  construction programs 60-1, 63, 65, 69-70, 302-3

dams in 60
economic crisis 68-9, 71, 73, 87(n4)
governments of 59, 61, 63, 65, 69-70
  Min. of Agriculture & Livestock (SAG) 61, 62
  Min. of Agriculture & Water Resources (SARH) 62, 64, 66-9, 76
  and CNA 79, 81, 82
  in planning process 70, 71
  reform of 73-4
  Min. of Water Resources (SRH) 61, 62, 63, 64, 65, 76
  presidential system in 61-2, 64, 86, 90(n30), 301
  Salinas' election campaign 77-81
hydraulic bureaucracy in 58-9, 76, 86-7, 302
  autonomy of 63-6, 80, 83-6
    loss of 66-74
  history of 59-66
  staff 65
IMT in 57, 71-2, 306
  issue in election campaign 77-81
industrial sector 74, 84
irrigable area of 89(n11)
irrigation districts 60-2, 67, 71-2, 88(n4)
  IMT of 73-5, 83-4
  Rio Yaqui 62, 71, 76, 78, 89(n16)
legislation
  Federal Rights Law (1983) 72
  Federal Water Law (1972) 62, 83
  Irrigation Law (1926) 60, 88(n4)
  Irrigation Law (1947) 62, 63
  water law (1929) 62
  water law (1992) 57
model 4, 291, 294-5
National Development Plan 74
National Irrigation Commission (CNI) 60, 61
National Water Commission (CNA) 57, 60, 68, 83-4
  created 79-80, 81
  income of 84-6, *85*
  legal structure options 79
  National Water Plan 69-71
  PNH Commission 69-70, 78, 81
  PRODERITH program 70-1
  water board 62
  Water User Associations in 51-2, 62, 71-3, 83, 85
  World Bank in 63, 65, 70, 71, 73-4, 77, 79
mining sector 12, 53-4
models
  of implementation 242-3, 256-7, 257-8, 308
  policy 264
  reform 4, 291, 293-6
MRIIS (Magat River Integrated Irrigation System) 110, 131, 132, 137
Mugabe, Robert 197
Muller, Mike 18
Musengezi scheme 190-7, *191*, 199

Naidu, Chandrababu 245
National African Farmers Union (NAFU) *26*, 27-9, 42, 52
NIA *see* National Irrigation Administration *under* Philippines
Njobe-Mbuli, Dr Bongiwe 36
non-governmental organisations (NGOs) 28, 39, 157, 158, 244, 256, 293, 305
Nyanyadzi scheme 180-90, *181*, *182*, 199, 202(n18)

OECF (Overseas Economic Cooperation Fund) *100-1*, *102*, 119, 139
oil prices 147
on-farm water management (OFWM) 210-11, 221, 222, 295, 310
OPEC *99-101*, *102*
operation & maintenance (O&M) 97, 225, 296

costs/budget 63, 83, 96, 104, 107-8, 112, 135, 155-7, 178
  breakdown of 192-7, *193*
  consequences of 209-10
  lack of resources for 120-1, 145
  deferred 124-5, 210
  efficiency programme (EOM) 148-9
  farmer participation in 212, 216-22, 287
  financial viability of 121-5, 138
  reforms 146-50, 151, 152, 154
  responsibility for 1, 111, 190
  support for 136-7
  turnover of 128, 190, 303
  water charges for 72, 84-5, 98, 105, *106*, 107, 183-4, 258(*n*9), 259(*n*15)
  WUAs and 246
Orive-Alba, A. 60
Ostrom, Elinor 307-8
Overseas Economic Cooperation Fund (OECF) *100-1*, *102*, 119, 139

paddy cultivation *see* rice
Pakistan 2, 9, 207-39, 295, 302
  Area Water Boards (AWB) 212, 228
  decentralization in 214
  Farmer Organisations 207, 212, 217-22, 230
    pilot projects 218-22, *219*, 227
  funding agencies in 208, 209, 210, 212, 222-5, 231, 304-5
    World Bank 211, 212, 214, 216, 221, 222, 224
  government 212-13, 215-16, 237(*n*43)
    provincial 225-7, 228
  IMT/PIM in 221-2, 223-4
  irrigation reform in 211-15, 222-3
    legal framework for 228-30, 231
  Kissan Board of (KBP) 213
  legislation 212-14
  Canal & Drainage Act (1873) 214, 224, 228, 234(*n*27), 237(*n*44)
  PIDA Act (1997) 215, 224, 226
  O&M in 207, 209, 211
  On-Farm Water Management 210-11, 295, 310
  privatisation in 212-13, 232(*n*13), 232-3(*n*15)
  Provincial Irrigation & Drainage Authorities (PIDAs) 207, 211, 212, 214-15, 218, 225
  schemes 230
    Indus Basin Plan (IBP) 208
    Lower Bari Doab Canal (LBDC) 212-13, 234(*n*25)
    National Drainage Programme (NDP) 212, 213, 222, 224, 228
    OFWM projects 210-11, 221, 222
    SCARP projects 208-10, 211
    Sindh pilot projects 218, *219*, 220-1
  stakeholders in irrigation reform 215-28
  Water & Power Development Authority (WAPDA) 208, 210, 295
  West Pakistan Irrigation Dept 208
  WUAs in 210-11
  see also Punjab
participatory irrigation management (PIM) 13, 96, 97, 105, 108-11, 222, 306
  impact of 240, 247-53
  institutionalisation of 110-13
  International Network for (INPIM) 309
  preconditions for 216, 219
  problems with 114, 223-4, 256-7
Partido Revolucionario Institucional (PRI) 59
Pazvakavambwa, S.C. 177
PDR (Process Documentation Research) 265, 286-8

Penaranda River Irrigation System
  see under Philippines
Philippines 2, 95-144, 222, 295, 298,
  305-6
  communal irrigation systems (CIS)
    in *126*, 140(*n*4)
    devolution of 96-7, 114, 118-19
      problems with 116, 117-18
    growth of *103*, 113
    PIM in 108-10, 111
    rehabilitation 120-1, 124-5
    turnover of assets 128
  Comprehensive Agrarian Reform
    Program (CARP) 119
  Dept of Agriculture 116, 117, 120,
    128
  IMT in 97, 127-8, 131-3
  irrigation rate change in *107*
  irrigation service fee (ISF) 97, 98,
    105, 106-7, 134-7, 139
    collection 114, 121, 122-3, *122*,
      135, 272, 279-81, 294, 295
    election issue 296
    engineer's stake in 283-4
    non-payment 124
    rates *134*, 135
  irrigation system types in 96, *107*
  Irrigator Associations in 108, 111,
    116, 120, 138, 279-80
    development 125-33
    ISF collection 283-4
    leaders 271, 272, 279
    National Confederation of
      (NCIA) 136-7
    role 306
    size 141(*n*10)
    structure 271-2
    turnover of CIS assets to 128
  legislation
    Agricultural & Fisheries
      Modernization Act (1997) 127-8
    Local Government Code (1991)
      115-16, 118, 120, 128
    NEDA Resolution 20 105, 107
    NIA Resolution AO 17 134-5

  Presidential Decree (PD) 552
    105-6, *106*, 108
  Presidential Decree (PD) 1702
    105, 107
  Republican Act (1964) 97-8
  Water Code 105
  loan/aid agencies in 95, 97-9, *99-101*, 102-4, 111, 120
    and reform 125, 127, 139-40
    support status quo 118-19
  Local Government Units (LGUs)
    116, 117, 120, 121, 128
  Magat River Integrated Irrigation
    System (MRIIS) 110, 131, 132,
    137
  model approach 291, 293-4
  National Irrigation Administration
    (NIA) 4, 8, 95-6, 269, 287
    budget/expenditure 101-2, 107-8
    bureaucratic resilience of 116,
      118
    charter 98
    construction projects 98-9, *99-101*, 102, 104-5, 302
    evaluated by zone 278-84
    growth & evolution 97-105
    management projects 110
    reform of 105-37
      administrative changes 114-15,
        115
      cost-cutting 121
      criteria for 119
      decentralization 113
    staff 104-5, 123, 126, 131-2, 137-8, 141(*n*11), 273
      and Irrigator Association 281
      threatened with violence 275,
        277-8, 285
    turnover programme 264, 271-84
    water scarcity guidelines 266-7
  national irrigation systems (NIS) in
    96, 120, 121, *126*, 140(*n*4)
    contracts 111-13, 129, 131
    growth of *103*, 139
    IMT in 129-30

performance *123*, 124
  PIM in 110, 111-13, 138
Penaranda River Irrigation System (PRIS) 274-84
  described 265-7, *266*, *274*, *282*
  PIM in 96, 97, 105, 108-13, 125, 306
  policy implementation, politics of 9, 263-90
    local elite in 263-4, 269-70, 277, 281, 300
    in management experiment 267-71
    machine model 242-3, 256-7
    politicians in 273-8, 281, 284, 285
    transactional model 242-3, 257-8
  resistance movement in 109
  self-sufficiency objective 104
  stakeholder interests, alignment of 102-5, 301
  UPRIIS programme 271-84
  World Bank in 111-14, 129-30, 132, 138, 271
PIM *see* participatory irrigation management
policy
  defined 167-8
  failure 168-9
politics 4-6
  defined 5-6, 95
  of policy implementation 263-90, 297-8
  reform policies dictated by 59, 61, 63-4, 312
pollution 12, *15*
population pressures 172, 176, 201(*n*9)
poverty 22, 35
  and drought 18
  policies to reduce 76
pricing of water 3
private sector 3, 26-7, 41, 61, 178, 315(*n*21)
  in water management 76, 214

Process Documentation Research (PDR) 265, 286-8
public consultation 8, 11, 19
public-private partnerships 20
pump systems 96, *107*, 108, *134*, 179, 181
  breakdowns in 182, 187, 188, 189-90, 195
  loans for 194-5
  maintenance of 192
Punjab 209, 212, *219*, 225, 227, 230, 232(*n*8)
  Irrigation & Drainage Authority 228
  Irrigation Dept 213, 215-16, 218, 225, 226-8
  PIDA Act (1997) 215, 226
  privatization in 232-3(*n*15)
  WUA Ordinance 232(*n*5)

quality control 250

rain-fed farms 125, 136, 192
rainfall 12, 27, 182, 188
  tax 21, 22
Ramesh, M. 167
Ramos, President 122
Rand Water 20, 49
reform models 4
rent-seeking interests 145-6, 307, 312
Repetto, Robert 145, 307, 312
research 7
  government funded 5
  politics of 8, 208-11, 292
  Process Documentation (PDR) 265, 286-8
  themes for further 8, 9, 291-318
reservoir systems 96, *107*, *134*
resettlement 61, 175, 200
  forced 70-1, 146
resource reallocation 17
Reynolds, N. 173
rice 107, 123, 135, 244, 265, 277
  Research Institute 109, 267, 269

self-sufficiency 34-5, 104, 114, 146
water risk factor in cultivation 268
Riddell Commission (1981) 175
Rio Conference 6
river basin management 3, 25, 102
run-of-the-river systems 96, 170
run-off water 21
rural development 22, 176, 196
  literature 58-9
rural industrialisation 176

SAAU *see* South African Agricultural Union
SADC (South African Development Community) 12, 44
Salinas, Carlos 68, 74, 75-6, 82, 86, 88(n7), 90(n30)
  election campaign 77-81
salinity 2, 208, 211, 314(n8)
sanitation 15, 51
Schreiner, Barbara 36-7
seepage 27
self-sufficiency, national 34-5, 104, 114, 146
SFRA (stream flow reduction activities) 21, 22, 27
SIDA (donor agency) 180
siltation 187, 189, 207, 251-2
Sindh 218, *219*, 220-1
siphons *182*
  illegal 251
small-scale irrigation 52-5, 60, 70
smallholders 22, 27-9, 40, 42, 52-4, 166-7
  management of 172-3
Smith, Ian 174
social power 298-302
socio-political perspective 5, 6-7
South Africa 2, 8, 295-6, 299
  agriculture/forestry sector 12, 14, 18, 19, 21-2
    commercial 22-7, 42, 51
    small-scale 22, 27-9, 40, 42
  apartheid system 12, 14, 17, 43, 56(n11)

Catchment Management Agencies (CMAs) 18, 20, 21, *25*, 51, 54-5
  and decentralisation 25-6, *26*
  constitution 13-17, *13*, *15-16*
  elections 13
financial sector 19, 20, 24, 28
government 36
  agriculture departments 35, 42, 46
  Agriculture, Water Affairs & Forestry Portfolio Committee (AWFPC) 23-4, 29, 33
  Dept of Water Affairs & Forestry (DWAF) 11, 16-18, 24, 29-31, 35
    in consultation process 37, 39, 46
    and National Water Bill 36-7, 41
    and small-scale irrigation 52-3, 54
    and trade unions 28-9
industrial sector 12, 17, 18, 19, 51
irrigation/water boards 20, 28, 51-2
legislation 23
  Irrigation Act (1918) 18
  National Water Bill/Act (1998) 11, 17, 25, *26*, 31, 51-2, 55
    consultation process 32-4, *32*, 37-44
    impact of 47-50
    provincial workshops 37-9, 53
    drafting 40-2
  review process 11-56
    commercial farmers in 22-7
    consultation 43-4, 301, 302
    institutional framework for 29-31, *30*
    and irrigation policy 34-7, 39-40
    issues raised 44-55
      by sectors 19-22, 48-9
    workshops 36, 39-40, 42-3
  Water Act (1956) 17, 41, 52
  Water Services Act (1997) 31
  white papers 39, 40
trade unions 19-20, 27-9, 39

agricultural 22, 23, 40, 42, 48, 53
water issues in 12-13
water supplier in 20
Water User Associations in *26*, 27, 34
South African Agricultural Union (SAAU) 23-4, 42, 55-6(*n*7)
   Irrigation & Water Affairs Committee 22
South African Development Community (SADC) 12, 44
South African Sugar Association 21, 24, 27, 48
sprinkler systems 170, 178, 190
stakeholder participation 20, 46-7, 95, 310
Stone, Ian 1
stream flow reduction activities (SFRA) 21, 22, 27
streambed cultivation 170
subsidies 68, 72, 84, 137, 167, 174, 181, 295
Sumatra 146
sustainability 226
   environmental 19, 33, 34
   financial 2, 302-3
   irrigation 159, 161, 230
sustainable development 15, 178

tail-end areas 253, 267, 276-7, 283
tank irrigation 258(*n*7), 259(*n*11)
tariffs 20
technology *see under* equipment
tenant irrigation 201(*n*2)
Thomas, J.W. 169
trade unions 19-20, 27-9, 39
   agricultural 22, 23, *26*, 27-9, 40
   training 125, 126, 132, 149, 180, 200, 257, 273
   models in 293
transparency 160-1, 212, 255, 300
tubewells 208-10, 211
Tungabhadra Canal irrigation system 240, 244, 247-53
   described 247, *247*

Turkey 4
turnover *see* Irrigation Management Transfer

United Nations High Commission of Refugees (UNHCR) 175
United States 43, 97, 145
United States Agency for International Development (USAID) 97, 98, 102, 111, 157, 210
   rehabilitation funds 175
United States Bureau of Reclamation (USBR) 97, 98, 102, 103, 105
UPRIIS programme 271-84

Valera, A. 270-1
Van der Ploeg, J.D. 168
village level development 147, 151, 153-4, 178
violence 273, 275, 277-8, 285, 301

Wade, Robert 285, 312, 316(*n*30)
Wade, Rovirosa 77, 78
Warwick, Donald P. 242-3, 256
water
   allocation *26*, 42
      tradability of 20, 21
   bailiffs 187
   boards *see under individual countries*
   charges 57, 63, 83, 89(*n*17), 187-8
   collection 255, 256, 260(*n*21), 279
      and cost recovery 173-4
   increased 246
   irrigation service fee (ISF) 97, 98, 105, 106-7, 114
   non-payment 177, 181, 183, 198
   for O&M 72, 84-5, 98, 105, *106*, 214
   payment in kind 106-7, 277
   see also Irrigation Service Fee
   drinking 19, 61, 82, 84, 85
   pricing 19, 20, 34, 80, 178, 180

quality 12, 15, 74, 161
rights *16*, 19, 28-9, 32, 37-8, 140(*n*4)
  and IMT 78, 84, 129
  and land ownership 21, 23, 24, 28
  legislation 211
  and PIM 251
scarcity 12, 16, 24, 55(*n*3), 178-80, 182, 220-1
  implementation problems 270
  planting controls to combat 266-7
tariffs 78
water demand management 26
water resource development 26, 41
Water User Associations (WUA) *26*, 27, 34, 51-2, 62, 87(*n*2)
  composition of 240, 246, 248-9, *248*
  elections 257, 259(*n*18), 261(*n*33)
  lack of resources for 157
  role of in management 71-3, 83, 85, 226, 293
  top-down approach to 256-8
  turnover to 148, 159-61, 178
  user organisations 2, 8
Water User Formations (WUFs) 212
water wholesaling 145
waterlogging 2, 208, 211
weirs 151-2, *182*, 243
wetland cultivation 170
Wickham, Tom 270-1
Wittfogel, Karl 173
women farmers *26*, 28, 40, 260(*n*30)
World Bank 3, 63, 65, 70, 73-4, 86, 99, *99-101*, 179, 258(*n*8)
  compared with other lenders 103-4
  IMT programme 131, 146
  and Indus Basin Plan 208, 209, 232(*n*13)
  influence on national policies 102-3, 147-8, 158, 159, 222, 224
  limits of 151
  Irrigation Operation Support Programme (IOSP) 271-2

Irrigation Subsector Projects (ISSP) 147-8, 152, 153, 154, 155, 162
Issues & Options document 211
overall lending figures 101
and privatisation 212-13
and reform 125, 127, 129-30, 139, 245
relationship with national bodies 111, 112, 113, 114
requirements/interests of 96, 97, 98, 111, 120, 132, 138, 313(*n*3)
role of 292, 295, 304-5
stops loans 71, 77
Water Resource Development Project (WRDP) 129-30
World Water Council 6
WUAs *see* Water User Associations
WUFs (Water User Formations) 212

ZANU PF party/government 174-5, 183, 195-6
Zimbabwe 2, 8-9, 166-206, 295, 297-8, 302
  African Nationalism in 173, 183, 201(*n*18)
  Agricultural Finance Corporation (AFC) 191, 192-3
  Agricultural Policy Framework (1996) 179-80
  Bank of (CBZ) 196
  colonialism in 170-1
  cost recovery in 173-4
  crops grown 170, 187, 188
  farmers rights in 173, 176, 177
  FARMESA project 180
  funding agencies in 179-80, 190-2, 193-4, 196, 199, 305
  government 191-2, 195-6, 198, 201-2(*n*10)
  Agricultural & Rural Development Agency (ARDA) 174
  Chavunduka Commission (1982) 175
  DWD activities 175-6, 179

privatised 187-8
election promises, impact of 296
Irrigation Policy Committee
  (1961) 174
Min. of Agriculture *172*, 175
  Agritex activities 175-6, 177-8,
  188-90, 192, 194, 196, 199-200
Min. of Internal Affairs *172*, 174
Min. of Lands, Resettlement &
  Rural Development, Derude
  activities 175, 176-8, 184, 198
Min. of Water & Natural
  Resources *172*, 175
post-colonial 174-6
Riddell Commission (1981) 175
structural adjustment programme
  (1991) 187, 188
IMT in 169, 174-80, 196-7
  case studies 180-97
  of financial responsibility 187-8
Irrigation Management
  Committees 183-4, 187, 198, 199
  politics of 188-90
irrigation sector structure *169*
  commercial farming *169*, 170,
  171, 175
  decentralisation of 178-9
  policy 176-80, 197-201
  smallholdings 166-7, *169*, 172
    donor assistance 176
    management regime 172-3, *172*

water charging for 183-7
state control of 171-3, *171*
legislation 172
  Control of Irrigable Areas
  Regulations (1970) 173
  Land Apportionment Act (1930)
  170
  Natural Resources Act (1941)
  170-1
  Rural District Council
  Amalgamation Act (1991) 179
  Water Act (1927) 170
  Water Act (1998) 179
  ZINWA Bill (1998) 188
National Farm Irrigation Fund
  (NFIF) 178, 191, 192
National Water Authority
  (ZINWA) 179, 187-8
O&M in 178, 183-4, 190, 191, 198
  breakdown of costs 192-7, *193*
policy discourse/practice in 167-9
Rural District Councils 179, 180
Save Valley 172, 180-90, 200
schemes 173
  Musengezi 190-7, *191*, 199
  Nyanyadzi 180-90, *181*, *182*,
  199, 202(*n*18)
WUAs in 178
Zimfreez activities 194-5, *195*
Zimfreez 194-5, *195*